Machines of the Farm

Theory and Practice

Machines of the Farm
Theory and Practice

Divaker Durairaj Chelladurai

ELSEVIER

ACADEMIC PRESS
An imprint of Elsevier

Academic Press is an imprint of Elsevier
125 London Wall, London EC2Y 5AS, United Kingdom
50 Hampshire Street, 5th Floor, Cambridge, MA 02139, United States

For accessibility purposes, images in electronic versions of this book are accompanied by alt-text descriptions provided by Elsevier. For more information, see https://www.elsevier.com/about/accessibility.

Books and Journals published by Elsevier comply with applicable product safety requirements. For any product safety concerns or queries, please contact our authorised representative, Elsevier B.V., at productsafety@elsevier.com.

Publisher's note: Elsevier takes a neutral position with respect to territorial disputes or jurisdictional claims in its published content, including in maps and institutional affiliations.

Notices

Knowledge and best practice in this field are constantly changing. As new research and experience broaden our understanding, changes in research methods, professional practices, or medical treatment may become necessary.

Practitioners and researchers must always rely on their own experience and knowledge in evaluating and using any information, methods, compounds, or experiments described herein. In using such information or methods they should be mindful of their own safety and the safety of others, including parties for whom they have a professional responsibility.

To the fullest extent of the law, neither the Publisher nor the authors, contributors, or editors, assume any liability for any injury and/or damage to persons or property as a matter of products liability, negligence or otherwise, or from any use or operation of any methods, products, instructions, or ideas contained in the material herein.

ISBN: 978-0-443-32906-7

For information on all Academic Press publications visit our website at https://www.elsevier.com/books-and-journals

Publisher: Peter Linsley
Acquisition Editor: Rakhshan Rizwan
Editorial Project Manager: Asmi Bharati
Publishing Services Manager: Deepthi Unni
Production Project Manager: Surya Narayanan Jayachandran
Cover Designer: Mark Rogers

Typeset by TNQ Tech
Printed in the United States of America
Last digit is the print number: 10 9 8 7 6 5 4 3 2 1

Contents

The book contains both a Companion and an Instructor's site.

Here are the URL to the two sites.
Companion: https://shop.elsevier.com/books/book-companion/9780443329067
Instructor: https://www.educate.elsevier.com/book/details/9780443329067

CHAPTER

Introduction to agricultural mechanization

1

Chapter Outline

Agriculture typically includes the basic operations of preparing the seed bed, sowing, weeding, management of an established crop, irrigation if needed, application of various inputs to enhance and upkeep crop health, protection of the crop from pests and diseases, harvesting of the matured crop, and, finally, handling of the harvested crop. Over the years, in history, these processes that were basically performed manually by hand were performed subsequently by tools and machines; hence, the evolution of modern agricultural machines started happening. A simple wedge tool such as a rudimentary plough was first used by humans to till soil, which evolved over the years into steel ploughs of various configurations pulled by tractors. The mould board plough was a pioneering development because it inverted the soil, thus controlling weeds as well pulverizing the soil for better soil aeration and crop growth. Seeds, which were broadcast by hand, saw the development of sophisticated precision planters in the modern era. Simultaneously, agronomy and plant science also evolved along with the search for new ways of planting so that better crop production could be achieved. Manual weeding between crops using hand tools evolved into high-production precision weeders that were used in a row of crop to perform inter- and intra-row weeding. Modern universal toolbars are now available that can perform weeding in crops over an extensive area in a short time. Crop harvesting done in the early days by hand tools, such as sickles and scythes, is done now by combines that perform the operations of cutting, separation, cleaning, and loading into hauling vehicles right in the field. Above all, mechanical power to perform these

Machines of the Farm. https://doi.org/10.1016/B978-0-443-32906-7.00001-3

operations was rendered in the early days by manual means, which progressed into the use of draught animals, steam-powered tractors, and, now, high-powered, four-wheeled tractors that can pull multiple plough bottoms and very large machines with ease. Modern tractors are designed to use high-tech engines, hydrostatic drives, and electro-hydraulic controls and have climate-controlled operator's cabs.

Agricultural mechanization can, hence, be defined as the application of equipment and machinery to improve the productivity of farm labour and land toward maximizing agricultural and food production. Mechanization is defined as the use of agricultural machines instead of traditional tools (Ulusoy, 2013). Agricultural mechanization is all about the design, manufacture, distribution, maintenance, repair, and utilization of machines (Anon., 2013). The level of agricultural mechanization in any region is indicated by the machine weight and numbers used per tractor, machine power utilized per area cultivated, and area cultivated per tractor (Akdemir, 2013). Mechanization helps in maintaining competitiveness toward lowering food/output prices.

The objectives of farm mechanization are:

- To save labour in agricultural operations
- To encourage better management practices in crop production
- To offer greater timeliness for critical operations such as harvesting
- To improve working conditions in jobs that are difficult or impossible to do manually
- To perform critical field operations in a shorter duration through the use of high-capacity machines
- To bring about machines suitable for small farm holdings as well to areas where conventional machines cannot be introduced
- To impart precision in agricultural operations toward increasing output efficiency
- To use modern science in the machines, improving the crop production system
- To bring about changes in the mechanized crop production system analogous to manufacturing industries
- To mechanize crop production and to address all farm-allied industries such as the livestock, fishery, poultry, and husbandry industries

The introduction of machines in agriculture has been the most outstanding development in world economics. The drudgery associated with farm work is thus reduced, with a corresponding enhancement in the output per worker. Moreover, mechanization has released millions of agricultural workers to other industries, allowing other industries to flourish. Since agricultural mechanization, only 3% of the US population is engaged in farming now, compared with 60% of US population that was engaged in farming in 1900. The use of modern agricultural machines has now made it possible for an North American farmer to manage more than 1000 hectares (ha) of farmland.

Mechanization in North America, Europe, and developed countries has made agriculture a major industry. Agricultural exports earning huge money are comparable to other lucrative industries such as the automobile industry. To keep pace with the world's rising population, which is expected to be 9.6 billion by 2050, the agricultural industry must innovate and introduce more tools that improve the timeliness of agricultural production operations, increase the efficiency of agricultural inputs, and aid modern agriculture in facing unpredictable climatic changes. The farm equipment industry, in particular, should support sustainable agricultural mechanization (SAM) in developing regions to contribute more to the world's food production.

As per recent trends, the agricultural machinery market is expected to increase by 5.4% in the next 5 years. This is because there is an increasing mechanization rate, particularly in developing countries. Shrinking land usage, water resources, and labour availability are continually increasing the need for farm machines in these countries.

1.1 Mechanization in developed countries

The farm mechanization market in North America and Europe is growing due to the increased demand for mechanization with advanced technology in the face of rising labour costs (Pawlak, Pellizzi, and Fiala, 2002). Mechanized farming is being found as more attractive with the use of geospatial positioning system (GPS)-enabled tractors and automated machinery systems. Most importantly, farmers demand more intelligent machine systems to reduce their labour costs. Moreover, the increasing demand for organic products is also driving the growth of automated machine systems in the farming industry. In the coming years, the development of machines for sustainable agricultural systems will be driven more by environmental concerns. This includes pieces of equipment that use less fuel, have electrical drives, and output less pollution, allowing for technology and strategies that conserve soil from chemical degradation. In Australia, there was an increase in the sales of small tractors. In 2020, the market for tractors under 40 horsepower (30 kilowatts) was up by 72% for the whole year. In Japan, where the use of tractors and combine harvesters is the lowest, fields are very small and the level of production per unit of agricultural-use area is high, with the inputs being significantly higher compared with North American and European countries. The average size of farms ranges from very small in Japan and Central, South, East, and Southeast Asia to very large in the United States and Canada. A farm's size is always negatively correlated with the number of people engaged in agricultural production, since mechanization is generally minimal in small holdings, with the exception of Japan, where small-machine usage is quite high. Developed, rich countries with the exception of Japan are intensely cultivating large areas, using modern machines and technology, whereas developing countries, especially poor ones, tend to use traditional tools and technology due to financial constraints. The numbers of those engaged in agriculture are steadily dropping due to urbanization in developed and many developing countries like India. The economically active population involved in agriculture is correlated with the level of mechanization, the intensity of agricultural production, the size of farms, and the status of the labour market.

1.2 Mechanization in developing countries

The status of mechanization is much different in Asian countries (Chen, 2018; Pawlak, Pellizzi, and Fiala, 2002) such as India, where 70% of the population relies on agriculture and most people farm on small and medium holdings, with sizes of 5 ha and less. Accordingly, the use of smaller tractors and machines is on the rise and, slowly, the use of animal-drawn implements of the old order is being phased out. All the developing countries of the world put together showed a 100% rise in demand for agricultural mechanization in the four and half decades from 1960 to 2005, whereas sub-Saharan Africa showed a negative demand for mechanization (Hazell and Wood, 2008). This happened in these countries due to aid programs that introduced Western mechanized agriculture, especially the tractor, to the world (Rijk, 1997).

Generally, Asian and South American countries have experienced significant mechanization using tractors compared with Africa, where human power is still prevalent due to economic reasons. Rice production is the major attributing factor in the mechanization of many Asian countries. However, many Asian and African countries like Malaysia, Pakistan, Sri Lanka, Ghana, and Nigeria have also used mechanization effectively in crops such as cotton, oil palm, cassava, cocoa, wheat, tea, and spices (Clarke and Bishop, 2002; Diao et al., 2014). In addition, the different characteristics associated with the geographical locations and terrains of most countries in Africa, Asia, and Latin America may also necessitate different types of machinery and equipment.

There is an abundant opportunity for enhancing agricultural production through large-scale mechanization in African countries where unutilized cultivable area is abundantly available. However, socio-economic reasons impede the progress of such large-scale mechanization (Clarke and Bishop, 2002; Rotimi, 2010; Viegas, 2003). In developing countries, a large number of rural people live under the poverty line (Boutayeb and Boutayeb, 2005); of those, most rely on either performing agriculture on small farms, with an area of 2 ha and less, or working as agricultural labour (Altieri, 2008). Agricultural mechanization in these regions targets small and reliable machines that are cost effective and less costly for the farmer. Zimbabwe (Rukuni and Eicher, 1994) and Latin America (Altieri, 2008) produce most of their food for consumption and export only through small farms. These small-scale farmers do not tend to enjoy the benefits of research in advanced mechanization. But, of late, many machines have been developed for small farms and are being adopted slowly but not extensively.

One of the examples of successful machines that proved themselves in rice-growing countries is the two-wheeled tractor or the power tiller, which was introduced in many Asian countries by Chinese and Japanese manufacturers in the

1960s. Many countries like India introduced their own brands of these tractors as well and are manufacturing these machines continuously. Many of these machines are also being imported and sold in many Asian countries. They are found to be very popular in the rice fields, especially for puddling. Another such machine that was successfully introduced in India and other rice-growing countries is the automatic rice transplanter, which is being imported from Japan and China.

Governments in Asia generally are in support of farm mechanization, have been framing policies on financing subsidies to farmers (Mondal and Basu, 2009), and encourage research, training, and programs for farm mechanization (Deininger and Byerlee, 2012; Viegas, 2003). Despite these efforts, the level of mechanization is growing only moderately, mainly due to the weak economic status of the different stakeholders.

In 2018, India and China accounted for a 45% to 55% increase in mechanization. The population in this Asian Pacific region is creating a rising demand for the mechanization of agriculture. In African countries, the farm machinery market including that of combines is also set to grow.

1.3 Global outlook of mechanization and its market

Pioneering firms such as Deere and Co., CNH Industrial, AGCO Corporation, Kubota, Mahindra and Mahindra Limited, Iseki & Co. Ltd, JC Bamford Excavators Limited, and Lindsay Corporation are in the fray and expanding their product bands. Recently in India, sugarcane harvesters, rice transplanters, and combines from these companies have found a market. Apart from these firms, very many small companies have emerged in the past decade, introducing customized machines suitable for specific regions and their needs. For instance, in India, we observed that there was a successful introduction of small power weeders in orchards and round balers and rice transplanters produced by such firms. In south India, there has been an exponential rise in firms making multigrain threshers and rotary tillers.

The mechanization scenario has evolved in leaps and bounds over the years by the use of electro-hydraulic controls on these machines and the improvement of precision through agro-electronics and sensor technology. Precision agriculture, which uses geographical information systems (GIS) and GPS, is now in vogue, prescribing agricultural inputs such as seeds and fertilizer as per the locationwise variability in soil fertility data of the field plot to achieve maximum production from the crops (Wang, 2001). Artificial intelligence and Internet of Things (IOT) are being incorporated appropriately to introduce more robotics into agricultural mechanization. The encouragement rendered by the respective governments is expected to boost sales and lead to the introduction of robotic, unmanned machinery on farms. The recent interest in drones for field spraying is one such instance. Without tractors, this kind of progress in suiting appropriate machines to crop production could not have taken place.

Simultaneously, advances in the biological sciences of agriculture have helped mechanization. Better plant breeds have been evolved to suit the mechanical harvesting of crops requiring multiple pickings and harvestings. Varieties of crops with uniform growth at maturity help make mechanization easier for machines. Planting and management practices for some crops have been modified to suit mechanization.

1.4 Challenges ahead in mechanization

Since the market is competitive, each manufacturer is putting effort into improving products to make profit. For instance, the cotton pickers that are popular in many of the world's cotton belts are still yielding output that is inferior to handpicked cotton because of trash inclusion in the lint. Constant effort is rendered by manufacturers to improve their products in this respect.

The mechanization of special crops such as plantation crops is still a challenge to engineers. The mechanization of these crops grown on difficult terrain is hard to incorporate. A lot more needs to be done in the field of fruits' and vegetables' production. The selective picking of fruits and vegetables continues to be a challenge. The robotic selection and harvesting of matured fruits are still being researched, although robotic machines are emerging slowly. The harvesting of flowers is still more complex because they are tender and difficult to handle.

The economic adaptation of certain types of machines to small farms is a problem that needs more attention before agriculture can become completely mechanized. On many farms, the acreage planted of a particular crop is too small to justify the ownership of expensive equipment. Joint ownership and contract or custom operations are alternatives to individual farmer ownership, but they have their limitations and disadvantages from the farmer's standpoint.

One of the most important differences between farm engineering and other engineering sciences is that the former involves biological consideration in their applications and the engineer needs to address the developments with the basic principles and practices of agriculture in mind. The cropping and processing practices of a particular crop pertaining to that region must be addressed to make a machine effective. Yet another challenge to the industry is that the manufacturing costs of farm machines must be at a minimum to keep the hourly usage cost of seasonal machines at an affordable level to the farmer. Machinery designs must be optimal and simple to make this possible. In spite of all these challenges, there is still demand for the development of newer designs of these machines toward improving existing machine efficiency. Management issues related to enhancing the efficient utilization of existing machines are also constantly explored.

Typically, manufacturing firms are profit conscious and put their effort more into the manufacturing and marketing of their machines that are already in market demand. They are generally cautious about introducing machines addressing entirely new problems. This role is typically taken up by research institutes, which are historically known for innovating methods and machines in these cases. Moreover, research studies on fundamental problems such as soil–tool–plant interaction in the context of machine development are taken up by these government-funded institutes. Additionally, engineering scientists work in collaboration with agricultural scientists such as agronomists and plant breeders, finding an ample scope of innovating and improving machine usage on a farm.

The farming scenario of each country is different and so diverse that the machinery required by each is varied depending on the farm holding size, the soil type, weather, cropping season, the crop grown, and the general economy of the farmers. The evolution and use of these machines are, hence, different temporally. For instance, in North American countries, machines for sowing have seen so many changes in the past five decades. Check row planters were out of vogue after the introduction of precision planters. Today, US farmers are going for high-speed sowing using air drills that provide immense coverage of an area in a short time. However, in Asian countries like India, the use of planters and seed drills of relatively smaller widths are slowly catching up. Backpack sprayers and mist blowers are more popular in these countries, unlike in North American and European countries, where boom and high-clearance sprayers of large swaths are being used. Similarly, machines used in Korea and Japan are mainly small and compact automatic machines and many are wonderfully made walk-behind machines that perform sowing, weeding, and transplanting. Furthermore, the methods and systems of using agricultural machines have also changed a lot over the years. One-way ploughs are now extinct in North America, Europe, and Australia and semi-mounted, reversible ploughs with more than eight bottoms are commonly used. Because of this, farmers in these countries are not familiar anymore with the old methods of ploughing systems such as ploughing in lands. But these systems are still relevant in Asian countries with smaller farm holdings.

So, an agricultural engineer should be trained to have a clear and precise understanding of all the aspects related to the development, use, economics, and management of farm machines as a whole.

References

Akdemir, B. (2013). Agricultural mechanization in Turkey. *IERI Procedia, 5*, 41–44. https://doi.org/10.1016/j.ieri.2013.11.067

Altieri, M. (2008). Small farms as a planetary ecological asset: Five key reasons why we should support the revitalization of small farming in the global South. http://twn.my/title/end/pdf/end07.pdf.

Anonymous (2013). Food and Agriculture Organization of the United Nations Statistics Division. Fao Stat: Rome, Italy. http://faostat3.fao.org/home/E.

Boutayeb, A., & Boutayeb, S. (2005). The Burden of non-communicable diseases in developing countries. *International Journal for Equity in Health, 4*(1), 1–8.

Chen, G. (2018). Mechanization of agricultural production in developing countries. In *Advances in agricultural machinery and technologies*. CRC Press.

Clarke, L., & Bishop, C. (2002). Farm power–Present and future availability in developing countries. *Agricultural Engineering International: CIGR Journal, 4*, 1–19.

Deininger, K., & Byerlee, D. (2012). The rise of large farms in land abundant countries: Do they have a future? *World Development, 40*(4), 701–714. https://doi.org/10.1016/j.worlddev.2011.04.030

Diao, X., Cossar, F., Houssou, N., & Kolavalli, S. (2014). Mechanization in Ghana: Emerging demand, and the search for alternative supply models. *Food Policy, 48*, 168–181. https://doi.org/10.1016/j.foodpol.2014.05.013. Available from http://www.elsevier.com/inca/publications/store/3/0/4/1/9/index.htt.

Hazell, P., & Wood, S. (2008). Drivers of change in global agriculture. *Philosophical Transactions of the Royal Society B: Biological Sciences, 363*(1491), 495–515. https://doi.org/10.1098/rstb.2007.2166. Available from http://rstb.royalsocietypublishing.org/.

Mondal, P., & Basu, M. (2009). Adoption of precision agriculture technologies in India and in some developing countries: Scope, present status and strategies. *Progress in Natural Science, 19*(6), 659–666. https://doi.org/10.1016/j.pnsc.2008.07.020. Available from https://www.sciencedirect.com/journal/progress-in-natural-science-materials-international/issues.

Pawlak, J., Pellizzi, G., & Fiala, M. (2002). On the development of agricultural mechanization to ensure a long-term world food supply. *Agricultural Engineering International: The CIGR Journal of Scientific Research and Development*. Invited Overview Paper.

Rijk, A. G. (1997). Agricultural mechanisation strategy. http://www.un-csam.org.

Rotimi, A. O. (2010). Measurement of agricultural mechanisation index and analysis of agricultural productivity of some farm settlements in Southwest Nigeria. *Agricultural Engineering International: CIGR.Journal, 12*(1), 125–134.

Rukuni, M., & Eicher, K. (1994). *Zimbabwe's agricultural revolution*. Harare: University of Zimbabwe Publication Office, Harare University of Zimbabwe Publication Office.

Ulusoy, E. (2013). Agricultural mechanisation in Turkey. *IERI Procedia, 5*, 41.

Viegas, E. (2003). Agricultural mechanization: Managing technology change. *Agriculture: New Directions for a New Nation East Timor, 113*, 32–44.

Wang, M. (2001). Possible adoption of precision agriculture for developing countries at the threshold of the new millennium. *Computers and Electronics in Agriculture, 30*.

CHAPTER

Primary tillage implements

2

Chapter outline

All the soil working tools that are used to open the soil after a cropping season or open a virgin soil are termed as primary tillage tools (Bell, 2005; Bernachi, Haman, and Kanafojski, 1972; Culpin, 2013; Hunt, 2008; Kepner, Bainer, and Barger, 1978; Klenin, Popov, and Sakun, 1970; Smith, 1965; Srivastava et al., 2006). The types of ploughs such as the mould board, disc, indigenous, and chisel ploughs are classified as primary tillage tools. The operations made after using these tools to manipulate the soil to a finer tilth, form the tilled land to our requirement, and do earthing-up operations are some of the secondary tillage operations. But many times the demarcation between the two operations is rather thin depending on the required soil tilth or landform desired by the farmer. Today, shallow ploughing with a powered rotavator is used extensively, though the direct use of this equipment is not advised as

Machines of the Farm. https://doi.org/10.1016/B978-0-443-32906-7.00002-5

a primary tillage operation. A rotovator is not a full-fledged rotary plough but rather a less-powerful piece of equipment developed for breaking soil clods after ploughing to bring forth a fine tilth. However, the farmer whose field soil is friable and loamy is using the rotavator as primary tillage equipment without using any other plough.

Objectives of tillage

The objectives of tillage are:

- to control weeds by exposing and burying them in tilled soil. Reasonably well-tilled soil not only kills weeds but also discourages weed germination, at least for some time.
- to thoroughly invert and incorporate the crop residues in the soil so as to decompose them into manure for the crops to be sown.
- to bring forth a desirable soil tilth suitable for a seedbed. A finer soil tilth with as few clods as possible allows good infiltration and retention of water, provides adequate air entrapment, and gives better seed germination and plant stand through the least resistance to root penetration. It also allows for forming the soil to requirements as well allowing further operations such as sowing to be effective.
- to disturb the soil adequately to bring trash such as crop residue from the previous season to the surface. This process provides mulch on the surface and reduces water evaporation in arid agriculture environments.
- to minimize soil erosion by following such practices as contour tillage, basin listing, and bunding.
- to form the land in proper slope and geometrical configurations to support irrigation. Making ridges and furrows is one such operation.
- to incorporate applied manure, fertilizers, and other soil inputs into the soil.

2.1 Mould board plough

Though there are many different types of ploughs that have been developed since the day humans began to till soil, the mould board plough is known to be "the" plough, owing to the advantages it offers. The first and foremost is that it offers a rectangular furrow and, hence, contributes to providing complete coverage of the land to be tilled in fewer passes. Moreover, it provides a relatively good inversion and pulverization of the soil being tilled. It is a suction-assisted and rather well-behaved plough even when it is drawn by animals. Since it is a plough used for opening up the soil after a cropping season, it is classified as a piece of primary tillage equipment. Although studies have shown that, in some crops and under certain climatic conditions, ploughing of soil does not provide any marked advantage, the mould board plough remains the most used primary tillage implement globally.

The basic unit of a mould board plough is the plough bottom. Figure 2.1 shows the construction of a mould board plough bottom. The share is the part that has a cutting edge and a share point. The share point is the first point that penetrates the soil and moves forward. It is thus susceptible to maximum wear. The cutting edge of the share is oriented at an oblique angle to the forward travel of the plough and, hence, progressively penetrates and cuts the soil to the full width of the plough. The throat width of a plough as measured perpendicularly to the travel direction is the projected distance between the share point and the wing of the share. When the share cuts this width of the soil and penetrates down to a working depth, it effectively cuts a rectangular section of the soil. The soil slice that is cut and flows over to the mould board is known as the furrow slice.

The size of a mould board plough bottom is the width of the furrow that it cuts. The most common sizes are 30 cm, 35 cm, and 40 cm. The mould board is aligned in flush with the share and the part that receives the furrow slice from the share and turns it. The mould board is the part of the plough that has an intricate curvature both perpendicularly as well as in the direction of travel, thus allowing the furrow slice to be turned upside down as well as bent and broken into smaller fragments by these curvatures, respectively. The turning action "inverts" the soil slice upside down before throwing it down, so that the crop residues and weeds on the soil surface are thrown upside down. This action is known to be the soil "inversion" of the mould board. Most of the inversion process is brought out by the upper region of the mould board. The soil traveling on the mould board due to the forward speed of the plough is, finally, thrown from the top of the mould board into the furrow made by the

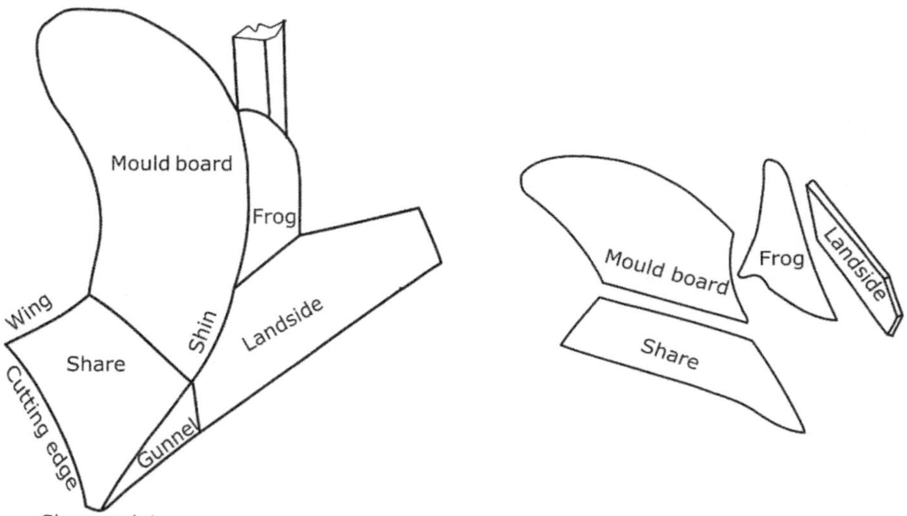

FIG. 2.1 Parts of a Mould Board Plough

Components of a plough body

No Permission Required

previous bottom or pass. The direction of the throw depends on the ploughing speed and the shape of the mould board.

Simultaneously, as the furrow slice runs on the mould board, it is bent and fragmented into finer particles, known as the "pulverization." Hence, it can be understood that the shape and curvatures of the mould board play a key role in the inverting and pulverizing performance of the plough. Different types of soils and soil conditions demand different mould board shapes to give the desired inversion and pulverization of soil. Mould boards are classified as sod, stubble, general-purpose, and high-speed types based on their configurations to deal with each type of soil. The shapes of mould boards can be very diverse to contend with each type of soil. However, the above classification is a broad one. The stubble type of a mould board (Figure 2.2) is curved more abruptly along the top edge for the slice to be thrown with a force on to the ground, thus pulverizing it towards loosening the stubbles or crop residues from the previous crop season. This is, hence, more suitable for the first ploughing of the soil cultivated in the last season. The sod bottom is most suited for soils that had been uncultivated for long periods of time with the surface soil held up firmly by the grass roots (sod). The shape of this bottom has deeper curvatures to bend and break the soil slice more. The general-purpose board is a combination of the sod and stubble types and is effective for both soils.

The high-speed bottom (Figure 2.2) has slightly less curvature than the general-purpose plough so that the speed of travel, rather than the curvature, breaks the soil more by the velocity of the furrow slice exiting the board. Slat mould boards (Figure 2.2) are specially shaped, with slots running over the surface. This reduces the sticking of soil to the mould board; they are used in sticky clay soils. A mould board for which the soil falls off the mould board easily without sticking is said to be "scouring" well.

For mould board ploughs without support wheels, such as animal-drawn ones, a vertical clearance is deliberately provided (Figure 2.3) at the joint between the share and the landside, so that the share point is oriented with a downward inclination to the soil surface. This clearance is called the vertical suction and allows the mould board plough to penetrate into the soil on its own. It is hence known to have suction-assisted penetration rather than weight-assisted penetration like the disc plough. Similarly, the bottom is also fashioned with a horizontal suction (clearance) between the furrow wall and the junction between the share and landside. This aids the bottom to penetrate sideways into the furrow wall easily. Tractor-drawn trailed ploughs and mounted ploughs do not have built-in suctions and are adjusted to have these clearances by the operator. The rear furrow wheel is raised or lowered to pitch the bottoms forward to provide for the vertical suction in trailed ploughs. In mounted ploughs, the top link of the three-point hitch is adjusted to yield this clearance. Generally, the vertical suction is about 5 mm to 10 mm and the horizontal suction is 10 mm to 13 mm.

The landside is a flat piece of metal that slides on the furrow wall while the plough is working. The mould board plough, by virtue of action, is an unsymmetrical soil working tool because it cuts the soil and throws it to one side. Since the soil

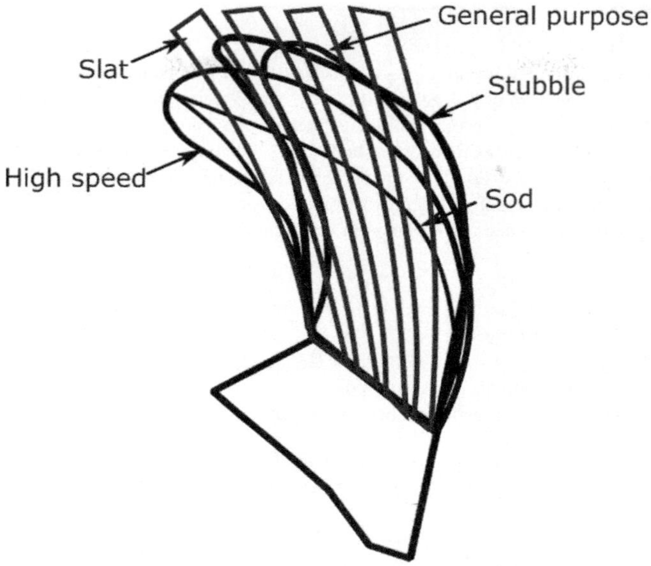

FIG. 2.2 Types of Mould Board Surfaces

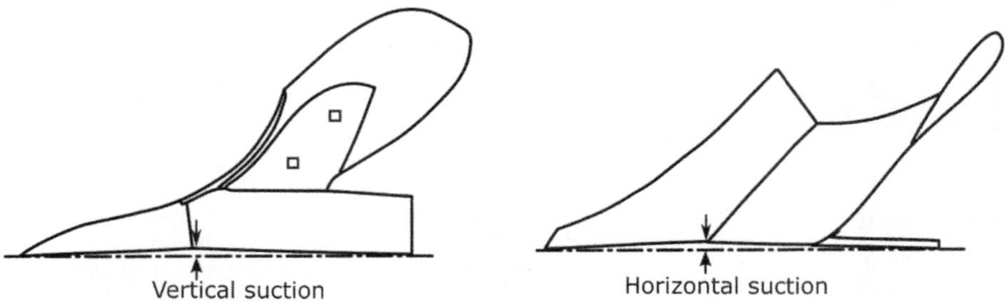

FIG. 2.3 Vertical and Horizontal Suction of Trailed Ploughs (without Support Wheels)

flows on the mould board in a lateral direction to travel, it pushes the mould board against the furrow wall. The landside hence rests on the furrow wall and counteracts this side thrust of the soil in a lateral direction to travel. The absence of a landside will hence not allow the plough to move forward in a straight line. However, this action of landside sliding on the furrow wall causes considerable frictional forces in the travel direction that adds up to the pull to be applied to the plough. Gunnel is that part of the share that is aligned with the landside and does the same job as that of the landside, except that it is pre-built into the share. In modern tractor-drawn ploughs, the gunnel is absent and instead the landside is extended forward accordingly. The frog is the part that fastens all the different parts together.

Box 2.1 Example 2.1

Example 2.1. A 25 cm mould board plough has a board, the mean path of which has soil flow, as seen in the plan, in the form of $y = ae^{bx} = 10e^{0.025x}$, with x being the coordinate in the plough's travel direction and y being the lateral coordinate, both of which are measured from the share point as origin. When the plough is being operated at 3 km h^{-1}· find out the exit velocity of the soil slice from the mould board and also how far it travels before falling into the next furrow. Assume that the soil exits the plough at a height of 35 cm from the share point.

The mean soil path as seen in the plan is in the form of:

$y = ae^{bx}$

$y = 10e^{0.025x}$

Differentiating $\frac{dy}{dt} = a\frac{d}{dx}\left(e^{bx}\right)\frac{dx}{dt} = ab\left(e^{bx}\right)\frac{dx}{dt} = by\frac{dx}{dt} = (0.025)y\frac{dx}{dt}$

i.e., $v_y = byv_x = 0.025yv_x$.

The speed of ploughing $= v_x = 3.0$ km h^{-1} = 83 cm s^{-1}

At the end of path where the soil is ejected,

y = 35 cm (substitution to be in cm, because the empirical equation is in cm).

Then, $v_y = 0.025 \times 35 \times 83 = 73$ cm s^{-1}

$\therefore$ The resultant velocity of soil ejection $= \sqrt{0.83^2 + 0.73^2} = 1.1$ ms^{-1}

The direction of the resultant = tan^{-1}(0.73/0.83) = 41 from the travel direction

$v^2 = u^2 - 2gs$

where v is the final velocity when it reaches ground = 0,

u is the initial velocity (the soil ejection velocity) = 1.1 ms^{-1}

g is the acceleration due to gravity = 9.81 ms^{-2}

and s is the distance that the object reaches, m

$0 = u^2 - 2gs$

$$s = \frac{u^2}{2g} = \frac{1.1^2}{2 \times 9.81}$$

$= 0.061$ m

Since the effective width of the plough is 25 cm and the path extends to 35 cm laterally from the share point, the soil exit point from the mould board is extending laterally into the preceding furrow on the right by 10 cm.

As per the calculations, the soil slice exiting the path travels 6.1 cm at a direction of 41° from the travel direction, implying that the lateral distance it travels = 6.1 sin(41) = 4 cm.

Then, the soil slice falls at 4 + 10 = 14 cm distance into the preceding furrow at this speed of ploughing.

The example problem in Box 2.1 shows how the shape of a mould board plough influences the soil throw from the board's end. The distance to which the soil is thrown affects the placement of the dug soil into the adjacent furrow being made by the plough.

2.1.1 Materials of construction of shares

Soil that has a very abrasive nature wears off the surface of the soil-working parts considerably. Especially, the cutting edge of the share and the share point are the spots of maximum wear. Soft-centre steel is the choice for clay and clay loams, providing good scouring, whereas carbon steel with a carburized surface provides better shock resistance when the share hits stones. Chilled cast-iron shares give the best wear resistance in sandy soils. Solid or soft-centred steel shares are sharpened by forging and grinding. But chilled cast-iron shares used in small animal-drawn ploughs are ground or replaced. Hard-faced alloy steels are used nowadays to reduce the rate of wear.

2.1.2 Attachments of mould board ploughs

A rolling coulter (Figure 2.4) is a circular device with a sharp cutting edge on the periphery. It is mounted on the plough frame such that its axis of rotation is right above the share point, with its edge a little bit outwards from the furrow wall that the plough will make. It makes a precut in the soil where the share point and landside travel. It makes a perfect furrow wall as well as cutting the trash along that line.

The depth of penetration of the coulter is usually set to be half the working depth of the plough. Coulter diameters usually vary from 30 cm to 40 cm. Sometimes a notched coulter is used; this can cut the trash better. A jointer (Figure 2.4), another accessory that looks more like a small plough bottom, is mounted in front of the plough to cut a narrow but shallow furrow slightly away from the furrow wall. It moves the surface trash and roots from the surface of the plough path, so that trash coverage by inversion is complete. Weed hooks are used to push away the tall weeds in front of the mould board to improve coverage. These are simple rods protruding from the plough frame to perform this action.

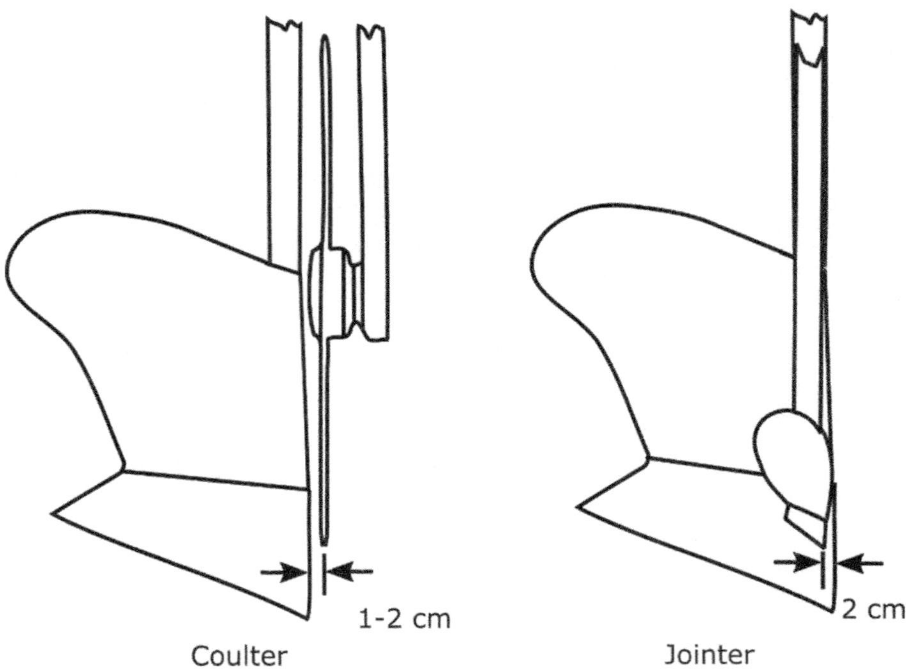

FIG. 2.4 Attachments of Mould Board Ploughs

2.1.3 Types of mould board ploughs

Generally, mould board ploughs are fabricated to throw the furrow slices to the right of the travel direction. These one-way bottoms are known as right-hand bottoms. The use of one-way bottoms will always end up in sets of dead and back furrows in the field; these need to be filled or broken, respectively. The meaning of dead and back furrows will be explained later in detail when learning about systems of ploughing with soil working tools.

2.1.4 Tractor-operated mould board ploughs

Based on the method of connecting (hitching) the plough to the tractor, the methods are classified as trailed, semi-mounted, and integrally mounted. Mounted ploughs have one to five such bottoms (Figure 2.5), and trailed ploughs may have many such bottoms depending on the power of the tractor used to tow them (Figure 2.6). Mounted mould board ploughs are totally supported by the tractor's three-point linkage on the rear. They can be lifted and lowered by the tractor's hydraulic system.

This system is detailed in the chapter on the vertical hitching of mounted mould board ploughs (Figure 5.16 and 5.17). The side draft or the lateral thrust of the soil flowing on the plough bottoms is fully taken up by the bottoms′ landsides. Trailed mould board ploughs are totally supported on the ground by a set of three wheels; they are pulled behind the tractor and connected by a drag bar to the draw bar of the tractor (Figure 2.6). The three support wheels are the two furrow wheels, one on the front and another on the rear, and a land wheel. This is detailed in the chapter on the horizontal hitching of trailed ploughs (Figure 5.8). The rear furrow wheel runs in the last furrow made in the present ploughing run. The side thrusts on the plough

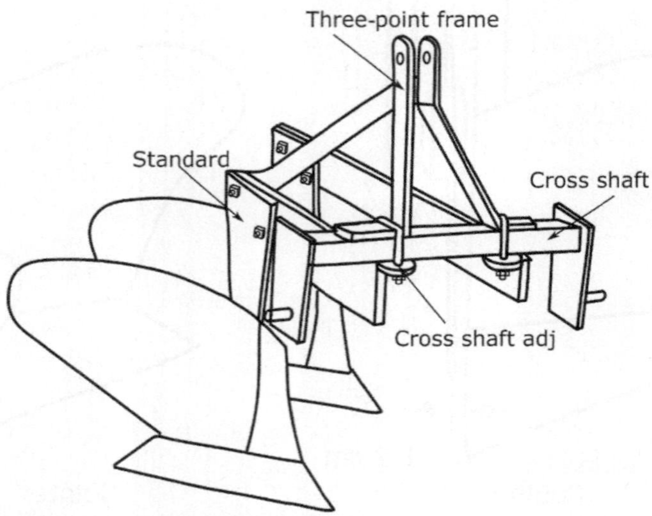

FIG. 2.5 Tractor-Mounted Mould Board Plough

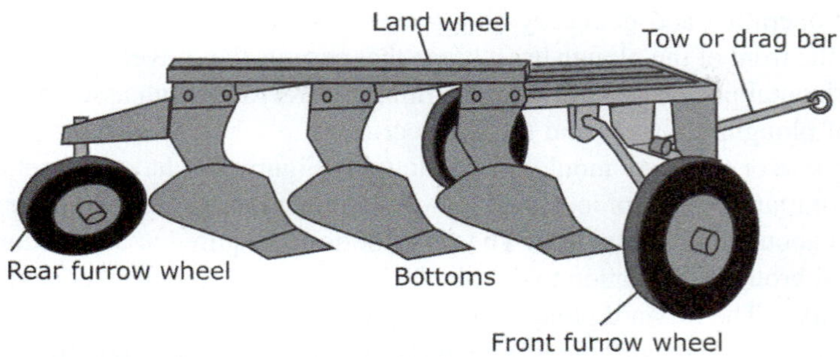

FIG. 2.6 Tractor-Drawn Trailed Mould Board Plough

FIG. 2.7 Tractor Semi-Mounted Reversible Mould Board Plough

New Holland Agriculture, CNH Industrial America LLC

bottoms are taken up by their own landsides. However, the rear furrow wheel is also slightly inclined towards the furrow wall so that it can help in taking up the side thrust of the whole plough more effectively. The land wheel rides on the unploughed soil (Figure 5.8) and is located slightly forward of the front furrow wheel. The front furrow wheel rides in the last furrow made in the previous adjacent pass made by the implement, thus aligning the present pass to the previous one. Adjustments are provided on all three wheels independently to adjust the depth of the ploughing, their vertical suction, and the level of the plough frame as a whole. When the rear furrow wheel's height is adjusted, the whole plough can be tipped forward or backward, thus changing the vertical suction of the bottoms. Similarly, all three wheels are raised or lowered to adjust the depth of the ploughing and the plough frame's horizontal level. Additionally, hydraulic cylinders remotely controlled from the tractor can lift and lower the plough bottoms out of the soil for transporting the trailed plough out of the field.

A semi-mounted plough is one where the front of the plough is mounted on the three-point linkage of the tractor and supported by the tractor only on the front. It is provided with a rear furrow wheel that support the tail-end of the implement (Figure 2.7). Generally, these ploughs have more than six bottoms, covering larger widths of operation and pulled by high-powered tractors. The three-point hitching frame on the front of the plough has a hinge that permits the swivelling of the plough in the horizontal plane. This allows for turning the tractor plough assembly easily at the end of ploughing passes and the field corners.

Reversible or two-way mould board ploughs (Figure 2.7) have one set of right-hand bottoms and one set of left-hand bottoms arranged on a common frame that can be rotated about a horizontal axis. The set of bottoms required at any instant can be rotated and brought into action to do the ploughing, while the other set is idle on the top of the axis. The forward ploughing pass is done with the right-hand bottoms and the return pass is done with the left-hand bottoms. The left-hand bottoms throw the furrow slice into the furrows already made by the right-hand bottoms. Then, the next forward pass with right-hand bottoms will once again throw the furrow slice into the furrows made by the left-hand bottom. This cycle of alternating the bottoms will give a level, ploughed field with all the furrows filled with loose soil and completely devoid of dead and back furrows. This process of ploughing with a "two-way" plough is known as "one-way" ploughing. Since two exclusive sets of opposing bottoms are used in a two-way plough, they are more expensive than one-way ploughs. The reversing action of the plough is actuated from the tractor through remote hydraulic rams provided on the plough. This system of ploughing is explained in more detail in Section 2.3.2.

2.2 Disc ploughs

The disc plough is an ingenious invention that was intended to replace the mould board plough in an attempt to reduce the draft requirement of a primary tillage tool. It has a concave disc rolling about a stub axle and is set at an inclination to the direction of travel as well to the vertical plane. The disc, while cutting and throwing the soil, rolls on its own axis on a set of anti-friction bearings; it was expected to reduce the draft requirement because of the reduction in friction with soil as compared with the mould board plough. However, unfortunately, the disc plough does not have adequate suction such that it can penetrate into the soil on its own. It is made to penetrate by the addition of adequate weight. This addition of weight offsets the advantage of the reduction in frictional forces and makes it into a tool requiring considerable draft. However, the disc plough is known to work in soils in which the mould board plough cannot perform well. It can outperform a mould board plough in the following soil conditions:

1. heavy, clay soil that is sticky, which does not scour well.
2. relatively dry and hard soil in which a mould board plough cannot penetrate by its suction.
3. soil with a lot of thick roots and stones in which the disc can ride over the stones and cut the roots.

4. peaty soils that are very loose and in which a mould board plough cannot invert the soil properly.
5. soil that is to be relatively ploughed more deeply than with mould board ploughs.

This is a plough highly suitable for soils of rain-fed land as well soils that are flooded and clayey (Kepner, Bainer, and Barger, 1978; Summer, Chen, and Bierl, 1983). However, by virtue of its shape, the furrow cut by a disc plough is not rectangular and is part of an elliptical section. This is why considerable overlap of the working widths of subsequent bottoms is provided: to reduce unploughed soil strips between furrows.

As already explained, a standard disc plough has its disc tilted backward from the vertical plane at an angle known as the "tilt angle" (Figure 2.8). The disc is also set with the plane of its face oriented at a horizontal angle, known as the "disc angle" (Figure 2.8), to the direction of travel. The disc angle alters the width of the furrow it makes. The rolling of the disc while cutting happens because of the disc angle making the furrow slice move on the rolling disc. The tilt angle is typically between 15° and 25°, and the disc angle is typically between 42° and 45°. An optimal tilt

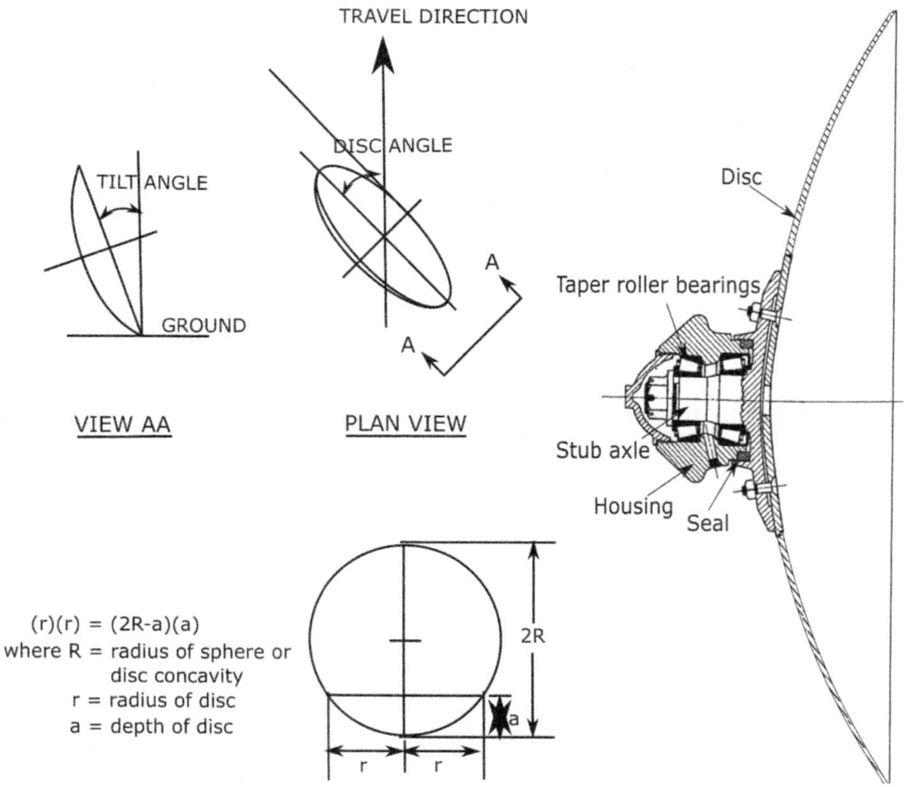

FIG. 2.8 Typical Disc Bottom Parameters and Assembly

Box 2.2 Example 2.2

Example 2.2. A 50 cm steel disc has a concave depth of 10 cm. Find the disc's radius of curvature.
Radius of the disc, $r = 50/2 = 25$ cm and $a = 10$ cm
We know that $(2R\text{-}a)\ a = r^2$
$(2R - 12) \times 10 = 25^2$
$2R - 10 = 62.5$
$R = 36.75$ cm.

angle aids the disc in penetrating with the assistance of weight. The diameter of a standard disc is between 600 mm and 900 mm. These rolled-steel discs are sections of a sphere with a radius of curvature ranging between 450 mm and 600 mm. The relationship between the radius of the curvature of the disc, R (i.e., the radius of the sphere from which the disc is a segment); the radius of the disc, r; and the depth of the disc is illustrated in Figure 2.7.

The example problem in Box 2.2 uses this relationship to calculate the radius of curvature of the disc blade.

Scrapers with a spoonlike surface are positioned inside the disc such that they can scrape off the soil from the inner surface of the disc as they cut and roll in the soil. A scraper is a regular part of the disc plough and helps not only in scraping but also in inverting the soil by impeding the flow of soil over the rotating disc.

As the disc blades cannot penetrate on their own, they are made to be heavy, are attached on heavy frames, and sometimes are added with extra weights in order to make them penetrate. As explained later, large thrust forces act on the discs and, hence, are mounted on tapered roller bearings (Figure 2.8) to bear the axial thrust as well the radial forces acting on them while they rotate on their axis and cut the soil. The housing that carries the stub axle and the disc is mounted to a standard or drop, which extends downward from the heavy plough frame. The housing is adjustable relative to the standard to adjust the disc and tilt angles.

2.2.1 Tractor-operated disc ploughs

Based on the method of connecting (hitching) the plough to the tractor, these methods are also classified as trailed, semi-mounted, and integrally mounted types similar to the mould board ploughs. Trailed disc ploughs are also totally supported on the ground by a set of three wheels and pulled behind the tractor (Figure 2.9). Arrangements are provided for adjusting the tilt and disc angles of the disc, which influence the penetration and width of operation, respectively. The three support wheels are the two furrow wheels, namely, the front and rear ones, and a land wheel. This arrangement is similar to the trailed mould board ploughs except that the front furrow wheel running in the last furrow of the previous ploughing pass is connected to the hitching system to guide and turn the plough. This is detailed in the chapter on the horizontal hitching of trailed disc ploughs. The rear furrow wheel runs in the last

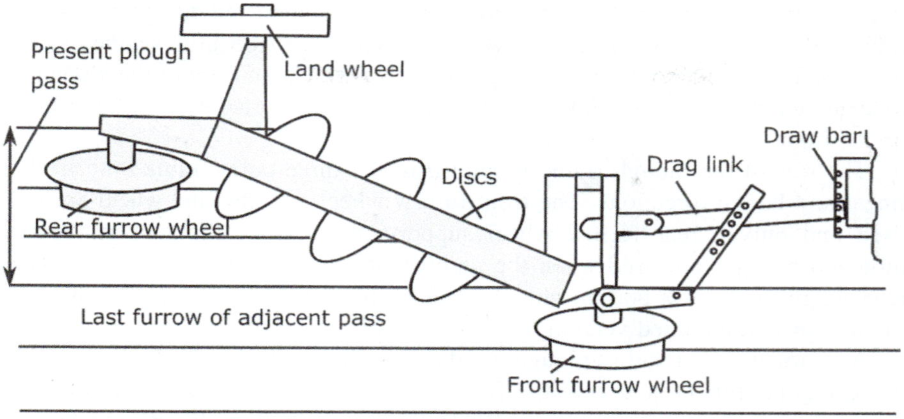

FIG. 2.9 Tractor-Drawn Trailed Disc Plough

FIG. 2.10 Three-Bottom, Tractor-Mounted Disc Plough

India Agrovision Implements Pvt. Ltd

furrow made in the present ploughing run. Both furrow wheels are more inclined than the mould board ploughs' furrow wheels since the side thrust on disc ploughs is to be compensated only by the furrow wheels, unlike the mould board bottoms, which have landsides to do that job. The land wheel rides on the unploughed soil and is located slightly forward of the front furrow wheel. Just like the trailed mould board ploughs, all three wheels are raised or lowered to adjust the depth of ploughing and the plough frame's horizontal level. Additionally, hydraulic cylinders remotely controlled from the tractor can lift and lower the plough bottoms out of the soil for transporting the trailed plough out of the field.

Integrally mounted disc ploughs (Figure 2.10) are totally supported by the three-point hitch on the rear of the tractor such that they could be lifted and lowered by the tractor's hydraulic system. A rear furrow wheel on the mounted disc plough set to ride in the last furrow being made serves to counteract the side thrust on the plough bottoms. The semi-mounted disc plough is one for which the front of the plough is mounted on the three-point linkage of the tractor and supported by the tractor only on the front.

The rest of the plough frame is hinged to the three-point frame both in the linear and lateral directions. The front furrow wheel and the land wheel are not used and only a rear furrow wheel supports the rear of the plough. Semi-mounted ploughs are used when the number of plough bottoms are more. The rear weight of a multi-bottom plough, which has more bottoms, may pitch back the tractor in a mounted version.

As learned earlier, a disc plough, unlike a mould board plough, does not make a rectangular furrow and cuts the furrows in the shape of a skewed ellipse's segment (Summer, Chen, and Bierl, 1983). So, a multi-bottom disc plough leaves out unploughed soil in between the bottoms. The example problem in Box 2.3 shows the extent of such unploughed strips left out between bottoms when using the disc plough and shows how the projected shape of the disc having disc and tilt angles is causing this.

Box 2.3 Example 2.3

Example 2.3. Draw the front view (projection) of a disc plough bottom of 60 cm diameter set at a 45° disc angle and 15° tilt angle. Graphically, find the area of furrow ploughed by the single-disc bottom operating at a 15 cm depth. (You may use graph paper to solve the problem or use a freeware 2D CAD such as the LibreCad to draw and solve.)

Draw an inclined line at 45° representing the disc angle. On that base line, draw the side view of a disc of 60 cm diameter after selecting an appropriate scale. The disc's face is to be oriented at 15° to the base line, representing the tilt angle with its bottom touching the base line. The disc's concave depth at the centre may be assumed to be about 15% of the disc diameter (this assumption will not influence the solution). Then, the line representing the disc's face may be divided into equal parts with three parts on either side of the disc centre, marking them as "a" to "g" with "d" as the centre. Draw projection lines that are perpendicular to the base line through these marks and extend them on both sides of the disc's sideview. You can see that the projected width from the top edge of the tilted disc to its bottom edge, i.e., the projected width "ag" perpendicular to the base line, will be disc diameter × cos(15°) = 58 cm. As illustrated in the example plot, the front view of the tilted disc, as seen perpendicular to the disc's face, will be an ellipse of a height of 58 cm (the projected width ag) and a width of 60 cm (the diameter of the disc). An ellipse of the said dimensions is drawn, as accommodated, between the projection lines drawn from "a" and "g." Any standard graphical method of drawing this ellipse may be used for drawing the ellipse.

After drawing the ellipse, project the points "a to g" over to the drawn ellipse and mark the corresponding intersections with the ellipse's half edge facing the base line. The projection lines may be extended to the opposite edge, but mark them with a different set of letters, viz., "m to q," which will form the other projected half of the final solution. (Only the projection lines of one half are shown in the example plot to make the explanation clutter free!) It may also be noted that the points b & q, c & p, d & o, e & n, and f & m overlap one over the other in the side view, since they lie on the same line.

Box 2.3 Example 2.3—cont'd

Now, project the points on the ellipse parallel to the base line (45° line) and turn them all perpendicular to the base line. Refer to the example plot on how this is done. The projection lines corresponding to the points "a to g" are drawn thus. Then, the points "a to g" on the sideview of the disc are projected parallelly to the base line. The points of intersections of the projection lines emanating from each corresponding point (line from "a" on ellipse and line from "a" on disc sideview) are marked one by one. These points, when joined by a smooth line, will form one symmetrical half of the tilted disc's plan (the half facing the disc's sideview). The other half can then be completed likewise as projected from the points "g, m, n, o, p, q, a." This would make the full plan of the tilted disc.

Since we proceeded with the projections relative to the base line representative of the disc angle, the plan of the tilted disc is formed parallelly to this base line. This profile is then to be projected at 45° downward to represent the view along the plough's direction of travel. This will be the front view of the disc vertically tilted at the given tilt angle and set at the given disc angle. To do it, the points "a to g" from the newly drawn plan are projected downward along the direction of travel. The points "a to g" on the disc's side view are once again projected downward and turned perpendicularly to the direction of travel. The intersections of the corresponding points ("a" to "a", "b" to "b", etc.) form one half of the front view of the disc set at tilt and disc angles. The points may be joined by a smooth line to get half the view. Similarly, the points "m to q" are also projected and the other half of the view is drawn.

Now, a horizontal line, representing the ground level, is drawn at a height of 15 cm from the bottom-most point (bottom of the furrow) of the disc's front view. The furrow area of the plough below this line is measured by counting the graph squares and applying the scale. (Display the grid in freeware to count the squares.)

Calculated furrow area = 364 cm^2.

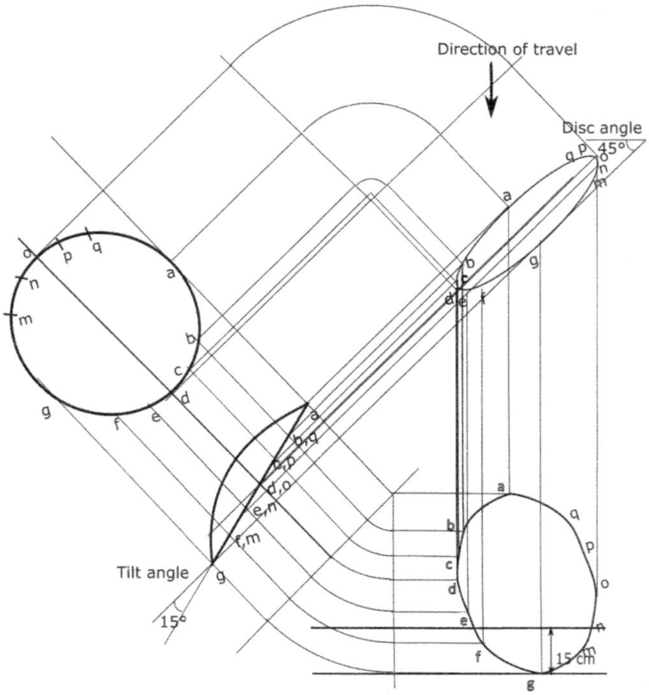

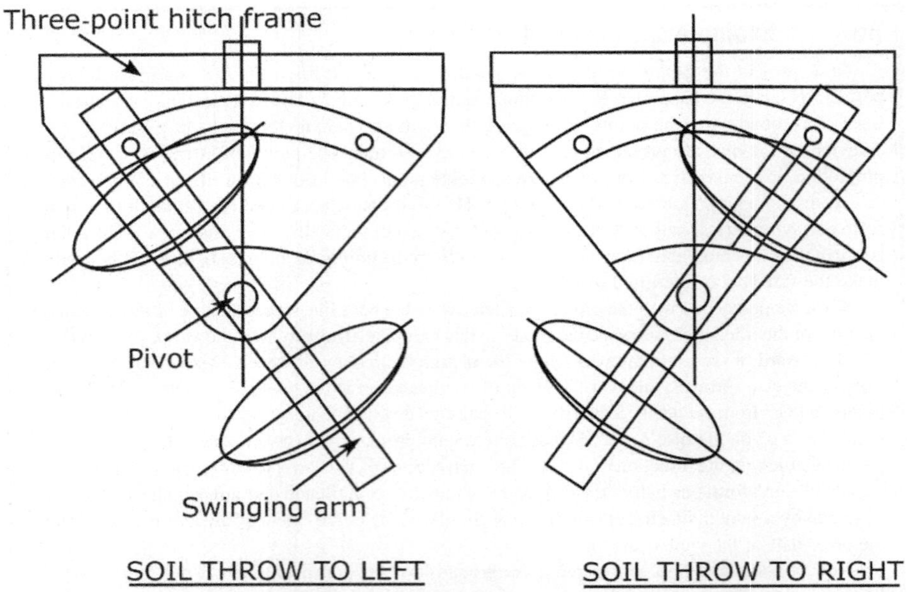

FIG. 2.11 Tractor-Mounted Reversible Disc Plough

Figure 2.11 illustrates an integrally mounted two-way or reversible disc plough. The two bottoms are mounted on a hinged frame that can swing the plough bottoms from one position to the other. The bottoms, which act as a right-hand bottom in one position, can be changed to left-hand bottoms by swinging the bottoms to the other side. A manual lever or hydraulic arrangement provides this sidewise movement on the go.

2.2.2 Accessories of disc ploughs

Disc plough bottoms are always provided with scrapers so that a moderate pulverization and inversion of the furrow slice is obtained. Additional weights, which can be mounted on the frame, are used sometimes in hard soils to help the ploughs penetrate better and to prevent the furrow wheels from riding out of the furrows.

2.2.3 Vertical disc ploughs

The vertical disc plough looks similar to a heavy-duty disc harrow (explained in the chapter on secondary tillage implements) rather than a standard disc plough. The discs are mounted on a common axle or gang bolt with spacing spools in between to provide uniform spacing between discs. The whole gang assembly locked by a gang bolt rotates as a single unit on support bearings. Since they are on a gang, the discs do not have any tilt angle. The discs of a vertical disc plough are smaller than that of a standard plough and larger than the disc harrows. Greater widths of

Box 2.4 Example 2.4

Example 2.4. A vertical disc plough, with six numbers of 60 cm discs spaced at 22.5 cm, is operating at a depth of 15 cm with a gang angle of 45°. Find the percent of effective furrow area cut by the plough relative to the theoretical furrow area. (You may use graph paper to solve the problem or use a 2D CAD means, such as the LibreCad, a freeware, to draw and solve.)

This can be found graphically by projecting the shapes of a pair of disc blades onto a plane inclined at 45° to the direction of travel.

Select an appropriate scale and draw a circle equivalent to a 60 cm diameter representing a single disc. Mark mutually perpendicular centre lines on the disc at 45° to horizontal. Draw three parallel lines to the first centre line of the disc, one at the bottom of the disc, the next one at 15 cm from the bottom corresponding to the working depth in soil, and the third in between the two. Mark the points of intersection of all four lines with the peripheries of the disc as 1, 2, 3, and 4.

Now, draw an inclined line of 45° (gang angle) to the side of the circle and project the disc onto this line, representing the plane of the disc blades as seen from the plan view of the gang. Draw the shape of the disc arbitrarily without changing the projected width. Mark the direction of travel. Draw a similar disc at 22.5 cm spacing from the first disc on the line, representing the projection plane. Draw four parallel horizontal lines below the projected discs, one corresponding to the bottom of the discs in the front view, the next one at 15 cm from the bottom corresponding to the working depth in soil, the third in between the two, and the fourth at 30 cm from the bottom, corresponding to the centre of the disc. Then, project the points 1, 2, 3, and 4 onto the projection plane and to the four lines at the bottom and name them accordingly. Then, connect the points 1, 2, 3, and 4 by a smooth curved line representing the disc edge as viewed from the front of gang.

Similarly, project the disc edges of the other half of the disc onto the bottom four lines, thus completing the bottom half of that disc.

Project the other disc also similarly or copy the edge profile of the first disc between the two vertical projection lines representing the second disc.

The area where the discs are not cutting the soil at all are marked (shaded in the figure) and measured by counting the graph squares and applying the scale. (Invoke the grid in freeware to count the squares.) The area marked on the left extreme in the demonstrative illustration (118 cm^2) is the uncut soil area at the extreme end on the gangs. The area marked next (11 cm^2) is the uncut soil between two discs on the gang.

Since there are six discs, there will be two such uncut areas on the extreme ends of the gang and five areas of uncut furrow in between the discs, making the total area of uncut soil–metal

$= 2(118) + 5(11) = 291\ cm^2$.

The theoretical rectangular furrow area of the gang with six discs set at a 45° gang angle

= (projected width of last disc + 5 times × projected width that is not overlapping with next disc) × working depth $= (60 \sin(45°) + 5 \times 22.5 \sin(45°)) \times 15 = 2587\ cm^2$

where 22.5 m is the spacing between discs on the gang.

∴ Effective soil cut area of the furrow made by the disc gang = (2587 - 291) / 2587 = 88 %.

cut are obtained using these ploughs. The example problem in Box 2.4 shows the geometrical projection of a vertical disc plough's bottoms, which do not have a tilt angle. The reader is encouraged to compare this to that of the earlier example of a regular disc plough.

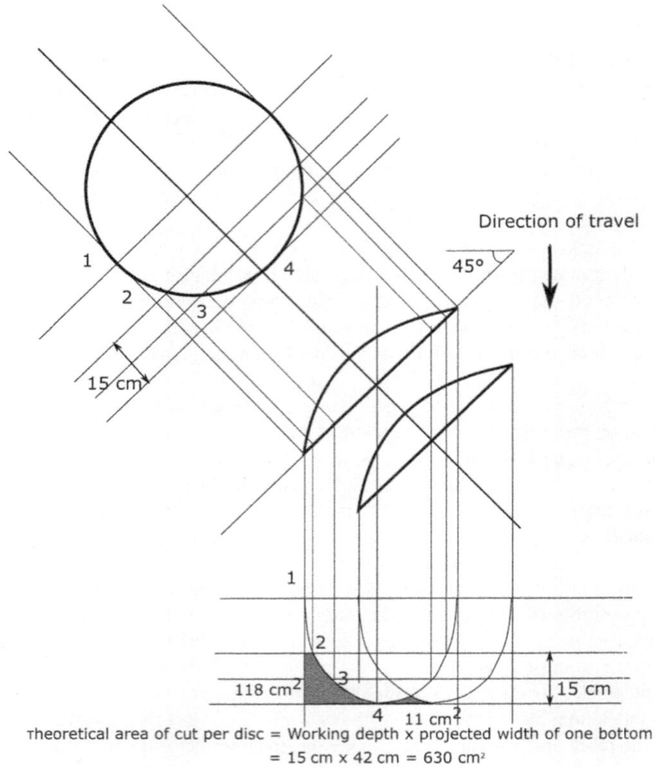

2.3 Systems of ploughing

2.3.1 Ploughing with a one-way plough

As already stated earlier, ploughing with a one-way plough will always result in a set of dead and back furrows . Let us learn why that is so. For instance, if we are using a right-hand mould board bottom, it digs a rectangular furrow and throws the furrow slice to the right of the cut furrow. Let us say the right-hand bottom is used to start the first pass from the centre of the field plot (Figure 2.12) and that the furrow slice falls on the unploughed land to the right (see the sectional view at the bottom of the illustration). At the end of the field (pass), the operator turns clockwise after raising the plough and running outside the field edge (in the headland, as explained in Section 2.3.3) to make a downward ploughing pass adjacent to the previous upward pass. However, the operator will not be able to plough the strip of unploughed pass on which soil from the previous pass was thrown. So, the operator leaves that strip and ploughs the adjacent strip, going downward, throwing the furrow slice on the unploughed strip already having soil on it. This implies that two layers of

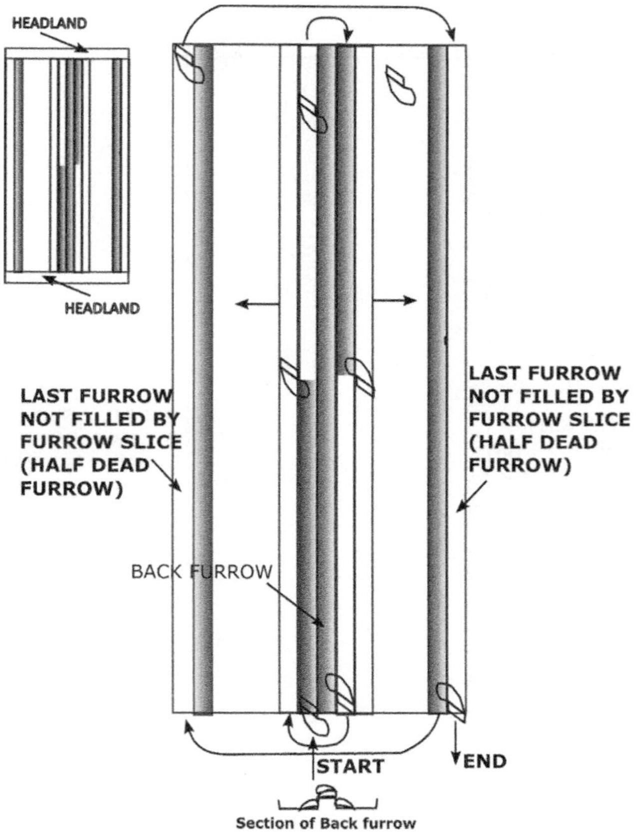

FIG. 2.12 Ploughing with a One-Way Plough: Starting at Centre

furrow slices are laid on the unploughed strip underneath. This strip is known as a back furrow and has unploughed soil beneath. Then, as the downward pass is over, the operator turns clockwise once again and ploughs the strip adjacent to the first upward pass. The furrow slice cut now is thrown into the furrow of the first upward pass, filling it with loose soil. At the end of this upward pass, the operator turns clockwise once again and ploughs the strip adjacent to the first downward pass. This process of starting from the centre of the field plot and going around in a clockwise direction fills all the furrows made in the last pass that were made in the same direction with loose soil. At the end of the process, the last furrow made will be empty, without being filled by loose soil. This can be viewed as half of a dead furrow, which is a side-by-side joining of two such empty furrows. This will be clear when we learn about the formation of a full dead furrow next. It may also be noted that the operator, while starting from the centre of plot, will not be able to turn anti-clockwise for ploughing the next pass because the ploughed furrow slice cannot fall into an empty

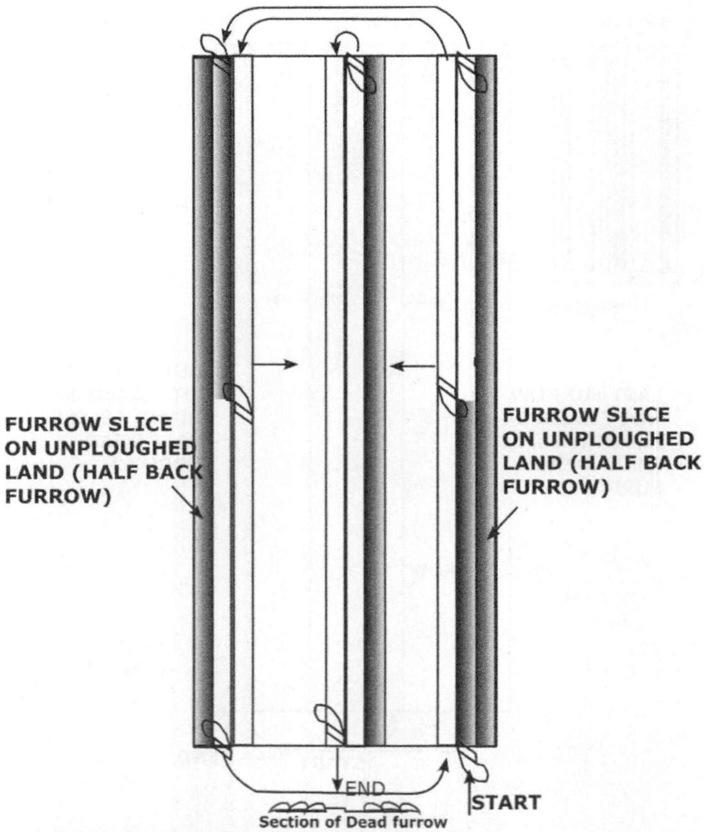

FIG. 2.13 Ploughing with a One-Way Plough: Starting at Side

furrow made by the previous pass in that process. Figure 2.12 shows the sequence of operations explained above as well the section of a back furrow formed at the centre of the plot. Most importantly, it should be noted that, when ploughing is done, unploughed land should exist on the left of the right-hand bottom to support the landside for compensating the side draft, making the plough proceed straight.

Now, let us see how a field plot can be ploughed by the same right-hand bottom in another way. Here, the operator proceeds (Figure 2.13) from the right-side edge of the plot and proceeds up. Here, the furrow slice is dumped on the unploughed strip outside the boundary of the field plot. At the end of the plot, the operator turns anti-clockwise, proceeds to the left-most edge of the field plot, and proceeds down along the left edge, dumping the furrow slice outside the boundary of the plot on an unploughed strip. When this downward pass is over, the operator goes to the right edge of the plot and ploughs the strip adjacent to the first upward pass. The furrow slice dug out now falls into the first furrow and fills it with loose soil. This ploughing process is continued in anti-clockwise direction until the operator reaches the centre of the field. Here, two empty furrows will be formed at the centre, adjoining each other. This pair of empty furrows is known as a "dead furrow."

While ploughing the very last of the furrow at the centre, the plough's landside is along the already ploughed furrow and may not get adequate support. Hence, the depth of ploughing is reduced to a three-fourths level so that the plough's landside can get a bit of support to go straight. It may be observed that, at the left and right edges of the plot, a back furrow, each with one layer of furrow slice, is formed. When the adjacent plots on the right or the left of this plot are ploughed using the above explained process, these back furrows will get another layer of furrow slice dumped over the previous layer. This makes the back furrow of width equal to one bottom but with two layers of soil on the unploughed strip. It may be noted that the dead furrow formed at the centre of the plot is twice the width of a bottom. Here, again, it may be noted that the operator may not be able to traverse in a clockwise direction starting from the left edge of the plot since the furrow slice will be dumped into the land to be ploughed at the centre. Figure 2.13 shows the sequence of operations explained above as well the section of a dead furrow formed at the centre of the plot.

The above processes are applicable for an implement with multiple bottoms. The effective width of a ploughing pass is then equal to the effective width of a single bottom multiplied by the number of such bottoms. Now, it is time for us to understand why the plough bottoms on an implement are offset linearly in the direction of travel. The reasons are:

- each bottom should have solid, unploughed land on the side of the landside to get support and compensate the side thrust of being an unsymmetrical tool;
- the furrow slice that is cut by the second bottom is to have sufficient time to fall into the furrow made by the first bottom and so on;
- the last bottom on the left will make an unfilled furrow in each pass.

Now, if we compare the two alternate processes of ploughing that were explained, it can be understood that the formation of dead or back furrows cannot be avoided at all when using one-way ploughs. It may be noted that, in the first system of ploughing from the centre to the side in a clockwise direction, the furrow slice in each pass seems to be "gathered" or thrown towards the centre of the plot. Hence, it is known as a ploughing system of "gathering," whereas, in the other system of traversing in the anti-clockwise direction, the furrow slice is thrown or "cast" away from the centre of the plot. Hence, it gets the name "casting." It may be noted that "casting" or "gathering" is meaningful only in the case of using one-way ploughs such as a mould board or disc plough and not for a symmetrical implement such as an indigenous plough or a chisel plough.

When using a left-hand plough instead of a right-hand plough, the sequence of operations will be quite opposite to that explained. When using a left-hand plough, the process starts from the centre and proceeds in an anti-clockwise direction to do the "gathering," and from the left side, proceeding in a clockwise direction to do the "casting." However, "casting" will always result in a dead furrow at the centre, and "gathering" will always yield a back furrow at the centre, irrespective of using a right-hand or left-hand plough.

It can be seen from above that, in all the explained systems of ploughing, the turning into the next subsequent pass gets tighter as it approaches the centre of the plot. The turning of relatively larger ploughs, such as three- or four-bottom ploughs, will be very difficult to negotiate as the ploughing passes approach the plot centre. Moreover, the idle run from one to the next pass outside the edge of the plot (in the headland) is long between the edges of the field, which makes the ploughing operation more time and energy consuming, thus reducing the overall work efficiency. This is very much true for relatively large fields, meaning that casting or gathering a large field as a whole is inefficient. So, a better "system of ploughing" is to divide the big field into strips of predefined widths, and then casting and gathering them alternately is more efficient. This system is known as "ploughing in lands" or the NIAE (National Institute of Agricultural Engineering, Silsoe, UK) system (Bell, 2003).

2.3.1.1 Ploughing in lands

Here, the relatively larger field plot is divided into lands of smaller width, so that the distance of transit between passes is reduced. The width, "w," of each land is to be multiples of the effective width of the plough being used and approximately 14 times.

Figure 2.14 illustrates how the field is divided into lands and also the sequence of operations of this system of ploughing. The field is laid out in such a way that the first and the last lands are three-fourths of "w" and the rest are of width "w." And, each of these lands is ploughed in parts of one-fourth of "w," implying that the first and last lands have three parts of one-fourth "w" each and the rest have four parts of one-fourth "w" each. Firstly, the operator starts with a right-hand bottom at the right edge of the first land. The operator does a "casting" operation (turning anti-clockwise) around the central strip of one-fourth "w" until completing the two strips of one-fourth "w" in the first land, leaving the central strip intact. The operator comes out of the extreme right edge of the first one-fourth strip and enters into the right-most edge of the central strip. The operator then does a "gathering" operation, turning clockwise around the already ploughed third strip, thus entering downwards the first strip of the second land, until completing the central strip of the first land and the first strip of the second land. It may be noted that in this process, at the place marked "df," a dead furrow is formed where open furrows at the right edge of the first strip and the left edge of the central strip adjoin. Simultaneously, a back furrow, "bf," is formed at the place between the third strip of the first land and the first strip of the second land. Altogether, by alternating casting and gathering, the first land and the first one-fourth strip of second land are completed. The advantage of ploughing this way provides an optimal distance of idle transit between subsequent passes. The distance is neither too small, making it difficult for the plough to turn, nor too large such that more energy and time are wasted in the idle run between passes. Since ploughing is done by a one-way plough, a pair of dead and back furrows were formed inevitably (Bell, 2003; Lovegrove, 1968).

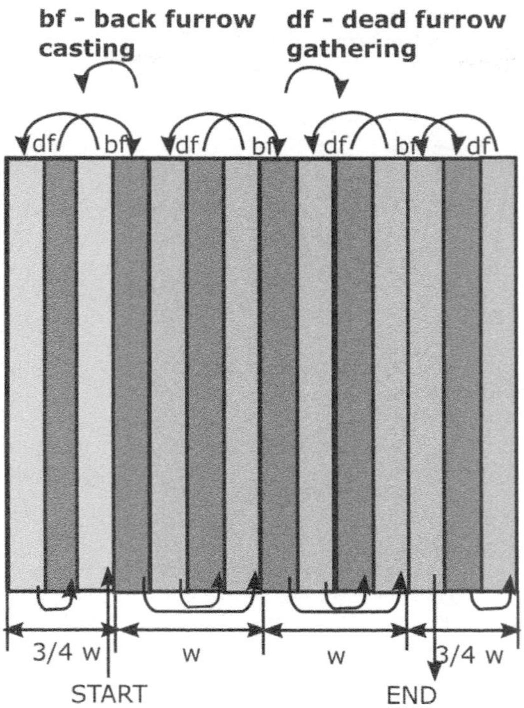

FIG. 2.14 System of Ploughing in Lands

It may be observed now that, after the first alternating casting and gathering sequence, the second land is now three-fourths of "w" and the subsequent land is of width "w," which is a similar status as in the beginning. So, once again, the same operation of alternate casting and gathering is repeated. In the final phase, the operator enters into the central strip of the final land of three-fourths "w" and does a gathering to complete the whole field.

The division of land and strips is sometimes marked by stakes to guide the operator. However, this division is not done by accurate measurements but by approximate eyeballing. An experienced operator can do the whole operation of marking and ploughing with considerable accuracy. If there is any mismatch of implement width to the remaining space in the strips, the operator tilts the implement accordingly to make one or two bottoms only to plough and match the remaining space.

It may be remembered that a dead furrow is a depression with no loose soil in it, and a back furrow is an unploughed land beneath a mound of loose soil on top. These, especially the back furrows, can hinder any secondary tillage and sowing operations later, if operations are carried out perpendicularly to these furrows. So, if needed, an operation of the splitting of back furrows or the partial filling of dead furrows is carried out after the whole field is ploughed. Splitting is to plough the back

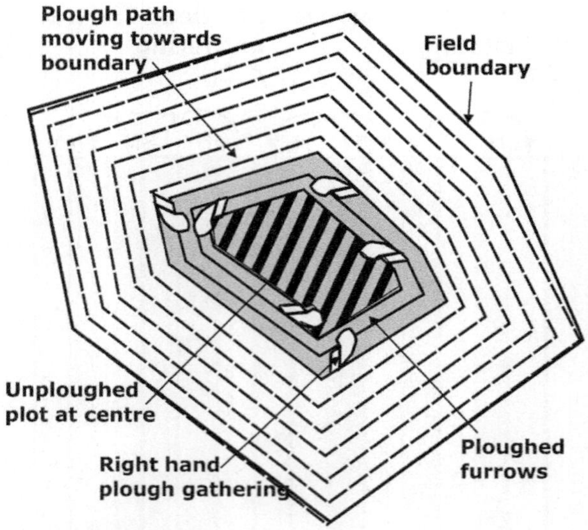

FIG. 2.15 Round-and-Round Ploughing

furrow along with the mound of loose soil, and the filling of dead furrow is pushing some soil into it from the ploughed land on the sides.

2.3.1.2 Round-and-round ploughing

This system of ploughing (Figure 2.15) with a one-way plough avoids the formation of back and dead furrows completely and can be used instead of using a two-way plough. Here, the one-way plough can be used by either starting at the centre of the field and going around and around until it reaches the field boundary or by starting at the boundary and ploughing around and around, going inwards until the centre of the field is reached (Bell, 2003; Hunt, 2008; Lovegrove, 1968). If it is a right-hand plough, then gathering is done by going clockwise from the centre of the field or casting is done by going anti-clockwise from the boundary to the centre of the field. In each case, the cut furrow slice falls into the furrow made in the previous pass (the inner or outer round of ploughing), thus avoiding the formation of dead and back furrows.

Since going around by turning will be difficult at the centre of the field, a small plot is marked first in the centre of the field, around which the ploughing runs are done. This plot is ploughed separately using linear passes.

2.3.2 One-way ploughing using a two-way plough

In earlier sections of this chapter, we learned about reversible or two-way ploughs that can switch from a right to left bottom by mechanical or hydraulic means at the will of the operator. If the ploughed field plot is large and regular, the operation of a two-way plough would be relatively easy, but not otherwise.

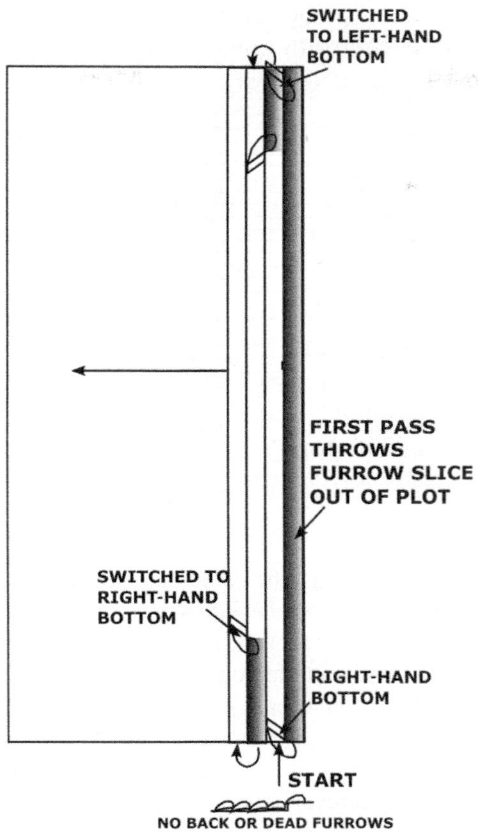

FIG. 2.16 System of One-Way Ploughing

Figure 2.16 illustrates the system of using a reversible or two-way plough. The process starts from the right-hand side of the field plot. All the upward ploughing passes are made using the right-hand bottom, and all the downward passes are made after switching to left-hand bottoms. This implies that the furrow slices dug in the present pass always fall into the furrow made during the previous pass. This system will never form any dead or back furrows. The field plot will be almost level after ploughing since all the furrows are filled with ploughed soil. Interestingly, the process of ploughing with a "two-way plough" is known as "one-way ploughing." The only disadvantage in this system is that the switching of bottoms needs to be done at the end of every ploughing pass. However, the availability of reversible ploughs that can do this switching hydraulically (Figure 2.7) has made this process easy, and two-way ploughs are now used more and more. But, obviously, these ploughs are at least two or three times more expensive than one-way ploughs. The process of dividing the whole field into smaller lands to do the ploughing with a one-way plough is not very appealing to a tractor operator presently, and the use of one-way ploughs has almost disappeared in nations where the holding

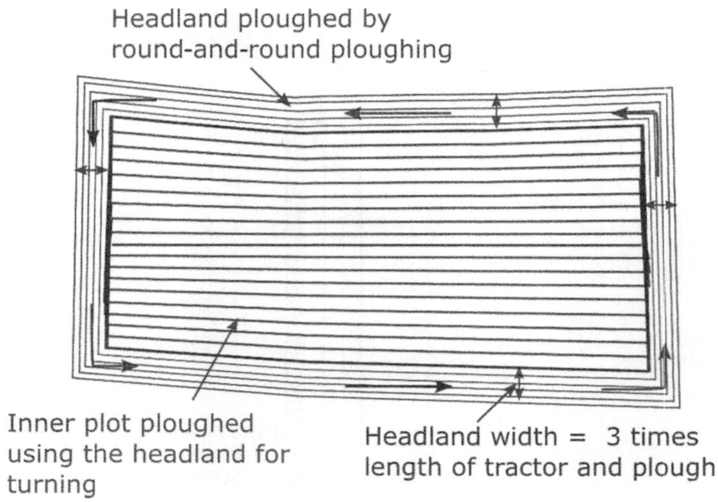

FIG. 2.17

sizes are very large. But, it may also be noted that an experienced operator can give an equally good ploughing performance with a one-way plough itself. Moreover, the use of one-way ploughs is still the preference in countries and regions where the majority of holdings are of small and medium field sizes. One-way ploughs are hence being manufactured in these countries. The understanding of the system of ploughing with one-way ploughs will give an enhanced perspective of the process of ploughing to the student in terms of how it affects the field efficiency of operation. Even in the process of using a reversible plough, it is necessary to keep the idle run on the headlands optimal by moving from one strip to another.

2.3.3 Headland

It is obvious that tractor ploughing with a hitched implement requires space in the field plot to take the turn and enter the next ploughing pass in many of the systems of ploughing explained so far. This turning space is known as the headland and is provided on all sides of the field plot, especially those perpendicular to which the ploughing passes are made. Its width is usually two to three times the overall length of the tractor and hitched implement (Figure 2.17). This space is planned and marked in the field plot before the ploughing process is started. After the desired system of ploughing is followed and the whole plot between the headlands is completed, the headland is ploughed separately by adopting a round-and-round system. Ploughing in lands requires the field plot to be marked with where the back furrows will be formed after the ploughing job. The back furrows as well the headland boundaries are marked by the implement itself by using the tip of the first plough bottom alone to make a cleavage on the soil surface. This is achieved by tilting the implement in the lateral direction to travel. The back furrow markings are done not by measurements but by eyeballing. This requires considerable experience on the part of the operator. This approximation will also demand considerable operator's skill to merge the passes at the dead furrow junctions.

The example in Box 2.5 shows how a field is divided into smaller lands and ploughed systematically.

Box 2.5 Example 2.5

Example 2.5. A field plot of 1 ha is to be ploughed by a tractor-mounted, three-bottom mould board plough, each with a 40 cm width. The north, west, and south sides of the plot are straight sided with the northwest and southwest corners right angled. The east side of the plot subtends 60° to the southern edge of the plot. The west and north sides are equal in length. Plan the sequence of ploughing it in lands with an illustration, and mark the lines on which the dead and back furrows will be formed.

The shape of the plot can be split into a square with the north, west, and part of the south sides as equal sides of length = a (say). Its area is hence a^2. The rest of the plot is a right-angled triangle with the base = b (say) and the height = a.

Since the angle subtended by the triangle between the east edge of the field (hypotenuse) and the base is 60°, a/b = tan(60);

the area of the triangular part of the plot = 0.5(ba) = 0.5(a cot60)a= 0.29 a^2

\ the total area = $a^2 + 0.29\ a^2 = 1.29a^2 = 10000\ m^2$

a = 88.04 m.

The width of the implement = 3 bottoms × 0.4 m = 1.2 m

Each land is usually taken as 14 times the implement width = 14 × 1.2 = 16.8 m.

Then, considering the square part of plot, it has to be divided such that the first and last lands are three-fourths the width of the land and the rest of the lands are of full land width,

i.e., (2(¾) + n)w = (1.5 + n) w = 88 m

Where w = land width and n = the number of full width lands.

By trial and error, assuming n = 4, w = 16 m, which is closer to the optimum land width.

The square plot is hence divided into six lands, with the first and last of ¾(16) = 12 m width and the four lands in between of 16 m width.

The number of lands in the triangle = b / 16 = a cot(60) / 16 = 50.83 / 16 = 3.17.

To round it off to three lands, the width of land can be = 50.83 / 3 = 16.95 m.

These three lands are ploughed by alternate casting and gathering, which forms the dead and back furrows at the centre of the strips.

The overall length of a 50 hp tractor with a three-bottom MB plough = 3.64 m + 1.2 m = 4.84 m

\ Headland width = 2 × 4.84 = 9.7 m

Headlands can be formed to a width of 9.7 m on both the regions for the tractor to turn and do the operation.

The dead and back furrows formed in the operation as well as the lands are shown in the solution.

Problems

1. A two-bottom disc plough with 60 cm discs, set at a 40° disc angle and 12° tilt angle, is ploughing at a 20 cm depth. The spacing between the discs, as seen on the front view of the plough, is 25 cm. Graphically calculate the percent of

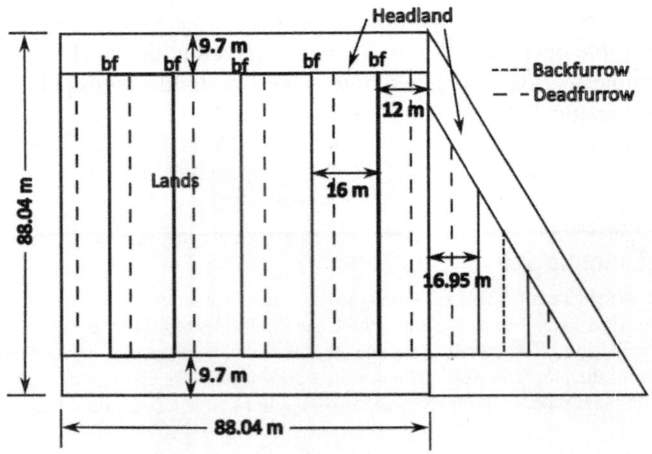

No Permission Required

effective furrow area cut by the plough relative to the theoretical furrow area. (Hint: Use the similar example problem on the single-disc bottom as a model to solve for the furrow area of a single-bottom set at the given angles. Then, copy a similar furrow at the given offset. The solution may be found from the intersection of the two furrows obtained.)

2. Draw the sectional profile of a rectangular plot ploughed by a tractor mounted to a two-bottom, left-handed mould board plough. Assume that the ploughing operation was by casting. Discuss the profile of how it was formed, particularly with respect to the formation of back or dead furrows.
3. A field plot of the given shape and dimensions is to be ploughed by a tractor-mounted, two-bottom mould board plough with 40 cm widths each. Plan the sequence of ploughing it with an illustration and mark the lines on which the dead and back furrows will be formed.

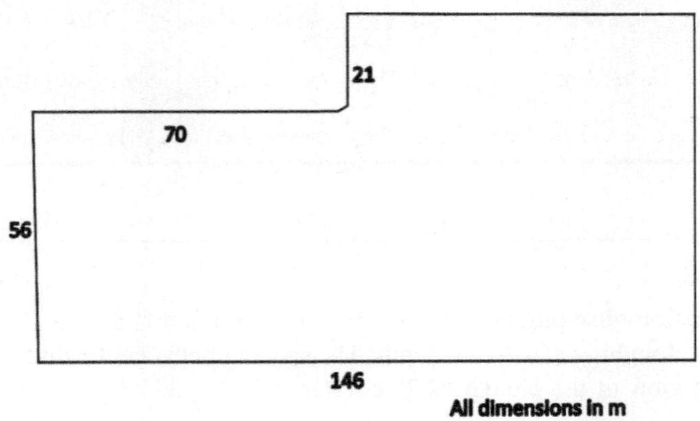

No Permission Required

4. A 40 cm mould board plough, as shown in the illustration below, has a board surface that can be expressed as a part of a sphere, whose centre is 65 cm away laterally from the share point on a line perpendicular to the landside (ploughing direction). The plough's throat width is 40 cm, and the extreme point of the mould board is at 50 cm, 45 cm, or 25 cm from the share point. Find the location at which the soil slice will be thrown from the extreme end of the mould board, when the plough is working at 5 km h^{-1}. (Hint: Use the equation of the sphere to find the partial derivative of the y coordinate with respect to x.)

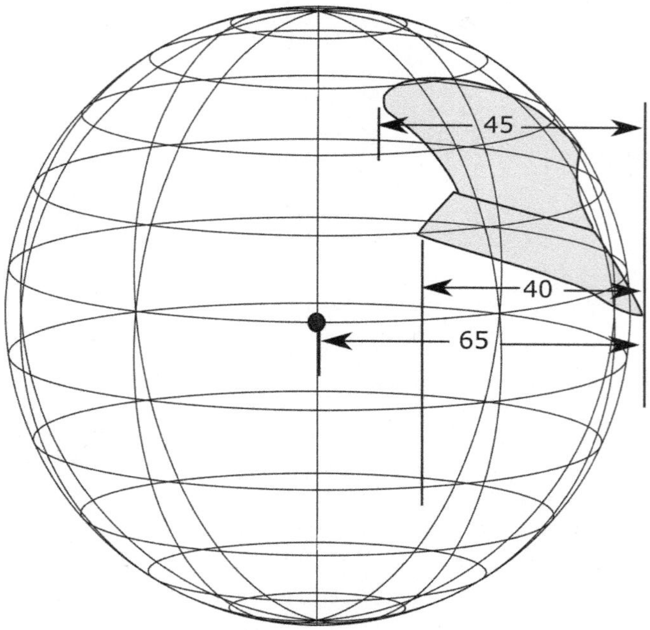

No Permission Required

References

Bell, B. (2005). *Farm machinery*. UK: Old pond Publ.

Bell, B. (2003). *The tractor ploughing manual*. UK: Old pond Publ.

Bernachi, H., Haman, J., & Kanafojski, C. Z. (1972). *Agricultural Machines, theory and construction* (Vol. 1). Warsaw: Scientific Publications Foreign Cooperation Center. Published for USDA and NSF by Foreign Cooperation Center of the Central Institute for Scientific, Technical and Economic Information.

Culpin, C. (2013). *Farm machinery*. Hesperides Press.

Hunt, D. (2008). *Farm power and machinery management*. Waveland press.

Klenin, N. I., Popov, I. F., & Sakun, V. A. (1970). *Agricultural machines.* USDA and NSF by Amerind Publishing Co. Pvt. Ltd.

Kepner, R. A., Bainer, R., & Barger, E. L. (1978). *Principles of farm machinery.* Avi Publishing Company.

Lovegrove, H. T. (1968). *Crop production equipment.* Hutchinson press.

Smith, H. P. (1965). *Farm machinery and equipment.* Tata McGraw Hill Pub. Co. Ltd.

Srivastava, A. K., Goering, C. E., Rohrbach, R. P., & Buckmaster, D. R. (2006). *Engineering principles of agricultural machines.* ASABE Publications.

Summer, M. S., Chen, S. H., & Bierl, J. F. (1983). *Disk blade performance. ASAE Paper* (pp. 83–1537).

CHAPTER

Materials of Construction and Heat Treatment

3

Chapter outline

3.1 Materials of construction

Engineering materials used for the fabrication of machines are classified as metals, such as iron, aluminium, and copper, and non-metals, such as plastics, asbestos, and wood. Metals are classified again as ferrous metals, such as cast iron, steel, and alloy steels, and non-ferrous metals, such as copper, aluminium, brass, and bronze, to mention a few.

3.1.1 Mechanical properties of materials

The strength of a material is its ability to withstand various forces to which it is subjected in a machine. Metals owe their strength to interactive forces residing between

Machines of the Farm. https://doi.org/10.1016/B978-0-443-32906-7.00008-6

their molecules. Each material has its own tensile, compressive, shear, and impact strengths. Elasticity is the ability to regain an original form after the removal of a load and the deformation it causes is removed. Young's modulus is a measure of the stiffness. Plasticity is yet another property that enables the formation of permanent deformation in a material. Ductility is the ability of a metal to withstand elongation or bending. Brittleness is the lack of ductility. Malleability is the property of a material to be hammered or rolled into thin sheets without rupture. This property generally increases in metals and alloys with an increase in temperature. Toughness is the strength of the material to oppose rupture. The area under the stress–strain curve of a material is indicative of its toughness. Hardness is the material's resistance to penetration. Hardness is a function of the elastic limit of the material. Fatigue is the phenomenon that leads to failure of the material when subjected to fluctuating or repeating loads. Strength under fatigue loads is very much less than strength under steady loads. Creep is the slow, plastic deformation of materials under constant stress or under prolonged loading usually at a high temperature. Machinability is the ease with which a material can be machined. Weldability is the capacity of a material to be welded to fabricate an assembly that can bear the intended load satisfactorily. Stainless steel, low-alloy steels, cast iron, and carbon steel are some of the weldable metals used for machine fabrication.

3.1.2 Metals and alloys

Metals are made of crystals differently oriented with respect to each other. They have high thermal and electrical conductivity. Pure metals generally possess low strength and are hence not used directly for fabricating machine parts. But alloys are made by melting or sintering metals with metals or metal with a non-metal. Alloys possess superior qualities compared with pure metals and are manipulated by changing their composition to acquire the required properties. For instance, steel is an alloy of iron and carbon. Chromium, nickel, and iron are alloyed to get stainless steel and so on.

3.1.2.1 Ferrous metals and alloys

Ferrous metals and alloys are iron-based metals that include all varieties of irons and steels. These metals go a long way in bringing prosperity to a country. Cast iron and steel are not pure irons but are iron alloys. Depending on the percent of carbon as an impurity in the iron, the iron is classified as pig iron, grey cast iron, and wrought iron. Pig iron is the most impure iron and wrought iron is the purest form, neither of which is used as manufacturing materials. Steel is an iron carbon alloy in which the percent of carbon is deliberately varied to obtain various grades of steel having differing properties. Its plasticity, both at room temperature and at red-hot temperatures, allows it to be formed as required. Carbon steel is still the most popular ferrous alloy that is used in the industry.

Cast iron is classified into the following types. Grey cast iron has a composition of C at 2.5–3.8 percent, Si at 1.1 percent to 2.8 percent, Mn at 0.4 percent to 1 percent, P at 0.15 percent, and S at 0.10 percent. It has a large portion of its carbon in the form of graphite and can be cast into a desired shape in a sand mould. It has better machinability than steel and has high resistance to wear. It is brittle and has low ductility and impact strength as compared with steel. It is strong in compression and, hence, cast as machine bodies, housings, and casings. White cast iron has a composition of iron at 94 percent and graphite carbon at 0.5 percent, with other materials being impurities. It is hard, brittle, and not machinable. It possesses very high resistance to wear. It is used for parts that are subjected to excessive wear such as implement wheels in agricultural machines.

Malleable cast iron is obtained by a lengthy annealing process on cast iron. It has high yield strength and can be hammered and rolled into different shapes. It is tough and easily machined.

Since the cast iron is brittle and cannot bear tensile or shear loads, it is mainly alloyed with nickel to impart these qualities.

Wrought iron was the form of iron used earlier for fabricaton before the advent of steel. It is virtually pure iron containing negligible carbon. Therefore, it cannot be hardened by quenching in water as steel could be. Carbon steels are classified into low-carbon steels or mild steels having 0.05 to 0.35 percent carbon, medium-carbon steels with 0.35 to 0.65 percent carbon, and high-carbon steels with more than 0.65 percent carbon. Mild steels are general purpose steels that are predominantly used for all the machine pieces that are exposed to tensile, bending, and shear loads. Medium-carbon steels are amenable for heat treatment such as hardening, carburizing, nitriding, and making soft-centred steels. These are used for fabricating wear-resistant parts of soil working tools. For instance, the mould board of a plough is made of soft-centred steel. High-carbon steels are very hard and are mainly used as tool steels to cut mild steel and cast iron.

When enhanced strength properties are desired in steel, elements such as nickel, chromium, manganese, vanadium, tungsten, and others are alloyed to the carbon steels to acquire alloy steels. Alloy steels, depending on their alloying element, can have improved case-hardening properties, elasticity, corrosion resistance, hardness, toughness, tensile strength, machinability, wear resistance, and ductility.

Nickel steel has improved toughness, forming properties, and special electrical and magnetic properties.

Chromium steel has properties of rust resistance, toughness, and strength.

Vanadium steel has improved response to heat treatment and is used for high-speed tools.

Tungsten steel remains hard and tough even at high temperatures and is used in tools and dies.

Silicon steel has high electrical resistance and magnetic permeability and is used in electrical machines. Molybdenum steel has enhanced corrosion and abrasion resistance.

More than one of the above elements can be alloyed to steel to obtain a combined enhancement of properties.

For instance, steel with Mn at 1 percent and Si at 2 percent has a very high elastic limit and is used for making springs. Steel with 0.6 percent to 1 percent silicon and 0.75 percent to 0.95 percent manganese is used in structures that demand a high load-bearing capacity. High-speed tool steels are used as cutting tools and contain 18 percent tungsten, 4 percent chromium, and 1 percent vanadium with a carbon content of 0.6 percent to 0.7 percent.

The stainless steels are generally known as nickel chromium steel that is non-magnetic and cannot be hardened. They are rust-proof and corrosion-resistant steels and are used a great deal in the food industry. The predominantly used 18-by-8 steel is used for making cutlery, cooking vessels, etc. and has 18 percent chromium and 8 percent nickel.

3.1.2.2 Non-ferrous metals and alloys

Copper

Copper is a reddish-brown metal and an excellent conductor of heat and electricity. It is malleable and ductile. Copper is not used in a pure form in machines except as an electrical conductor.

Copper Alloys

The two principal copper alloys are brass, an alloy of copper and zinc, and bronze, an alloy of copper and tin.

Brass has 5 to 45 percent zinc alloyed to copper and has good mechanical properties, corrosion resistance, and appreciable machinability. Manganese brass has a composition of copper at 60 percent, zinc at 38 percent, tin at 1.5 percent, and manganese at 0.5 percent. It is tough and resistant to corrosion by sea water. Because of its high strength, it is used in pump rods, hydraulic rams, tubes, and propellers.

Iron brass, known also as delta metal, is another common alloy, with Cu at 60 percent, Zn at 37 percent, and Fe at 3 percent. It, too, is strong, tough, and easily cast. It is also corrosion resistant and strong in tension. It is used in place of mild steel in corrosive environments.

Tin brass is an alloy of copper, zinc, and tin at 2 percent. Tin provides a higher tensile strength and hardness to the alloy. Leaded brass has lead at 1 percent to 2 percent, making brass machinable.

Bronze is an alloy of copper and tin. Bronze has good tensile strength and has a maximum of 20 percent tin. Bronze is very ductile when it has 5 percent of tin and amenable for rolling. The commonly used bronze has 8 percent to 11 percent of tin. It was also known as gun metal in earlier days. Bronze can be cast into machine parts. Bronze with 25 percent of tin is very hard and brittle; it is used for making bells.

Zinc bronze has Cu at 88 percent, Sn at 10 percent, and Zn at 2 percent. Zinc prevents gas cavities while the bronze is being cast. It also increases the strength and ductility.

Phosphor bronze is a de-oxidized bronze in which phosphorous is added to molten bronze and oxidized phosphorous is present only at 0.1 percent. Phosphor bronze has excellent tensile strength, elasticity, and fatigue resistance. It can, hence, be forged, drawn, cold-rolled, and cast. It is used in place of steels in corrosive environments.

Aluminium

This silver-coloured metal is ductile and malleable. Its electrical conductivity and thermal conductivity are high. It is easily corroded by sea water. Metallic contact with other metals under moist conditions causes corrosion. Pure aluminium is used to cast light machine enclosures such as seed metering units. Rolled aluminium sheets are used to make hoppers in planting machinery.

Aluminium Alloys

Duralumin is an aluminium alloy of 94 percent Al, 4 percent Cu, and 0.5 percent each of Mg, Mn, Si, and Fe. It has a high tensile strength with high electrical conductivity and can be cast, forged, and stamped. It is widely used for making electrical cables, sheets, tubes, forgings, rivets, nuts, and bolts.

Y-alloy has 92.5 percent Al, 4 percent Cu, 2 percent Ni, and 1.5 percent Mg. It has a higher strength than pure aluminium, even at 200 °C.

Magnelium is an alloy with 85 percent to 95 percent Al, 0 percent to 25 percent Cu, 1 percent to 5.5 percent Mg, 0 percent to 1.2 percent Ni, and 0 percent to 3 percent Sn. It is light in weight and has good tensile strength. It is machinable and weldable but cannot be cast. It is used for making machine housings and handles.

3.1.3 Organic materials

Organic materials are derived from carbon and chemically combined with hydrogen, oxygen, or other non-metallic substances. They have complex chemical structures. Some of the popular organic materials are plastics and synthetic rubber, which are extensively used in agricultural machines. These are known as polymers and are made by the process of polymerization. PVC (polyvinyl chloride), PTFE (polytetrafluoroethylene), and polythene are some common plastics used in agricultural machines.

3.1.3.1 Plastics

Plastics are non-metallic materials and polymers that can be moulded and extruded to shape and also coated as a film. Most plastics consist of organic polymers composed of hydrogen, oxygen, carbon, and nitrogen. Plastics are mainly classified into two types, namely, thermoplastic and thermosetting. Thermoplastics are those that are softened by heat, with or without pressure, but require cooling to set them to shape. They can be heated and cooled any number of times. They are highly plastic and are easy to mould or shape. They possess a low melting point and are lesser in strength than thermosetting plastics. Polythene, polyvinyl chloride (PVC), and polytetrafluoroethylene (PTFE) are some popular thermoplastics.

Thermosetting plastics require heat and pressure to mould them into shape. They cannot be re-softened after setting and hardening. A part requiring rigidity, strength, and relatively better resistance to heat is made of thermosetting plastic. Phenolics such as Bakelite, polyesters, epoxies, and silicones are some popular thermosetting plastics.

Laminated Plastics

Laminated plastics are formed by impregnating sheets of fibrous materials such as paper, linen, canvas, or silk with a phenolic, urea formaldehyde, or vinyl resin and then compressing the sheets together with the application of heat. These sheets are light, strong, and machinable with resistance to wear, acids, and alkalis. They are water proof and oil proof with a high dielectric constant. They are used as electric insulation and for making silent gears, water-lubricated bearings, pulleys, and pump parts.

Fibreglass-Reinforced Plastics

Glass fibres are mixed with synthetic resins to form a composite. The fibres provide stiffness and strength while the resin provides a matrix to transfer the load to the fibres. They are light in weight, easy to repair, durable, and corrosion resistant. They are used for making water-storage tanks, roof sheets, domes, and structural sections.

3.1.3.2 Ceramic materials

Ceramic materials contain compounds of metallic and non-metallic elements. Earthen pottery is an example of ceramics in which clay is shaped, dried, and fired into a permanent structure. Ceramics are also formed by injection moulding, sintering, and hot pressing. Ceramic materials are classified as abrasives such as alumina and carborundum, pure oxide ceramics such as MgO, Al_2O_3, and Si_2O_2, fire-clay products such as bricks, tiles, and porcelain, inorganic glasses such as window glass and lead glass, cementing materials such as cement and lime, and refractories such as fire bricks.

3.2 Heat treatment of metals and its purpose

Heat treatment combines the heating and cooling of a metal or alloy in the solid state to specified temperatures and at certain heating/cooling rates for the purpose of obtaining the desired changes in properties. This occurs as a result of the microstructural changes in the material produced by the heat treatment operations. The desired changes are intended to improve machinability, ductility, strength, hardness, and toughness as well to relieve internal stresses. Some heat treatment processes can also enhance corrosion resistance and heat resistance and modify electrical and magnetic properties. Some of the important heat treatment processes are explained next.

3.2.1 Annealing

Annealing is done to soften the steel so that it becomes machinable. It also restores ductility and toughness and relieves internal stresses. It is also a step for preparing it for subsequent heat treatment processes. Full annealing does a complete phase recrystallisation and cleans the previous structure.

It is done by heating the steel slightly above the critical point and holding it at this temperature for a longer period. The retention of temperature is achieved by burying it in sand, lime, or ashes. This slow cooling enables austenite to decompose into a pearlite dominant structure.

Process annealing is a process that lessens the increased hardness due to cold working processes and makes it suitable for plastic deformation. It is accomplished by heating the steel to a temperature close to the critical point and holding it at this temperature to cool it very slowly. It is performed only on low-carbon steels. Isothermal annealing is carried out on alloyed steels in which the cooling rate is to be extremely slow. The steel after heating above the critical point is cooled rapidly down to a temperature of just 100 degrees less and held at a constant temperature for a prolonged time.

3.2.2 Normalizing

Normalizing is a heat treatment process to remove the internal stresses due to working. It is intended to strengthen medium-carbon steels to a certain extent and improve the machinability of welds on low-carbon steels. It is the final treatment for equipment that is expected to operate at high stresses. It is performed by heating the metal to normalizing temperatures, which are usually 40°C to 50°C above the critical point, holding it at that temperature for about 15 minutes and then cooling in air. Normalized steels have a higher yield point, tensile strength, and impact strength, with a slight loss of ductility and machinability.

3.2.3 Hardening

Hardening is a heat treatment applied to all tools and machine parts made of alloy steel. Hardening resists wear, especially abrasive wear in soil working parts. However, tempering is a must after hardening if strength, elasticity, ductility, and toughness are to be enhanced to transform it into load-bearing or cutting parts. The process is to heat the steel above the critical point temperature and cool it rapidly by quenching in a liquid or a current of air. The resulting properties depend on the steel's composition, the carbon and alloy content of the steel, homogeneity and grain size of the austenite, heating rate and time, and the quenching medium and rate.

Hardenability is the property of a steel that determines the depth and distribution of hardness after quenching. Alloy steels are more effectively hardened by quenching in oil than in water.

There are many variations in the process steps; these classify the hardening into several types.

Martempering or interrupted quenching is a hardening operation that produces martensite. Here, the quenching is done in a salt bath at a temperature of 150 °C to 300 °C, where martensite starts to form. It is done to avoid quenching cracks and the warping of treated parts.

Austempering or isothermal quenching is a hardening process in which the part is quenched in a molten salt or lead bath usually at 300 °C to 350 °C. It makes the part tougher and more ductile than other quenched and tempered steels. Tempering is not usually needed after this process.

3.2.4 Case-hardening

Case-hardening is a heat treatment process in which only the surface of the steel or alloy steel is treated to make it hard, leaving the inner core soft. The treated part then has an extremely hard surface on the outside and a soft ductile core. This is, hence, known as soft-centred steel. It is predominantly used in the non-load-bearing parts of soil working tools such as plough mould boards. The different types of case-hardening processes are detailed next.

3.2.4.1 Flame-hardening

Flame-hardening is accomplished by heating the steel with an oxy-acetylene flame and then quenching it in water. This process is used when localized hardness to a limited depth is needed. The process is rapid and less expensive.

3.2.4.2 Induction-hardening

Induction-hardening is also used prevalently for the surface hardening of machine parts susceptible to wear. It is accomplished through an extremely rapid heating and quenching of the surface. A 2,000 Hz voltage is applied to an inductor block; this induces a current on the steel part that is in the vicinity, thus heating the surface rapidly to 750 °C. The induction-heated surface is then quenched by water sprays. A hardened case of 3 mm is thus obtained. Medium-carbon steel is amenable for this cheaper option of the rapid case-hardening of parts.

3.2.4.3 Carburizing

Carburizing is employed on low-carbon steel, of less than 0.15 percent carbon, because this type of steel does not respond to other heat treatments. Only the outer layer is converted into a high-carbon steel of 0.9 percent to 12 percent carbon. This process is performed to heat mild steel to a red-hot temperature in contact with a carbonaceous material such as wood or charcoal. The carbon thus enters the metal to form a solid solution with iron and converts the outer surface into a high-carbon steel. The part is then heated to the critical point temperature and then cooled to refine the core structure. It is then reheated to a point above the transformation range of the case (shell) and quenched to produce a hard, fine structure. Alloy steels are usually quenched in oil, and carbon steels are usually quenched in water. This results in a hard, fine-grained shell with a core of low-carbon steel.

3.2.4.4 Cyaniding

Cyaniding is a case-hardening method in which a hard case is obtained on the surface of low-carbon or medium-carbon steels by immersing the steel in a molten salt bath at 800 °C containing cyanide and then quenching it in water or oil. Usually, a hard case of less than 0.8 mm is obtained.

3.2.4.5 Nitriding

Nitriding is performed to obtain a hard surface on alloy steels only. The steel is heated in an atmosphere of ammonia gas at temperatures of 500 °C to 650 °C. The nitrogen in ammonia combines with steel to form nitrides that give hardness on the surface, up to a 0.4 mm depth. It is performed on machine parts such as pump shafts, dies, and gears.

3.2.4.6 Diffusion coating

Diffusion coating or metallic cementation is performed to impregnate the surface of steel with aluminium, chromium, silicon, boron, beryllium, and other elements. It is achieved by heating the steel parts in a solid, liquid, or gaseous environment of the elements to be impregnated. The surface, hence, becomes more like an alloy steel and is wear resistant.

Further reading

Martin, J. (2006). *Materials for engineering materials for engineering*. United Kingdom: Woodhead Publishing Limited. https://doi.org/10.1533/9781845691608. Available from http://www.woodheadpublishing.com/9781845691578.

Rajput, R. K. (2008). *Engineering materials*. S. Chand and company Ltd.

CHAPTER

Secondary tillage implements

4

Chapter outline

Secondary tillage is that which succeeds the primary tillage operation of opening up the field soil for cultivation. Generally, secondary tillage tools (Bell, 2005; Culpin, 2013; Hunt, 2008; Kepner, Bainer, and Barger, 1978; Lovegrove, 1968) penetrate and stir the soil to relatively lesser depths. These tools are used to make the seedbed and form the required field configuration for planting and other such operations.

The general objectives of secondary tillage are:

- to provide more pulverization to the soil ploughed deeply by the primary tillage tools;
- to shred and incorporate the crop residue and uprooted weeds into the topsoil;
- to form the ploughed soil into the desired field configurations such as ridges and furrows, wide beds, compartmental bunds, etc.;
- to bring about a friable soil of good tilth suitable for the seeds to germinate and develop into plants of good root hold;
- and to uproot and kill shallow rooted weeds.

Machines of the Farm. https://doi.org/10.1016/B978-0-443-32906-7.00004-9

Harrows, field rollers, tools for residue mulching, land-forming tools such as a ridger or buster, a bund former, and wide-bed former are some of the types of secondary tillage equipment.

4.1 Harrows

A harrow is a tool that cuts and crushes the bigger soil clods formed during the primary ploughing. Simultaneously, it stirs the soil and buries the weeds on the surface of the soil. Since harrows provide good stirring action, they are used to cover seeds that are broadcast on the soil surface. This practice is performed in rain-fed land to keep the sowing operation cost effective. Three commonly used types are the disc harrow, spike-toothed harrow, and spring-toothed harrow.

4.1.1 Disc harrows

These harrows consist of a series of concave discs assembled on a common axis to form a gang of discs. This gang does not have a tilt angle but is set at an angle to the travel direction. Since the whole gang has the same disc angle as assembled on a single axis, the disc angle of the gang is known as the gang angle. The gang of discs, similar to a disc plough, cuts the soil and rotates on its own axis. The gang penetrates by the action of its weight. The gang cuts into the soil clods and pulverizes them thoroughly. Such disc gangs are arranged one behind the other to provide repeated pulverization of the clods to bring about a fine tilth of soil. Generally, if the soil has a lot of grass roots and residue in the clods, a notched disc gang runs in the front of the implement cutting and shredding the roots embedded in the clods. Gangs of plain-edged discs follow; these shred and pulverize them further. Next to rotary tools, disc harrows are the best tools to offer a good seedbed. The disc harrow on operation brings about a relatively good level of soil surface, too.

Types. Disc harrows are available in sizes suitable for any size of tractor. Just like other tools, disc harrows are also classified based on the mode of hitching them to the tractor, namely, trailed and mounted disc harrows. Trailed disc harrows are hitched to the draw bar of the tractor and pulled by the tractor. The whole weight of the harrow is supported by the soil. Trailed disc harrows are usually mounted on a set of trailing wheels relative to which the whole implement can be lifted by hydraulic cylinders for on-road transport. Mounted disc harrows are usually small and are totally carried on the three-point linkage of the tractor.

Based on the arrangement of the gangs of the harrow, it is classified as a:

- Single-action disc harrow. This consists of two disc gangs placed end to end (Figure 4.1A). The gang angles of the two gangs oppose each other and throw the soil in opposite directions from centre to side. This causes a slight depression in the centre of the cutting width or swath. The cutting width of the disc harrow is the projected width along the direction of travel and ranges from

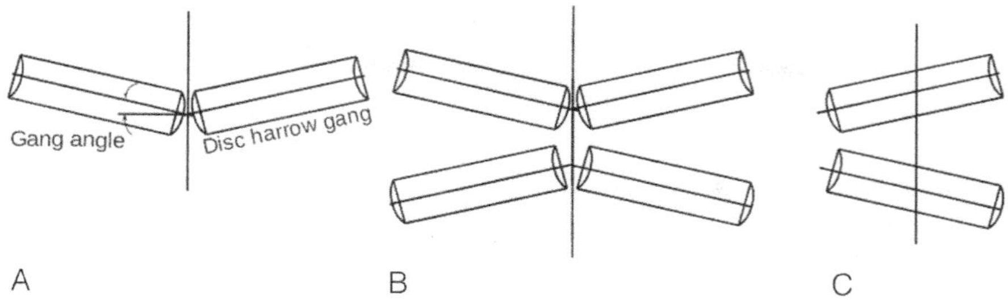

FIG. 4.1 According to Gang Arrangement as (A) Single Action (B) Double Action and (C) Offset.

1.2 m to 3.6 m. Harrows, which are more than 3.6 m in width, have end sections that can be folded over the mainframe to transport them easily.

- Double-action disc harrow. This is also known as a tandem disc harrow and is constructed with two sets of single-action gang assemblies attached back to back (Figure 4.1B). The rear set of gangs are arranged such that they throw the soil from the side to centre. The gangs on the front throw the soil in an outward direction from the centre line and the rear gangs throw the soil in the opposite direction. This action of the front and rear gangs together levels the tilled soil.
- The offset disc harrow. This is constructed as a lateral half of the tandem harrow. The front and rear gangs throw the soil in opposite directions and provide a levelled bed of tilled soil. The gang angles of the front and rear are set differently. The offset disc harrow is basically a trailed disc harrow that can be operated at various offsets from the centre line of the tractor while being trailed. By changing the gang angles of the disc gangs or by shifting the hitch point on the tractor, a side thrust on the implement is achieved; this pushes the disc harrow to one side of the travel line.

This offset can be altered by the tractor driver from the operator's seat using remote hydraulic cylinders or simple ropes. The operator has control over the implement's offset while traveling between obstacles such as trees. This harrow is most suitable for orchards; the tractor and harrow can be maneuvered easily between the rows of trees. Mounted offset disc harrows cannot be operated at an offset as trailed harrows since they are mounted on a three-point link of the tractor. However, they offer levelling of tilled soil.

A disc harrow consists of the disc gangs, a mainframe, standards or drops, bearings, bumpers, scrapers, and weight boxes. The disc gang is an assembly of discs on a single axle, namely, the gang bolt (Figure 4.2). The gang bolt is a square sectioned bolt on which a row of steel discs is assembled. The square holes in the centre of the discs key the discs to the gang bolt so that the gang of discs can rotate together with the gang bolt. Cast iron spacer spools with square holes are used to space the discs apart on the gang bolt so that there is sufficient space available between consecutive discs. At the end of the convex side of the gang bolt, a disc-shaped convex part,

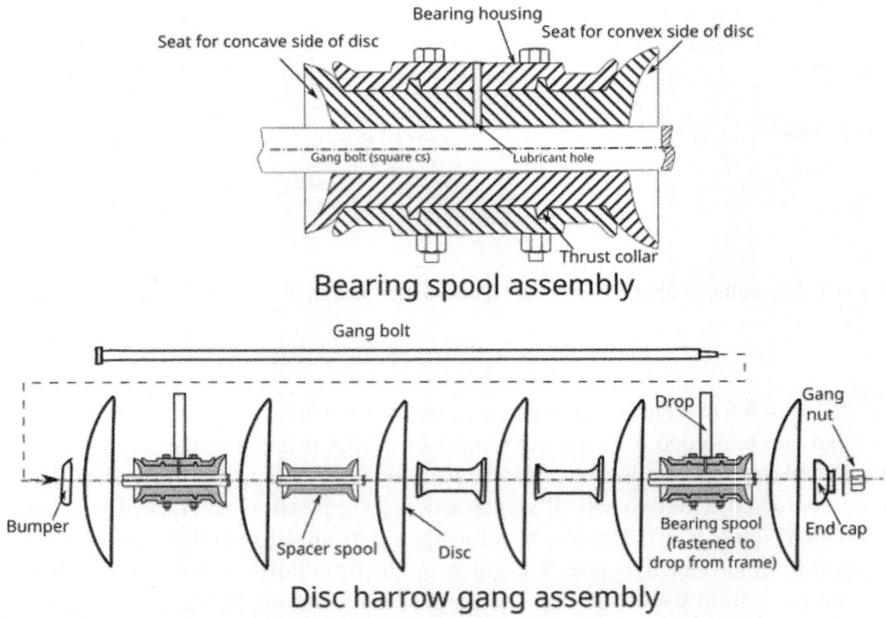

FIG. 4.2 Gang Assembly Construction.

known as the bumper, retains the discs in position. On the other side, a smaller cast iron disc is used to fasten all the discs and spacers together on the gang. Two of the spacer discs, one on either side of the gang, are known as bearing spools, which possess a bearing surface of chilled cast iron that can accommodate the bearing shells on which the whole gang can rotate. These bearing spools have ridges or thrust collars to bear the thrust load due to the side thrust on the gang. The bearing shells and the bearing housing are assembled over these bearing spools, and the bearing housing is fastened to the mainframe through a drop or standard. This arrangement provides for the whole gang assembly to rotate freely in the bearing assembly relative to the implement frame. The drops are fastened to the mainframe, and they can be adjusted to alter the gang angle. Since the speed of rotation of the gang at the harrowing speed is relatively slow, either wooden or chilled bearing shells are used as bearing material or the bearing housing is provided with a chilled layer. Grease nipples are provided on the bearing housing to lubricate the bearing adequately. Tapered roller bearing assembles are also used as alternatives to the explained journal bearings.

Round, smooth-edged steel discs of 450 mm to 600 mm diameter are used on most disc harrow gangs. Cutaway, notched, or scallop-edged discs are used on the front gangs for better cutting of residue in the tilled soil. The mainframe is generally a rugged rectangular or tubular frame and also has a weight tray on the top surface to add weights for improving penetration.

In the single-action and double-action disc harrows (Figure 4.3), the two opposing disc gangs in the front are set to throw the soil from the centre to the side. The axial thrust acting on these two gangs pushes the gangs against each other at the centre. The bumpers are made as heavy, half-moon-shaped cast iron cups fastened to the gang bolt on the convex side of the gang end. The bumpers on the opposing gangs butt against each other and absorb all the thrust instead of passing it on to the frame. The heavy bumpers also help to add weight on the convex side to maintain the gang level, as explained in Chapter 5. They are, hence, also used on the outer (convex) ends of the rear gangs of a double-action disc harrow. Similar to the scrapers of disc ploughs, scrapers are placed inside the concave side of the discs to remove the sticking soil.

4.1.2 Spike-toothed harrows

A spike-toothed harrow (Figure 4.4) has closely spaced, straight, long spikes resembling teeth that stir the soil beneath the frame. It is known by different names such as a peg-toothed harrow, drag harrow, section harrow, or smoothing harrow. It is used to stir the soil, break the clods, and level the soil directly after ploughing. Unlike a disc harrow, it does not effectively break hard and dry clods. Loamy and sandy soils that do not have much clay are amenable for these harrows since the clods remaining after a primary tillage can easily be broken, stirred, and levelled. This can, hence, be an alternative to a disk harrow in light soils. It can till the soil to a depth of about 50 mm. The tool is useful in covering seeds after they are broadcast on the soil surface. It can also be used to cultivate row crops at early stages of growth by altering or removing spikes to leave room for the rows of crops and stir the weeds in between the rows.

Since this equipment requires a lighter pull per unit width as compared to a disc harrow, it is usually made much wider than the tread width of the pulling tractor. The tool is, hence, made of lateral sections that can be folded over the

FIG. 4.3 Trailed Tandem Disc Harrow (Case IH True-Tandem™ 375, CNH Industrial, USA).

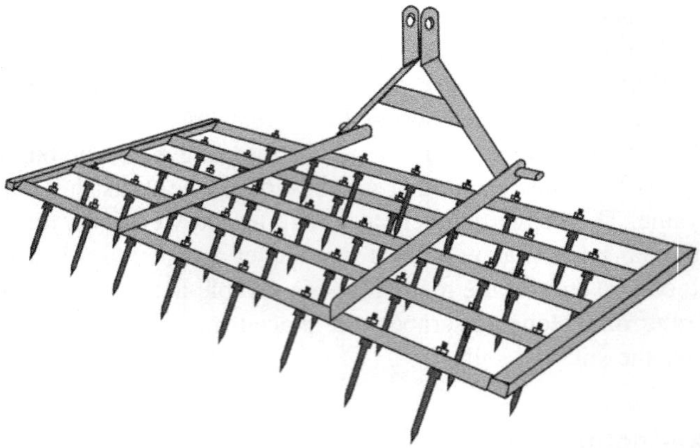

FIG. 4.4 Mounted Spike-Toothed Harrow.

others for easy transport. In the field, these sections are unfolded for operation. These sections are usually 2 m in width with about 20 to 35 teeth in each. Several such sections are attached to the tool frame and hitched to get a wide swath. The number of such sections will depend on the power of the tractor used. These are mostly trailed pieces of equipment, owing to their large width, and can give substantially greater coverage. They have on-road wheels for easy transport on roads. These wheels are retracted above the frame while harrowing.

4.1.3 Spring-toothed harrows

Spring-toothed harrows (Figure 4.5) are similar to spike-toothed harrows except that, instead of spikes, curved 'S'-type spring teeth are used. These teeth are wide, flat, oil-tempered, spring steel bars, curved in a partly spiral shape. One end of the tooth is fastened rigidly to a bar, and the other end has a penetrating point. The depth of penetration into the soil can be controlled by adjusting the approach angle of the teeth. This is done by swivelling the bar on which the teeth are fixed through levers and locks. On tilling the soil, these teeth flex back and forth. So, they are well suited for use in rough and stony ground. They are used to till previously ploughed soil to form a seedbed for small seeds. The spring teeth penetrate more deeply than those of spike-toothed harrows and are capable of riding over obstructions such as stones. A spring-toothed harrow is also used directly as a ripper in grassy land to kill the grass since the teeth penetrate and rip out the grass roots.

Spring-toothed inter-cultivators, which are basically down-the-row weeders, are also used interchangeably as harrows in lighter soils for clod breaking. More and

FIG. 4.5 Spring-Toothed Tines (Case IH Vibra Tine 265 S, CNH Industrial, USA).

more types of harrows are becoming available on the market that are suitable for different work requirements.

4.2 Land rollers and pulverisers

Land rollers or pulverisers are tools used for a finer preparation of the seedbed after harrowing. They crush the smaller clods on the surface, further bringing about a finely levelled seedbed with loose soil on top and firm soil beneath it. Based on the configuration of the roller periphery, they are of various types, such as 'V'-shaped roller-cum-pulverisers and sprocket-wheel pulverisers. A 'V'-shaped pulveriser has a gang of wheel sections loosely revolving on a common shaft. The periphery of these rollers are in a 'V' shape and form a narrow furrow in the loose soil. Each wheel is 300 mm to 450 mm in diameter and is made hollow and light. When a gang of wheels roll over the soil, individual rollers move up and down due to their loose fit on the shaft and follow the ground terrain precisely. They roll and leave small ridges and furrows on a level soil surface. Another gang trailing behind the first with a slight offset makes these ridges still finer and the surface finer. These are combination pieces of equipment available with front and rear roller gangs acting in tandem with a spring-toothed harrow in between. Sprocket-wheel pulverisers have alternate 'T'-shaped wheels and sprocket wheels, as assembled on a common shaft (Figure 4.6). 'T'-shaped wheels pulverize the soil clods and sprocket wheels provide a mulching effect, leaving loose soil on the surface. Figure 4.7 illustrates a trailed-roller array that pulverizes the soil and levels it to a certain extent.

4.3 Rotary tillers

Rotary tillers or rotavators are commercially available in many forms. Power tillers (Figure 4.8) and power weeders (Figure 4.9), which are self-propelled, two-wheeled

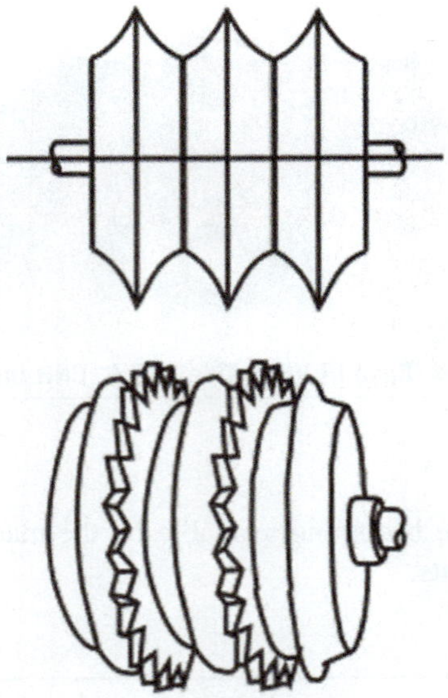

FIG. 4.6 Field Rollers with or without Sprocket Rollers.

FIG. 4.7 Trailed-Roller Arrays (Actiroll, Kverneland Group, Norway).

tractors, are provided with a rotavator as a standard attachment. They are popularly used in rice fields as well orchards for cultivating and weeding between crop rows. In rice fields, they are also used for puddling the soil water sludge. Puddling is an operation of stirring the soil sludge continuously to bring out the fine, colloidal particles of soil that settle at the bottom layer, thus sealing the downward water movement.

FIG. 4.8 Power Tiller with a Rotavator (Sharp Garuda Farm Equipments Pvt. Ltd., India).

FIG. 4.9 Power Weeder with Rotavator (Sharp Garuda Farm Equipments Pvt. Ltd., India).

FIG. 4.10 Tractor-Mounted Rotavator (4-u-series, India Agrovision Implements Pvt. Ltd).

This causes the water to stagnate in the rice paddy. Rotavators are also used to cut and shred crop stubbles after a harvest. They are available as tractor-mounted units (Figure 4.10) of varying widths, from 1500 mm to 2400 mm. The tractor-mounted rota-tillers are driven by the rear PTO shaft (power take-off) of the tractor. Rotary tillers are very effective in preparing seedbeds after a primary tillage. Rotary tillers break the soil clods thoroughly, even better than disc harrows, and bring about a bed of fine tilth. However, if the soil is softer, they are also used directly on unploughed soil to prepare nurseries for vegetable crops and lawn planting. The rotor consists of a power-driven shaft on which rotary tines are fitted and rotate as a gang. The drive to the shaft is through a heavy chain and sprockets and, in some machines, by cascaded gears. The whole mechanical drive is oil immersed. Since there is considerable shock and impact loads on the rotating tines, machines with hydraulic drives are also available now. Rotor speeds are generally around 200 rpm to 300 rpm. Many types and shapes of tines or knives are available, none being ideal under all conditions.

The important knife configurations are the ‘C’ and the ‘L’ types. The ‘L’ types are used for cutting the clods and stubbles and are not suitable for deep rotavation. The ‘C’ type is used for a deeper cut but is susceptible to clogging by excessive trash. The tines are made of a special grade steel to bear the much higher dynamic forces during soil cutting. Tine wear is a serious problem with these devices; it requires the replacement of a tine set after a period and will depend on the soil in which it is used.

The tines are spaced around and along the axis of the rotavator shaft and are angularly staggered around the shaft’s periphery. The arrangement is so designed to provide incremental cuts of soil and distribute the torque on the machine more uniformly without substantial impact (Smith, 1965). As the tines revolve and the machine moves forward, each tine cuts a soil slice, as illustrated in Figure 4.11. The cross-sectional area of the soil cut by each tine will depend upon the depth of tillage,

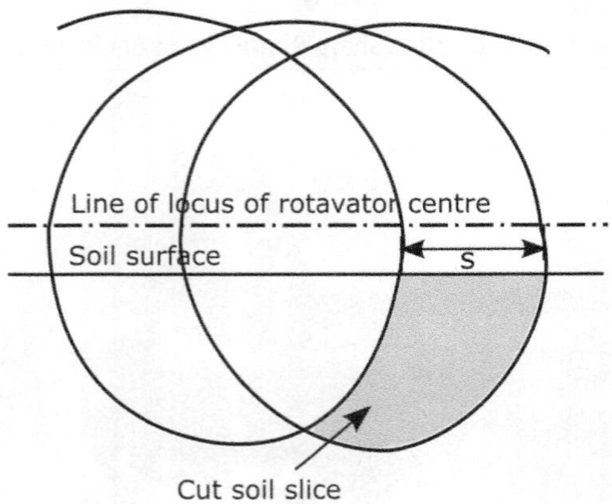

FIG. 4.11 Locus of Rotavator Tip.

the effective diameter of the rotavator tine assembly, and the forward travel made per cut, as shown in the illustration.

The locus of the rotavator tine tip can be expressed by the following equations. (Bernachi, Haman, & Kanafojski, 1972; Klenin, Popov, & Sakun, 1970)

$$x = \frac{V_t}{\omega} + R\cos\theta$$

$$y = R\sin\theta$$

where V_t is the travel speed, ω is the angular velocity of the rotating tine, R is the radius of rotation of the tine tip, and θ is the angle through which the tine tip rotates at any instant. The equation is applicable for the axis of rotation moving along the x-axis. The $\frac{V_t}{\omega}$ in the equation is the linear travel the axis of rotation makes per radian,

Box 4.1 Example 4.1

Example 4.1. Plot the locus of a single rotavator blade tip when the rotary speed of the tine is 200 rpm and the forward speed of travel of the machine is 3 km h^{-1}. The radius of revolution of the blade tip is 25 cm. Find the area of cross-section of the soil slice cut in each rotation when the depth of operation is 20 cm. Also, find the ratio of the tip peripheral speed to the forward speed of the rotavator.

Forward speed = 3 km h^{-1} = 83 cm s^{-1}

Rotary speed of the rotavator = 200 rpm = 3.33 rev s^{-1}

Forward travel per revolution of rotavator = 83 / 3.33 = 24.9 cm rev^{-1}

Forward travel per degree of rotation = 0.0691 cm deg^{-1}

Implying that the centre of rotation moves by 0.691 cm every 10° of rotation.

Before trying to calculate and plot the locus, let us recollect how one can plot a circle. It may be observed that a circle is the locus of the blade tip if there is no forward movement of the machine.

Knowing the radius of the blade tip, the x and y coordinates of the locus (circle) for every 10° rotation can be found as:

x = 25cos(θ°) and y = -25cos(θ°) where θ is from 0° to 360° in steps of 10° from the positive x-axis. The negative sign in the y coordinate is intended to move the locus down in clockwise rotation.

Use spreadsheet software, such as Excel, to calculate the x and y coordinates and plot them to draw the cycloidal locus. Table 4.1 shows a part of the Excel calculations. Or calculate by hand, tabulate, and plot the values on graph paper.

Once that is done, visualize the locus of the blade tip that is rotating with its centre moving at a constant rate of 0.691 cm every 10° of rotation. So, all that is to be done is to add the incremental movement of the centre to the x coordinates calculated for the circular locus. Then, if the coordinates are plotted, the locus of the blade tip with its centre moving forward is obtained. The calculation, if it is continued for another 360° rotation, of the intersection of the locus below the soil surface gives the area of the cut soil slice. Setting the X and Y grids in the spreadsheet chart accordingly, as shown below, or hand-plotted on graph paper, the area can be calculated by counting the number of squares inside the slice area.

Here are 17 (appr.) 25 cm^2 squares in the cut soil slice = 425 cm^2 = area of cut soil slice.

Now to calculate the ratio:

Peripheral speed of tip = 2.61 m s^{-1}

Forward speed = 3 km h^{-1} = 0.83 m s^{-1}

Tip speed to forward speed ratio = 2.61 / 0.83 = 3.14.

Table 4.1 Port of the EXCEL table of calculations.

Rotation, deg	Forward distance, cm	X for circle, cm	X for blade tip, cm	Y for blade tip, cm
0	0.00	25.00	25.00	0.00
10	0.69	24.62	25.31	-4.34
20	1.38	23.49	24.88	-8.55
30	2.07	21.65	23.73	-12.49
40	2.76	19.16	21.92	-16.06
50	3.46	16.08	19.53	-19.14
60	4.15	12.51	16.66	-21.64
....	....	....	....	....
170	11.75	-24.61	-12.87	-4.38
180	12.44	-25.00	-12.56	-0.04
190	13.13	-24.63	-11.50	4.30
....	....	....	....	....
350	24.19	24.61	48.79	4.42
360	24.88	25.00	49.88	0.08
10	25.57	24.62	50.19	-4.34
20	26.26	23.49	49.75	-8.55
....	....	....	....	....
350	49.06	24.61	73.67	4.42
360	49.75	25.00	74.75	0.08

and $R\cos\theta$ is the x coordinate of the tip on its rotation on the axis. The example in Box 4.1 throws more light on the locus of the rotavator tip. The above equations were not directly used in Excel in the solution to impart explicit understanding of the concept.

As the forward travel made per cut (per revolution of tine) is decreased, the pulverization of the soil increases as the cut slice is narrower. (A tutorial software is available with the book that can demonstrate the variations in loci of the rotor blades for various rotor and travel speed combinations. This can impart more insight to the student). Accordingly, the power requirement increases. The energy requirement to cut a unit volume of soil increases by four times when the forward travel made per cut is increased by 5.5 times. An analysis of the tine tip's locus showed that the ratio of the distance of tine travel in the soil to the depth of tilling is a minimum when the diameter of the pitch circle of tine tip rotation is about 1.25 times the depth. Thus, the power utilization of the rotavator can be improved by reducing the rotor diameter to approach this optimum ratio.

4.4 Subsoilers and chisel ploughs

Subsoilers and chisel ploughs are tools that do heavy work in unploughed soil but are classified as secondary tillage tools since they prepare the soil for the next season, before the summer rains and the ensuing crop season. Both types of tools are used to break the hard pan formed in the subsoil region due to soil compaction caused by excessive field traffic in previous crop seasons. This hard pan is generally formed in soils with more clay content and is a hard, impermeable layer of soil that impedes water infiltration to the soil. These tools are hence classified as conservation tillage tools, too, since they aid in moisture conservation by way of collecting the scanty rain in the soil cracks they form, especially in arid and semi-arid regions. Sometimes, the chisel plough is also viewed as a primary tillage tool used in place of conventional ploughs. It is used along with disc harrows on a common tool frame for stubble mulch tillage, as explained in Section 4.6.

The subsoiler (Figure 4.12) is usually a single-tined, deep tillage tool that can reach about 50 cm depth, shattering the hard pan and forming lateral cracks in the subsoil. It is used in dry conditions so that these cracks are formed, which helps with water infiltration at the time of rain. The subsoiler is run such that there is a minimum of one tread width of the pulling tractor between consecutive passes. This prevents the tractor from compacting the formed cracks underneath. The chisel plough (Figure 4.13) is a multi-tined implement that is ruggedly made and looks more or less like a rigid tined cultivator. This tool can also reach the subsoil, at 50 cm depth, but is pulled by a higher-powered tractor than what is used for conventional tillage. It is used at depths below the normal ploughing depth, when the hard pan layer is relatively thin and at a lesser depth.

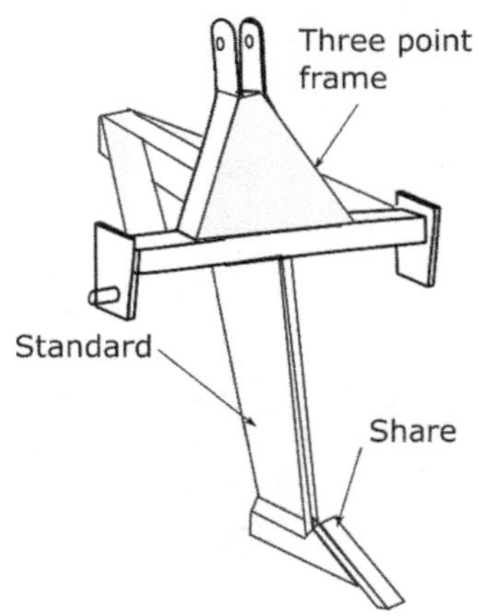

FIG. 4.12 Subsoiler.

FIG. 4.13 Chisel Plough (India Agrovision Implements Pvt. Ltd).

4.5 Ridgers

A ridger (Figure 4.14) is basically the same as two opposed, mould board plough bottoms placed side to side, with the landside eliminated, so that the soil is thrown to both sides. However, the mould boards have convex configurations rather than conventional shapes. They are used to form ridges and furrows in well-tilled seed-beds to grow irrigated row crops such as cotton and maize. In India and many Asian countries, ridgers are mainly used to form the ridges and furrows in which the

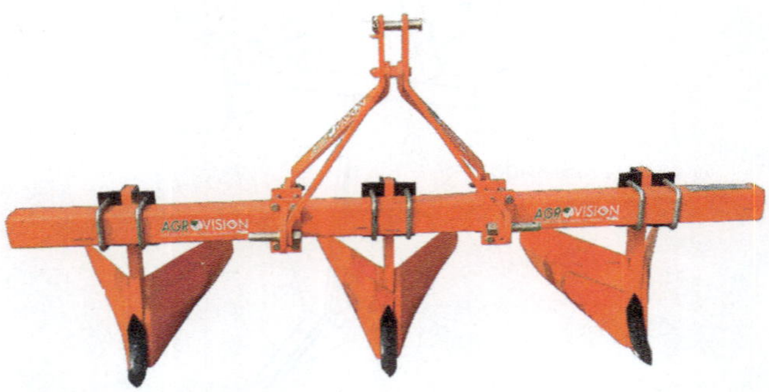

FIG. 4.14 Mounted Three-Bottom Ridger (India Agrovision Implements Pvt. Ltd).

planting is manually done on the sides of the ridges. The furrows receive irrigation for the crops to get established. But in many other countries, crops such as maize, soybeans, and cotton are planted with ridgers, known also as listers, having planter attachments that place the seed at the bottom of the furrow. Conventional planting devices are attached at the back of the ridger bottoms to achieve this. The planting is done in the furrow sole because that is the zone where the soil is firm enough to establish proper seed-to-soil contact. Once the crop establishes itself to a nominal height, the ridger is used once again to rework the ridges into furrows so that the plants are positioned on the ridges. Various sizes (measured between the outside wing tips of the ridger boards) are available, with 35 cm being a common size. The lateral components of the useful soil forces balance each other in a ridger, and the resultant useful force lies in a vertical plane in the direction of travel, making it into a symmetrical tool requiring a pull in the direction of travel. Ridger bottoms are mounted on trailed tool carriers but more often are attached to tractor toolbars. Depth wheels to control the depth of ridge formation are frequently employed in tractor-mounted arrangements.

4.6 Implements for stubble management and subsurface tillage tools

Tools and harrows for stubble management (McCalla and Army, 1961) are combination tools with various forms of secondary tillage devices that cut and incorporate

FIG. 4.15 Stubble Mulch Cultivation (CASE IH VT Flex 435 Vertical Tillage Tool, CNH Industrial, USA).

the stubble in the soil. They can also bring about a layer of shredded stubble to the top layer, providing a mulch cover and impeding the evaporation of soil moisture. Figure 4.15 is an example of such a tool that has a rugged, curved, tandem disc harrow with curved disc edges, 'S'-type spring tines preceding the discs, and paddle rollers or crumblers to sift and settle down the pulverized stubble and soil. These are very wide implements that have hydraulically controlled depth and gang angles. Subsurface tillage tools till the soil beneath the soil surface and under the crop residue left after a previous crop. The objective is to disturb the subsurface soil, leaving the crop residue undisturbed on the soil surface. This method of farming is highly suitable for dry land farms, where soil moisture already available in the soil is to be conserved for crop growth. Stubble mulching and crop residue management are among the nomenclature used to describe this method. Soil mulching is another concept in which bigger soil aggregates are manipulated to the soil surface so that the evaporation of subsoil moisture is impeded by the clods on the surface. Machines are available that sow the seeds beneath the crop residue without disturbing the soil surface. These tools and methods can also be called conservation tillage practices. The broader objectives of using these tools are to increase the capacity of the soil to absorb and retain moisture and reduce surface moisture evaporation, runoff, and erosion. Sweeps are tools that meet these objectives.

Subsurface tillage sweeps are set almost without any vertical suction on conventional cultivators; this allows crop residue to flow around and between the tool shanks and remain in place on the soil surface. However, the sweep blades skim the uppermost soil layer under the residue cover. Crop residue is not to get collected on the tines. Wider sweeps are used along with notched coulters to prevent crop residue from getting tangled on the cultivator tines. The construction of sweeps is explained in Chapter 7, which discusses row crop cultivators and weed control.

Minimum tillage, strip tillage, and zero tillage are all conservation tillage practices that use varied magnitudes of tillage. Zero tillage is the practice in which literally no tool is used to till the soil but seeds are planted with minimum soil disturbance. Minimum tillage, as the name implies, uses a very minimal tillage before planting. Strip tillage is tilling the land in strips in which the planting is to be done without disturbing the other areas of soil. Tool combinations appropriate for each system are used by selecting from the already available array of primary and secondary tillage tools.

4.7 Machines for formation of raised or wide beds

Raised-bed or wide-bed cultivation (Sayre and Hobbs, 2004) is practiced in many situations, including conservation agriculture, vegetable farming, and farming under waterlogged conditions. A wide bed is a land formation in which beds of 1 m and above with loose soil are formed with furrows on their sides. The formation of wide beds is performed by machines of different configurations. The simplest of

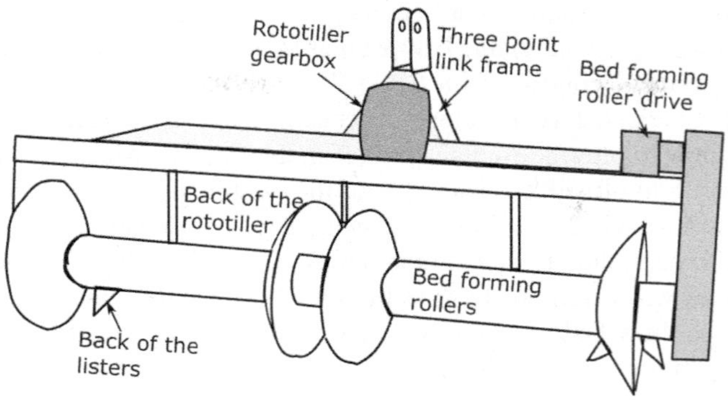

FIG. 4.16 Wide-Bed-Forming Machine.

the methods is that the land is prepared as usual and harrowed or rota-tilled to make the soil friable, after which listers or ridgers, spaced at the required bed width, are used to form the beds. The tractor's wheel tread is adjusted to match the width of the bed so that they can ride in these furrows without disturbing the beds. There are wide-bed forming machines, which are combination tools of wide rotavators with listers or ridgers set to form the wide beds (Figure 4.16). Bed-shaping rollers in the form shown are usually hydraulically driven pieces of equipment that shape the beds with sloping sides and compact them softly. These machines are also used with plastic mulching attachments that automatically lay a thin plastic sheet over the bed, covering the bed. Punch seeders or dibbers are used later on these beds to sow the seeds or plant the seedlings on these beds. Versions are available that can perform these operations inside greenhouses or poly-houses for raising vegetable crops or for raising crops in open fields.

The advantages offered by raised beds are:

- The beds are soft and loose and kept without any traffic on them, allowing the crop to establish with better root depth. This is because all the ensuing operations of sowing, weed control, plant protection, and the application of other inputs are performed by a tractor riding only in the trenches or furrows on the sides.
- These beds are amenable to plastic mulching and, hence, can save soil moisture that can control crop growth, leading to better productivity.
- This technology is useful in arid land for the conservation of available soil moisture due to scanty rains.
- It is also advantageous in water flooded areas, where the furrows allow for the draining of excess water to cross drains and thereon to the main drain channels off the field.
- It offers a systematic way of crop cultivation that can guide all the farm operations in the formed land configuration and usually is very efficient if planned properly.

- Wide beds are made permanent in certain farms and the same lanes are used for on-farm traffic to avoid undue compaction by heavy machinery. This is generally termed "controlled traffic".
- The beds are oriented usually along the existing slope of the field, and the slope is made shallow to prevent excessive runoff in the furrows. Mostly, wide beds are formed at 1.8 m of width to suit most of the existing tractors and machines. In many farms, wide beds are drip- or sprinkle-irrigated. However, irrigation is also achieved through the furrows in other crops. The beds, being soft and undisturbed, draw water from the furrows by sidewise water movement. Rows of crop are sown on these beds, and conventional weeding and plant protection operations are performed using tractor-mounted toolbars set to match the crop rows.

The following problems in Box 4.2 and Box 4.3 compare the energy requirements of secondary tillage implements.

Box 4.2 Example 4.2

Example 4.2. A farmer needs to decide whether to use a rotavator for preparing a seedbed of 15 cm depth or to go for conventional ploughing and harrowing to the same depth. The farmer wants to compare the energy cost of rota-tilling the soil to that of a sequence of ploughing with a mould board plough, followed by using a disc harrow twice, and finally finishing with a spike-toothed harrow. For the rotary tiller method, take the forward travel per soil slice as 22 cm with an energy requirement of 125 (kNm) m^{-3} of soil tilled. The unit draft of the mould board plough in the same soil is about 48 kPa at 5 km h^{-1}. The draft per unit width of a disk harrow is 2 kN m^{-1} and that of a spike-toothed harrow is 0.5 kN m^{-1}. Discuss the results.

For rotavator:

Depth of ploughing = 15 cm

For rota-tilling a width of 1 m at a depth of 15 cm with a forward travel of 1 m,

the energy requirement = 125 (kNm) m^{-3} × volume of soil tilled

= 125 × 1 × 1 × 0.15 = 18.75 kNm

For mould board plough:

Unit draft = 48 kPa = 48 kN m^{-2}

For ploughing to a depth of 15 cm to a forward travel of 1 m,

Area of ploughing swath = 1 m × 0.15 m = 0.15 m^2

Draft = 48 kN m^{-2} × 0.15 m^2 = 7.2 kN

Energy requirement = Draft × Linear run of 1 m = 7.2 kNm

For harrowing a width of 1 m and to a forward travel of 1 m,

Draft per unit width = 2 kN m^{-1} = Draft

Energy requirement = Draft × Linear run of 1 m

For spike-toothing a width of 1 m and to a forward travel of 1 m,

Draft per unit width = 0.5 kN m^{-1}

∴ Energy requirement = Draft × Linear run of 1 m = 0.5 kNm

Total energy requirement for ploughing with mould board, disc harrowing twice, and spike-toothing = 7.2 + 2(2)+ 0.5 = 11.7 kNm

Comparing the energy cost between the two operations, the energy cost of rota-tilling is 60% more than the conventional ploughing and harrowing. But, the quality of tilth offered by rota-tilling is superior and, hence, the farmer may prefer it over conventional harrowing.

Box 4.3 Example 4.3

Example 4.3. In the last problem, calculate the maximum forward speed at which a 1.5 m rotary tiller could be operated if the tractor has a PTO rating of 33 hp and can be loaded to only 80 percent of this rating. Find the maximum forward speed if the travel per soil slice is reduced to 10 cm, which increases the energy demand to 220 (kNm) m^{-3} of soil tilled. What is the drawbar horsepower required to pull a three-bottom 40 cm mould board plough at 5 kmh^{-1} at a unit draft of 48 kPa?

Energy requirement of the machine = 125 (kNm) m^{-3} of soil tilled
Width of the machine = 1.5 m
Depth of tilling = 0.15 m
Volume of soil tilled for per m run = 1.5 × 1 × 0.15 = 0.225 m^{-3}
Energy requirement of the machine per m run = 125 × 0.225 = 28.125 (kNm) m^{-1}
Let forward speed of machine = S ms^{-1}
Energy requirement of the machine = 28.125 × S = (28.125 S) $kNms^{-1}$ = (28.125 S) kW
PTO rating of tractor = 33 hp = 24.3 kW
Available power is 80 percent of the rating = 24.3 × 0.8 = 19.4 kW
Equating the energy demand to the available power,
28.125 S = 19.4
The forward speed permissible S = 0.68 ms^{-1} = 2.47 kmh^{-1}
When the forward travel per soil slice is reduced to 10 cm,
Energy requirement of the machine per m run = 220 × 0.225 = 49.5 (kNm) m^{-1}
Now, similarly equating the energy demand to the available power,
49.5 S = 19.4 kW
The maximum forward speed permissible S = 0.39 ms^{-1} = 1.41 kmh^{-1}
Effective width of MB plough = 0.4 × 3 = 1.2 m; depth of ploughing = 0.15 m
Area of furrow = 1.2 × 0.15 = 0.18 m^2
Unit draft = 48 kPa
∴ Draft = 48 × 0.18 = 8.64 kN
Forward speed = 5 kmh^{-1} = 1.38 ms^{-1}
Drawbar power required = Draft × Speed = 8.64 × 1.38 = 11.92 kW

Exercise problems

1. Assuming that there are two rotavator tines in line on the rotavator axis, plot the locus of the two blade tips. The rotary speed of the tine is 200 rpm, and the forward speed of travel of the machine is 4 km h^{-1}. The radius of revolution of the blade tip is 27 cm. Find the area of cross-section of the soil slice cut in each rotation when the depth of operation is 15 cm. Use the equations of the tip locus to solve the problem in Excel.
2. The rotary speed of a rotavator tine is 200 rpm, with the forward speed of travel of the machine being 3 km h^{-1}. The radius of revolution of the blade tip is 25 cm, and the depth of operation is 20 cm. Find the area of cross-section of the soil slice cut in each rotation by integration.
3. Find the rise in total energy demand if the energy demand per unit volume of soil is increased by 80 percent when the forward speed is reduced by 30 percent. Assume the depth of operation is 15 cm in both cases. Calculate the maximum

forward speed at which a 1.5 m rotary tiller could be operated if the tractor has a PTO rating of 33 hp and can be loaded to only 80 percent of this rating.

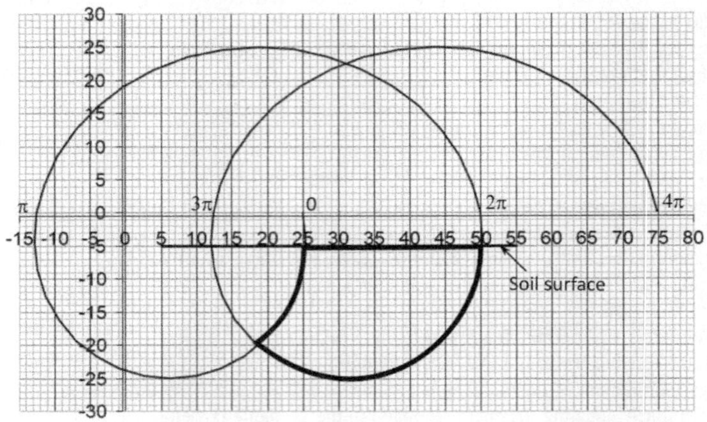

References

Bell, B. (2005). *Farm machinery.* UK: Old pond Publ.

Bernachi, H., Haman, J., & Kanafojski, C. (1972). *Agricultural machines, theory and construction* (p. 1). Warsaw, Poland: Published for USDA and NSF by Foreign Cooperation Center of the Central Institute for Scientific, Technical and Economic Information.

Culpin, C. (2013). *Farm machinery.* Hesperides Press, 144370301X, Hesperides Press. ISBN 144370301 :X.

Hunt, D. (2008). *Farm power and machinery management.* Waveland press.

Kepner, R. A., Bainer, R., & Barger, E. L. (1978). *Original from the university of michigan.* Avi Publishing Company.

Klenin, N. I., Popov, I. F., & Sakun, V. A. (1970). *Agricultural machines . USDA and NSF by Amerind.* Publishing Co. Pvt. Ltd, USDA and NSF by Amerind Publishing Co. Pvt. Ltd.

Lovegrove, H. T. (1968). *Crop production equipment.* Hutchinson press.

McCalla, T. M., & Army, T. J. (1961). Stubble mulch farming. *Advances in Agronomy. 00652113, 13*(C), 125–196. https://doi.org/10.1016/S0065-2113(08)60958-X.

Sayre, K. D., & Hobbs, P. R. (2004). *Sustainable agriculture and the international rice-wheat system. The raised bed system of cultivation for irrigated production conditions* (pp. 337–355). CRC Press. Sustainable agriculture and the international rice-wheat system. CRC Press, Sustainable agriculture and the international rice-wheat system.

Smith, H. P. (1965). *Farm machinery and equipment.* Tata Mc-Graw Hill Pub. Co. Ltd.

Srivastava, A. K., Goering, C. E., Rohrbach, R. P., Buckm, D. R., & ster. (2006). *Engineering principles of agricultural machines.* ASABE Publications.

CHAPTER

Mechanics of field machinery and their hitchings

5

Chapter Outline

Machines of the Farm. https://doi.org/10.1016/B978-0-443-32906-7.00005-0

An engineer who desires to learn about the description and working of agricultural machinery, especially tillage tools, must possess a deep understanding of the mechanics of the soil working parts of the machine. This does not pertain only to tillage equipment but also to crop cultivation, weed control, and sowing and harvesting machinery too. To find out a machine's power requirement as well to know how its proper hitching to the prime mover influences its performance, the forces acting on the implements, particularly the tillage tools, should be well understood by the engineer. Moreover, the structural design of such agricultural machinery as well the choice of standard machine parts in their assembly, such as fasteners and anti-friction bearings, require a thorough knowledge of the forces acting on them. An inquisitive student should see every nut and bolt of the machine and ask why it is there and how it has been chosen to be of that specification.

5.1 Forces on tillage tools and their analyses

In preparation for analysing the forces acting on a soil working tool, we need to understand how the soil reacts to a cutting tool entering the soil. This was explained by Sohne's model (Sohne, 1956) (Figure 5.1), which gives an insight into what are the influencing factors that contribute to the soil reacting to a simple inclined tool entering the soil.

When a simple inclined tool having a rake angle, α (the angle between horizontal and the tool surface), penetrates and moves forward in the soil, the tool pushes the soil in the front and causes a shear plane along the line "qr," breaking the soil and allowing it to move on the inclined tool face. The shear plane is oriented at the shear angle, β, which is assumed as 45° to the horizontal.

Soil, being a cohesive medium, obeys Coulomb's law (Gill and Berg, 1968), expressed by the equation $\tau = c + \sigma \tan \varphi$, where τ is the shear stress, σ the normal stress, c is the cohesion, and φ *is* the internal soil friction angle. This is in the form of a linear equation where c is the intercept and $\tan \varphi$ *is* the slope relating τ and σ. The soil failure along the shear plane obeys Coulomb's law expressed above.

The soil slice that moves over the tool surface is opposed by the adhesive force, A, due to adhesion and soil-tool friction on the interface between the soil and tool

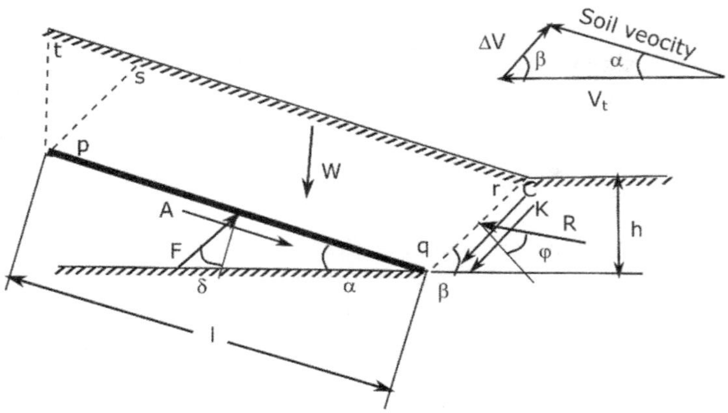

FIGURE 5.1 Sohne's Model of a Simple Inclined Tool

No Permission Required

surface. The soil-tool frictional stress generated is governed by an adhesion equation (Gill and Berg, 1968) analogous to the Coulomb's equation, which is of the form $\epsilon = a + \sigma \tan \delta$, where ϵ is the frictional stress, σ *is* the normal stress, a is the adhesion, and δ *is* the soil-metal friction angle.

The soil that is sliding on the inclined surface has a weight that is acting vertically downward and, for simplicity, it is assumed to be a single chunk of soil over the whole tool surface. Actually, the soil will be broken into sheared segments while riding over the tool.

Once the soil slides over the tool, it is accelerated up over it due to the forward velocity of the tool, V_t. The velocity vector of the soil sliding over the tool is a vector sum of V_t, and the differential velocity vector, ΔV, as shown in the vector diagram (inset of Figure 5.1).

$$\Delta V = V_t \frac{\sin\alpha}{\sin(\alpha + \beta)}. \text{ (by sine law)}$$

The acceleration or inertial force that is one force component of how soil reacts to the tool interaction can be calculated from ΔV. The soil slice that is accelerating during a time interval, Δt, yields a mean acceleration of $\frac{\Delta V}{\Delta t}$. The soil mass that is being accelerated at this time interval equals $\rho b h V_t \Delta t$, where b is the blade width, ρ is bulk density of the soil, and h is the working depth of the tool. The inertial force, K, is then of the form:

$$K = \rho b h V_t \Delta t \frac{\Delta V}{\Delta t} = \rho b h V_t^2 \frac{\sin\alpha}{\sin(\alpha + \beta)}$$

It may also be noted that the direction of this acceleration force is in the opposing direction of ΔV, as indicated in the vector diagram of the figure (inset).

Now, $C = \frac{hc}{\sin \beta}$, where C = cohesive force, c = cohesion, and h = the cutting depth of the tool,

$A = la$, where A = adhesive force, a = adhesion, and l = the blade length.

Also, W, the weight of the soil on the blade considered as a single chunk, is:

$$W = \rho blh \frac{\sin(\alpha + \beta)}{\sin \beta} + \frac{1}{2} \rho g h^2 \frac{\sin(\alpha + \beta) \cos \beta}{\sin^2 \beta \cos \alpha}$$

where the volume and thereby the weight are calculated for the soil segment as a sum of the parallelogram, "pqrs," and the triangular region, "stp."

So far, we have derived for the component forces opposing the tool movement, viz., weight of the soil chunk, the adhesive force, the cohesive force, and the acceleration force on soil. The vectorial sum of all the opposing forces will be in equilibrium with the blade force comprised of the unknown *F,* the resultant of the normal and its accompanying frictional force that the blade exerts on the soil. Apart from these, there is *R,* the reaction of the still undisturbed soil on the soil front opposing tool motion, which is unknown.

Now, equating all the vertical forces and horizontal forces, we get

Summing forces in the vertical direction,

$$W + A\sin \alpha + (C + K)\sin \beta = F\cos (\alpha + \delta) + R\cos(\phi + \beta)$$

Summing forces in the horizontal direction,

$$A\cos \alpha - (C + K)\cos \beta = F\sin (\alpha + \delta) - R\sin (\phi + \beta)$$

These two equations help us solve for F and R, which are both unknown. The resolution of F, the blade force, will then yield the horizontal and vertical reactions of the soil to the cutting tool. The following example problem in Box 5.1 will throw more light on the procedure.

Any soil working tool has soil reactions likewise. A symmetrical tool such as this simple inclined tool has only horizontal and vertical reactions, whereas an unsymmetrical tool such as a mould board plough has a lateral soil reaction also, perpendicular to the travel direction due to the soil being moved in that direction.

Before learning about the force dispositions on tillage implements, we need to define some more commonly used terms in this context. "Pull" is the net force applied on a farm implement by a power source, viz., tractor, power tiller, or animal. The horizontal component of pull parallel to the line of travel is known as the "draft" and the horizontal component of pull perpendicular to the line of travel is known as "side draft." The pull also has a vertical component in it. The drawbar horsepower of an implement is the power actually needed to move the implement at a uniform speed. It is understood that the following three forces are to be in equilibrium when the tool is in operation:

(a) weight of the implement:
(b) pull applied on the implement; and
(c) total soil reaction as offered by the soil opposing movement and weight.

Box 5.1 Example 5.1

Example 5.1. A 30 cm wide and 15 cm long flat blade with a rake angle of 30° is operated at a depth of 20 cm in a cohesionless soil having an internal friction angle of 35°. The tool is operated at 4 kmh^{-1} in this soil of 1200 kg m^3 density. The soil-metal frictional coefficient is 0.3. Neglect the effect of adhesion forces and calculate the horizontal and vertical soil reactions of this simple raked blade. The soil failure plane may be assumed to be at 45°.

Now from the problem given:

Rake angle, $\alpha = 30°$;

Blade width, b = 30 cm;

Blade length, l = 15 cm;

Depth, h = 20 cm;

Angle of soil internal friction, $\varphi = 35°$;

Angle of shear failure plane, $\beta = 45°$;

Coefficient of soil-metal friction = 1.3; angle of soil-metal friction, $\delta = \tan^{-1}(1.3) = 17°$

Soil density, $\rho = 1200\ kg/m^{-3}$;

Tool forward velocity, $V_t = 4\ kmh^1 = 1.11\ ms^{-1.}$

According to the Sohne's model equations, the acceleration force of the soil on the blade is

$$K = \rho bhV_t^2\frac{\sin\alpha}{\sin(\alpha+\beta)} = 1200 \times 0.30 \times 0.20 \times 1.11^2\frac{\sin 30}{\sin(30+45)} = 46N.$$

The weight of the soil slice considered as a single chunk is calculated as:

$$W = \rho blh\frac{\sin(\alpha+\beta)}{\sin\beta} + \frac{1}{2}\rho gh^2\frac{\sin(\alpha+\beta)\cos\beta}{\sin^2\beta\cos\alpha}$$

$$W = 1200 \times 0.3 \times 0.15 \times 0.2 \times \frac{\sin(30+45)}{\sin 45} + \frac{1}{2}1200 \times 9.81 \times 0,20^2\frac{\sin(30+45)\cos 45}{\sin^2 45\cos 30} = 395N.$$

Summing the forces in the vertical direction:

$W + A\sin\alpha + (C+K)\sin\beta = F\cos(\alpha+\delta) + R\cos(\phi+\beta)$.

Summing the forces in the horizontal direction:

$A\cos\alpha - (C+K)\cos\beta = F\sin(\alpha+\delta) - R\sin(\phi+\beta)$.

C and A = 0 since cohesion c and a = 0, being a cohesionless and adhesionless soil. Substituting the other values, the equations become:

$395 + 46\sin 45 = F\cos(30+17) + R\cos(35+45)$

$428 = 0.68F + 0.17R$ ———————— (1)

$-46\cos 45 = F\sin(30+17) - R\sin(35+45)$

$-33 = 0.73F - 0.98R$ ———————— (2).

Solving equations for F:

F = 595N

Horizontal reaction, L = 595 sin(30+17) = 435 N,

Vertical soil reaction, V = 595 cos(30+17) = 405 N.

The total soil reaction is classified into "useful" and "parasitic" reactions (Clyde, 1937; Kepner, Bainer, and Barger, 1978; Smith, 1965). Useful soil reactions are those that the tool overcomes to cut, manipulate, and move the soil. Parasitic reactions are the associated reactions including sliding and rolling friction that act on the implement to support and stabilize it. The plough sole and landside counteracting the downward and lateral reactions and forces, respectively, offer parasitic reactions. The implement designer as well as the operator has some control over these parasitic reactions by way of adjustments on the soil working tool.

It is fairly easy to measure the forces on a symmetrical trailed tillage tool, which does not have any soil reaction laterally perpendicular to travel in the horizontal plane. The implement is hitched to the power source by a chain or cable through which the line of pull passes. The position and angle of the tool is adjusted until the tool operates at the desired depth, width, and approach angle. The pull is measured by a load cell or a hydraulic dynamometer mounted in line with the pulling chain or cable. After measuring the tool's weight and its centre of gravity separately, the total soil reaction required for equilibrium is determined by the parallelogram law of forces from the known weight passing through the centre of gravity and measured pull passing along the line of pull. However, finding out the exact useful and parasitic reactions from the total soil reaction using this method is a difficult task.

As an instance of a symmetrical tool, let us look at how the indigenous plough used in India is hitched to a pair of bullocks (Figure 5.2). This plough is simply a symmetrical wedge type of tool, which cleaves the soil to a certain depth, making a triangular furrow. The beam of the plough, which is firmly attached to the plough body, is made to rest on the yoke being pulled by the pair of bullocks. The plough operator moves the animals back and forth relative to the plough before hitching it, so that the plough bottom has the right vertical suction angle for proper penetration. Then, the rope that is secured to the plough bottom at the hitch point (HP) is tied to the yoke firmly with a loop on the yoke for the beam to slide a little bit. The bullocks pull the plough with the rope as from the centre of the yoke, which is the centre of pull (CP). The plough penetrates into the soil and moves forward, cleaving the soil. HP is usually a small ring located on the plough body to which the rope is tied.

As already explained, when a dynamometer or load cell is connected directly in the line of pull, (i.e., on the rope), the total pull is measured. The draft is then calculated as a horizontal component of pull after measuring the angle of pull from the horizontal.

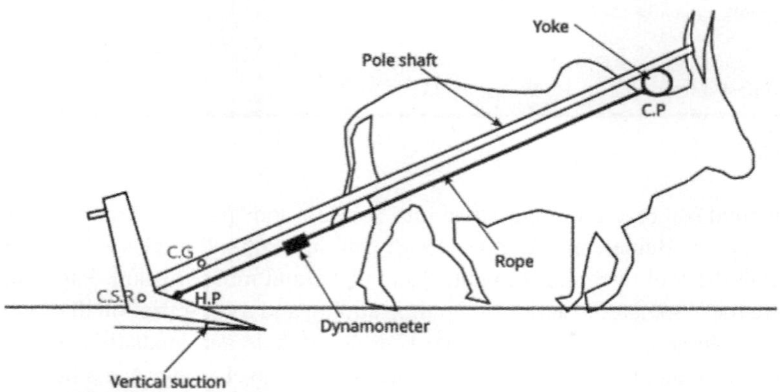

FIGURE 5.2 Hitching of a Bullock-Drawn Country Plough

No Permission Required

Table 5.1 Typical unit draft values for different soils

	Type of soil	Unit draft, kPa
1	Sand at 5 percent moisture	13 to 20
2	Sandy loam at 10 percent moisture	43 to 58
3	Clay at 15 percent moisture	61 to 77

No Permission Required

The total power required of any implement is a function of draft and not the total pull. Unit draft is defined as the draft of the implement per unit area of furrow cross-section it cuts and is expressed in kPa. The general range of unit drafts for many implements in different types of soil is given in Table 5.1 for reference.

5.2 Measurement of useful soil reactions on a soil working tool

A six-cell dynamometer (Clyde, 1936a, 1937a; Godwin *et al.*, 1993) is a device on which the tillage tool can be mounted and pulled to measure the complete set of soil reacting forces including the couple acting on a soil working tool. Forces are thus measured in six degrees of freedom, and these six forces are resolved into the desired system of force representation (Clyde, 1936a). Sophisticated tillage tool dynamometers are available today with computerized instrumentation and data loggers. Telemetric devices make it possible to measure the forces from a remote location well away from a tillage tool operated in the field.

Such a six-cell dynamometer (Durairaj and Balasubramanian, 1997) is shown in Figure 5.3. It can be of many designs, but what is shown consists of an "L" frame, that is carrying the plough bottom on a vertical drop. The L frame is fully restrained in six degrees of freedom by six load cells. The indicated locations of the force vectors, L_1, L_2, L_3, V_1, V_2, and S, are where the load cells are attached to the frame and, in turn, mounted to another fixed frame carried by the tractor. This implies that each load cell will measure the force exerted on it between the L frame and the tractor in the direction shown. It may be noted that the mounting frame on the tractor is not shown in the illustration for improved clarity. Of the six load-measuring devices, three are measuring the horizontal forces acting on the plough (three degrees of freedom restrained), two devices are measuring the vertical forces (another two degrees of freedom restrained), and one is measuring the lateral force (another one degree of freedom restrained), making the L frame carrying the plough completely restrained in all six degrees of freedom. This means that all the forces acting on the plough as soil reactions will be totally represented by these six forces at any instant, viz, three horizontal, two vertical, and one lateral forces. However, it is

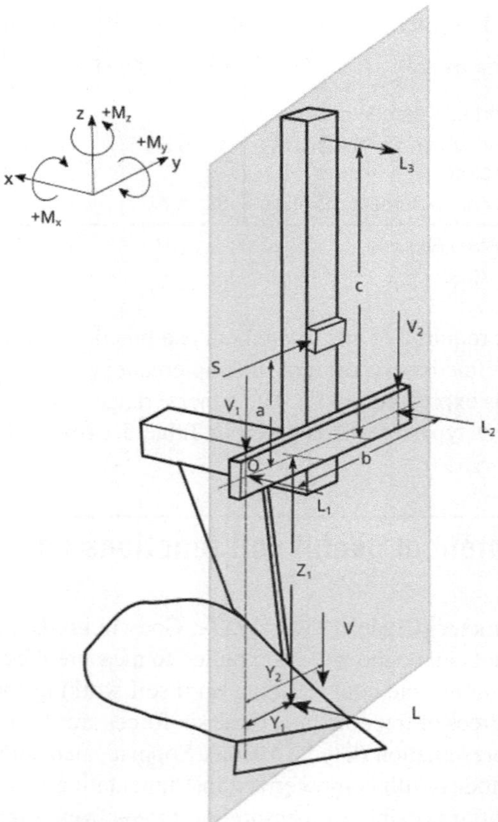

FIGURE 5.3 Forces Measured by a Six-Cell Dynamometer

No Permission Required

better if the forces can be represented in a more concise form, so that they can be useful for any design and analysis of the plough forces.

Parasitic forces can be partially eliminated from the measured soil reaction by removing supporting parts such as the landside. The landside of the plough is hence removed during measurement so that the lateral soil reaction perpendicular to travel is not compensated and can be measured by the dynamometer. The implement weight is eliminated from the force measurements on the tillage dynamometer by taking initial load readings with the tool just suspended on the dynamometer. The forces due to the weight of the plough when not in operation can hence be deducted from the measured vertical forces so as to compute the soil reactions alone exclusive of the weight. Now, let us consider a set of measurements thus observed (Table 5.2) on a single mould board plough bottom (Figure 5.3) as mounted on the measuring L frame of the dynamometer.

Table 5.2 A sample of measured forces on a single mould board plough bottom

Load cells	Measured force	Load cells	Measured force
L_1	1350 N	V_1	300 N
L_2	1000 N	V_2	150 N
L_3	1300 N	S	200 N

All the points of measurement on the tillage meter are conveniently constructed to be on a plane (shown in grey) perpendicular to the travel direction and passing through point "O." Point "O" is selected to be the point at which L_1 and V_1 are measured. The linear dimensions of the dynamometer are a = 0.1 m, b = 0.3 m, and c = 0.4 m. The axes of measurement as well as the convention for positive directions of the moments as per the righthand thumb rule are also indicated in the figure.

We know that:

L is the soil reaction along the travel direction = $\Sigma F_x = L_1 + L_2 - L_3 = 1350+1000-1300 = 1050$ N;

V is the soil reaction perpendicular to the ground = $\Sigma F_z = V_1 + V_2 = 300+150 = 450$ N;

S is the soil reaction perpendicular to the travel direction and parallel to the ground = $\Sigma F_y = S = 200$ N.

Taking moments about point "O" at which L_1 is measured.

Moment about the x-axis = $\Sigma M_x = V_2 \times b + S \times a = 150 \times 0.3 + 200 \times 0.1 = 65$ Nm;

Moment about the y-axis = $\Sigma M_y = L_3 \times c = 1300 \times 0.4 = 520$ Nm;

Moment about the z-axis = $\Sigma M_z = L_2 \times b - L_3 \times (b/2) = 1000 \times 0.3 \times 1300 - 0.15 = 105$ Nm.

Then, Y_1 is the horizontal offset of L from point "O" = $\Sigma M_z / L = 105 / 1050 = 0.1$ m;

Z_1 is the vertical offset of L from point "O" = $\Sigma M_y / L = 520 / 1050 = 0.495$ m, meaning that L passes through coordinates 0, 0.1, −0.495 m with reference to point "O";

and Y_2 is the horizontal offset of V from point "O" = $M_x / V = 65 / 450 = 0.144$ m, meaning that V passes through coordinates 0, 0.144, 0 m with reference to point "O".

We see that S (at the point of measurement itself) passes through coordinates 0, 0.1, 0 m with reference to point "O".

We have thus reduced the six measured non-intersecting forces on the dynamometer into three non-intersecting forces acting on the plough. Figure 5.4A shows these forces on the plane passing through point "O" with reference to the plough. It may be noted that these are the forces of resistance that the soil imposes on the plough. And, they can hence be called the soil reactions to the plough operation. L is the force due to cutting of the furrow slice and pushing it in the travel direction; V is the force due

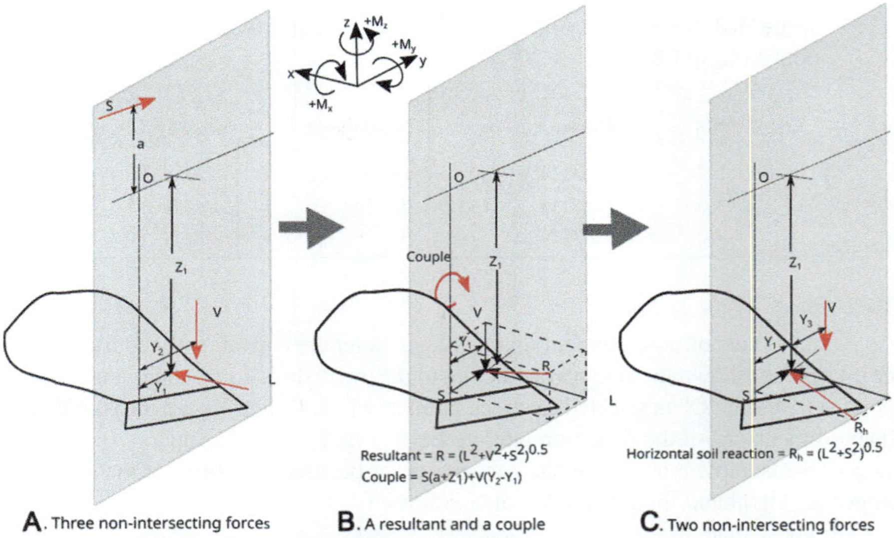

FIGURE 5.4 Different Representations of the Soil Forces Acting on the Plough

No Permission Required

to the vertical suction of the soil, making the plough to penetrate plus the weight of soil moving over the mould board; and S is the lateral soil force imposed on the mould board in a direction perpendicular to travel due to its movement on the mould board. Moreover, the acceleration of the soil on the mould board also manifests as forces in these three directions. Although the forces are fully represented, it would be convenient if the force representation is more concise.

The force S at a vertical offset of "a" from point "O" can be shifted down along the vertical plane to the point where L is positioned at a vertical offset of "Z_1" from point "O." By doing so, it would introduce a positive couple about the x-axis (CW, as seen from the front):

$$= S \times (a+Z_1) = 200 \times (0.1 + 0.495) = 119 \text{ Nm to bring the system to the original state}$$

Similarly, the force V at a horizontal offset of Y_2 from point "O" can be shifted horizontally on the vertical plane to the point where L is positioned at an offset of Y_1 from "O." By doing so, it would introduce another positive couple about the x-axis:

$$= V \times (Y_2 - Y_1) = 450 \times (0.144 - 0.1) = 20 \text{ Nm to bring the system to the original state}$$

1. The total couple that was introduced due to the shifting of V and S to the location of L = $S \times (a+Z_1) + V \times (Y_2-Y_1) = 119 + 20 = 139$ Nm.

Now that the two forces, S and V, have been brought to the location of L at 0, 0.1, −0.495 m with reference to point "O," they can be vectorially added into:

2. A single resultant as R $= \sqrt{(L^2 + V^2 + S^2)} = \sqrt{(1050^2 + 450^2 + 200^2)} =$ 1160 N passing through 0, 0.1, −0.495 m with reference to point "O" and at an angle of $\tan^{-1}$(V/L) = $\tan^{-1}$(450/1050) = 23.2° to the horizontal plane and at an angle of $\tan^{-1}$(S/L) = $\tan^{-1}$(200/1050) = 10.8° to the vertical plane along travel.

Figure 5.4B shows this new representation of the same set of three non-intersecting forces in the form of a single resultant (1) and a couple (2), as calculated above.

Sometimes, as in the case of a disc plough, it may be desired that the forces are to be represented as two non-intersecting forces. Figure 5.4C shows this mode of representation. This can be brought about from the last representation of forces and has

- the horizontal soil reaction R_h, the resultant of L and S $= \sqrt{(L^2 + S^2)} = \sqrt{(1050^2 + 200^2)} = 1068$ N passing through 0, 0.1, −0.495 m with reference to point "O" and at an angle of $\tan^{-1}$(S/L) = $\tan^{-1}$(200/1050) = 10.8 ° to the vertical plane along travel.
- The vertical soil reaction V is shifted away from the location of L along the lateral direction to equal the couple in the last representation (Figure 5.3B).

The total positive couple was S × (a+Z_1) + V × (Y_2−Y_1) = 139 Nm then, and V is now shifted through {S × (a+Z_1) + V × (Y_2−Y_1)} / V = 139 Nm / 450 = 0.31m from the location of R_h, i.e., V is passing through 0, 0.41m with reference to point the "O."

The above two are the two non-intersecting forces that express the same disposition of reactions on the plough.

So, a single force plus a couple being placed in a vertical plane perpendicular to the direction of travel is one convenient expression of the force disposition (Figure 5.3B). If the tool is symmetrical, then the couple will be zero and the net force is just one along the direction of travel. One could also express the net forces as two non-intersecting forces, as shown in Figure 5.3C. This is the choice of the analyst based on the requirements.

5.3 Soil reactions on a mould board plough

As from the above, the reactions can be expressed in any of the three methods according to the requirement. Because a mould board plough is a suction-assisted plough, the soil pulls down the plough, making it penetrate; so, the vertical force, V, is acting downward. Being unsymmetrical in action, it also has a couple, as expressed in Figure 5.4B. The centre of the soil resistance on the plough bottom, the point through which the resultant soil reaction passes, is approximately at a distance equal to about one-fourth the width of the cut on the face of the bottom to the right of the shin and near the joint between the shin and mould board (Clyde, 1937). The example problems in Box 5.2 and Box 5.3 illustrate the manner of dispositions of the soil reactions of a mould board plough.

Box 5.2 Example 5.2

Example 5.2. The line of pull of a trailed single-bottom mould board plough is at 15° to the horizontal plane and is at 10° to a vertical plane parallel to the direction of travel. The downward suction force of the plough is 500 N and acts at an offset of 15 cm to the right of the centre of the soil resistance as seen from the plough front. Neglecting the effect of plough weight and parasitic reactions, calculate the horizontal and lateral reactions and pull on the plough as well the couple. Find the drawbar horsepower required to pull the plough at a speed of 5 kmh^{-1}.

Let us shift the vertical force (suction force) to the centre of the soil resistance without altering the system of forces.

This would result in the CW couple being exhibited = 500 N × 0.15 m = 75 Nm.

Now that all the soil reactions are at the centre of soil resistance,

R_h is the resultant of L, the horizontal reaction (draft), and S, the lateral reaction (side draft). The overall resultant soil reaction is the resultant of R_h and V, the vertical reaction. The angle between the pull and R_h (on horizontal plane) is 15°.

Then $R_h = V / \tan(15°) = 500\ N / 0.27 = 1852\ N$.

The angle between R_h and the horizontal soil reaction L is 10°.

Then, $L = R_h \cos(10°) = 1852 \times 0.984 = 1824\ N$

and $S = R_h \sin(10°) = 1852 \times 0.17 = 322\ N$.

The pull, P, to be applied will be equal and opposite to the resultant of the horizontal, vertical, and lateral soil reactions (neglecting weight and parasitic reactions, as assumed in the problem).

Then, $P = (1824^2 + 322^2 + 500^2)^{0.5} = 1919\ N$.

Drawbar power = Draft × Speed of travel

$= L \times Speed = 1824\ N \times 5\ kmh^{-1} = 1824\ N \times 1.39\ m\ s^{-1} = 2535\ W = 3.44\ hp$.

Box 5.3 Example 5.3

Example 5.3. The pull applied on an animal-drawn, right-hand mould board plough was measured as 1500 N, the line of pull being at 14° to the horizontal plane and 11° to a vertical plane parallel to the direction of travel. Neglecting the effect of the plough weight and parasitic reactions, find the horizontal, lateral, and vertical soil reactions acting on the plough.

No Permission Required

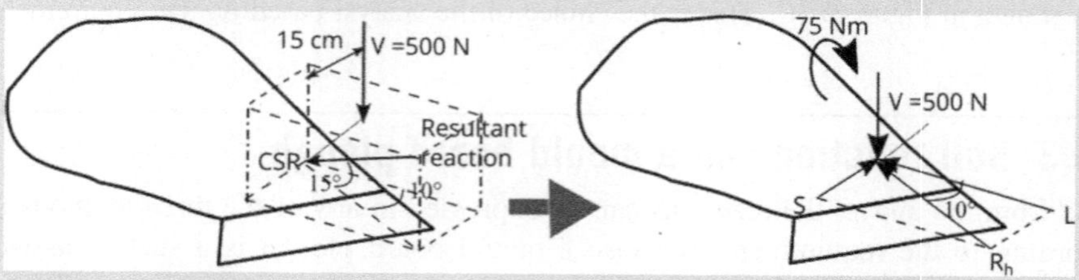

The applied P pull will be equal and opposite to the resultant soil reaction, which is the resultant of the horizontal, vertical, and lateral soil reactions (of course neglecting weight and parasitic reactions).

Since R_h is the resultant of L, the horizontal reaction (draft), and S, the lateral reaction (side draft), this overall resultant soil reaction is then the resultant of R_h and V, the vertical reaction. The angle between the pull and R_h (on horizontal plane) is 14°. The angle between R_h and the horizontal soil reaction is 11°.

Then, $R_h = P \cos(14°)$.

$L = R_h \cos(11°) = P \cos(14°) \cos(11°) = 1500\ N \times 0.97 \times 0.98 = 1426\ N$.

$S = R_h \sin(11°) = P \cos(14°) \sin(11°) = 1500\ N \times 0.97 \times 0.19 = 277\ N$.

And, $V = P \sin(14°) = 1500\ N \times 0.24 = 362\ N$.

For verification, $P = (L^2 + V^2 + S^2) = (1426^2 + 362^2 + 277^2)^{0.5} = 1498\ N \approx 1500\ N$.

5.4 Soil reactions on a disc plough

Unlike a mould board plough, a disc plough does not have a downward suction reaction from the soil (Clyde, 1937) and is made to penetrate by adding weight. It is hence a weight-assisted plough, and the weight counters the upward vertical soil reaction to make it penetrate. Figure 5.5A expresses the soil reactions on a 600 mm vertical disc tool operating at a 200 mm depth (Gordon, 1941). The force system is expressed as two non-intersecting forces, similar to Figure 5.4C, of the mould board plough. All the forces are in kN, and the linear dimensions are in mm. A vertical disc plough having a 38° disc angle and no tilt angle was considered to make the analysis simple (Kepner, Bainer, and Barger, 1978; Bernachi, Haman, & Kanafojski, 1972; Klenin, Popov, and Sakun, 1970). The two non-intersecting forces are:

1. R_h, the horizontal component of the soil reaction passing through the centre of the soil resistance as shown;
2. V, the vertical soil reaction acting upward (against penetration) at an offset from the soil centre of resistance.

Similar to a mould board plough, a disc plough also throws the soil to one side and hence is unsymmetrical in action. Figure 5.5B expresses the same state of soil reactions as another set of two non-intersecting forces: 1) a thrust T perpendicular to the plane of the disc and 2) a force U parallel to the plane of disc. The thrust force T perpendicular to the plane of the disc and a force U acting radially on the disc can be used conveniently to select the taper roller bearing taking up the thrust and radial

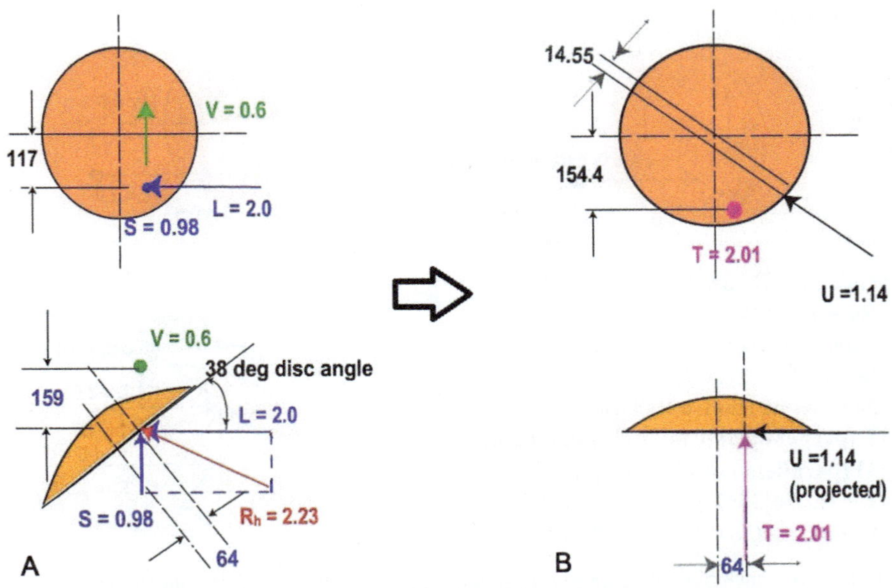

FIGURE 5.5 Soil Reactions on a Vertical Disc Bottom Expressed in Two Ways

loads on the disc while in operation. This is a typical example of how the same system of soil reactions can be expressed differently for our requirement. It may be seen that the radial force on the disc is not passing through the centre of the disc but is slightly offset from it. This offset provides the disc a torque to rotate continuously while ploughing the soil. This rotation of the disc makes the whole peripheral cutting edge of the disc interact with the soil. The centre of the soil resistance on the disc plough bottom, the point through which the resultant soil reaction passes, is as shown in Figure 5.4A, at a 20% distance from the disc centre toward the soil surface and at an offset of one-tenth of the diameter from the centre of the disc.

The following sequence of illustrations (Figure 5.6) shows how one system of two non-intersecting forces, R_h (the resultant of L and S) and V, can be converted into another system of expression having a thrust force T and a radial force U. All the forces are in kN, and the linear dimensions are in mm.

An animation of the above example is available for the student.

5.5 Soil reactions on disc harrow blades

The analysis of soil reactions on disc harrow blades (McKibben, 1926) is similar to the vertical disc plough bottom. Unlike a standard disc plough, the disc angle or the gang angle of disc harrows does not exceed 25°. Generally, they are between 19° and 22°. There is no tilt angle for the disc harrow blades. The combined soil reactions of

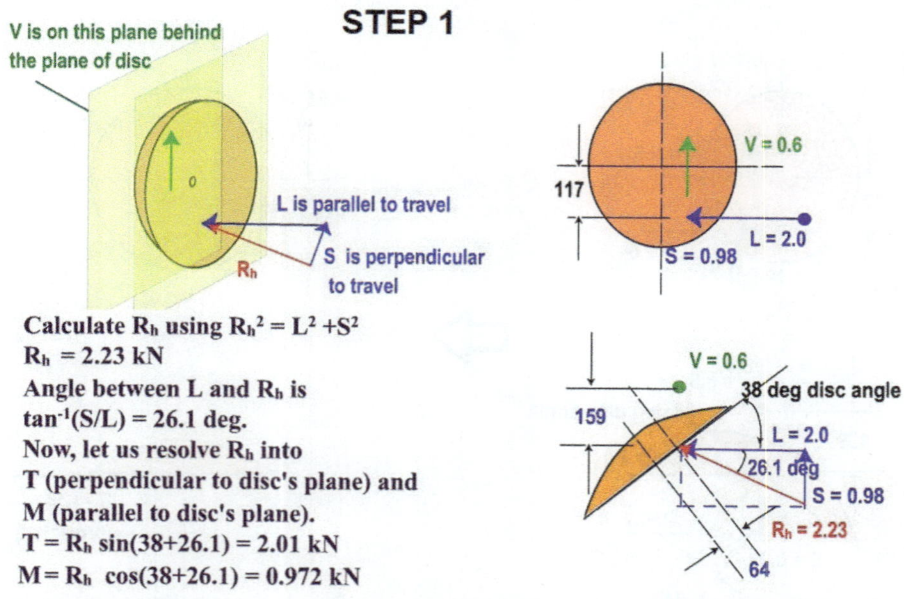

FIGURE 5.6 Conversion of Force Representation System in Vertical Disc Plough Bottom

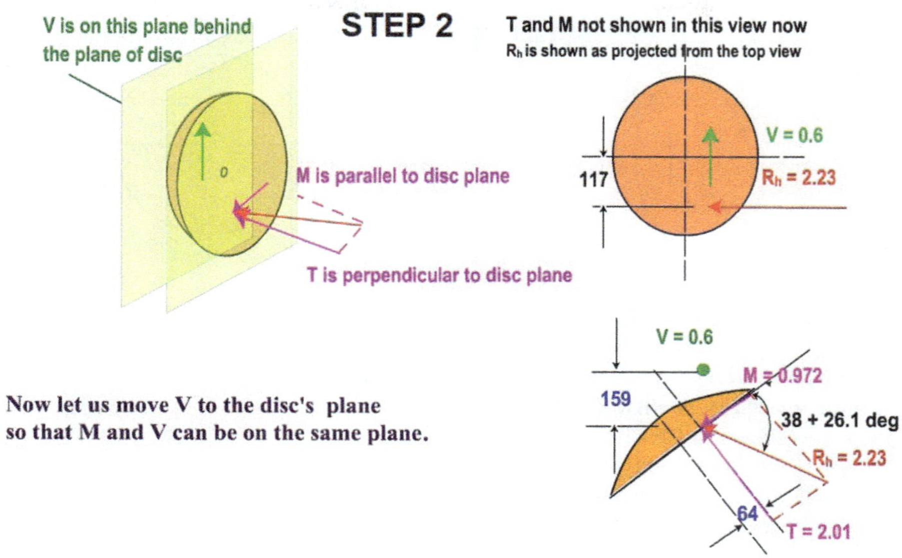

FIGURE 5.6B Cont'd

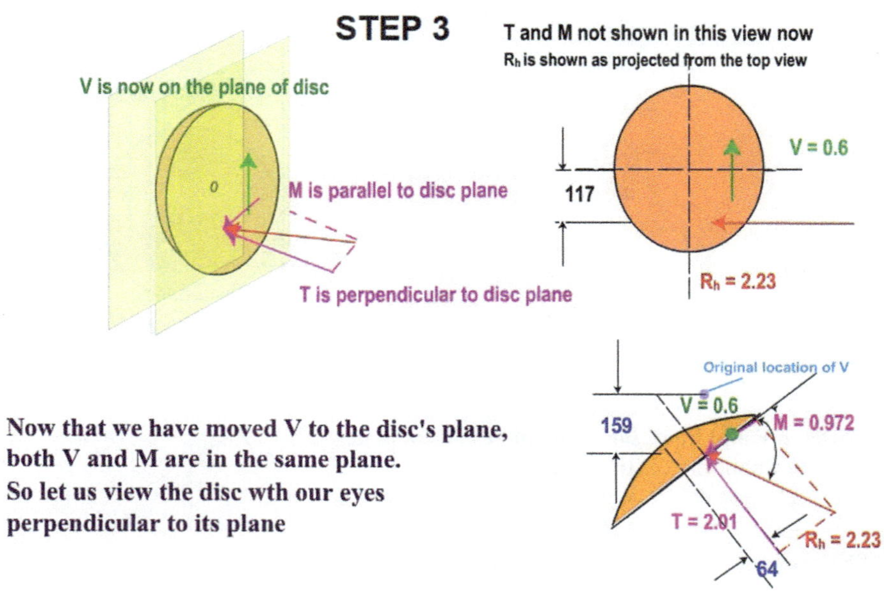

FIGURE 5.6C Cont'd

STEP 4

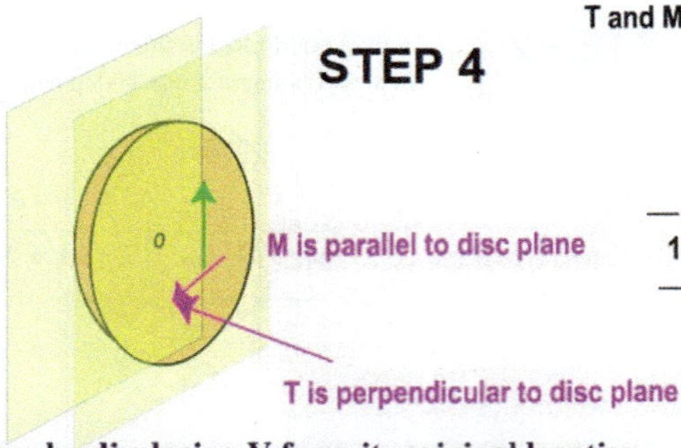

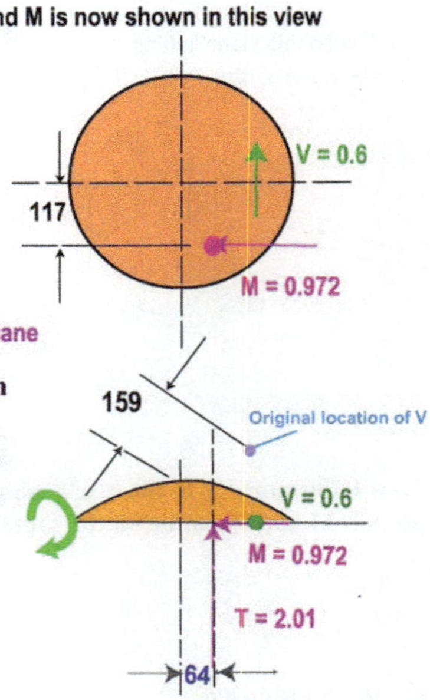

However by displacing V from its original location through 159 cos(38) = 125.29, we have disturbed the original force disposition and have cancelled a couple = V X 125.29 = 0.6 X 125.29 = 75.17 kNmm acting as shown in GREEN.

Hence to add that couple into the system again, let us shift T down by distance 'a'.

We know that T.a = 75.17 kNmm

2.01 X a = 75.17

a = 37.4 mm.

Therefore the new position of T is

(117+37.4) = 154.4 mm.

FIGURE 5.6D Cont'd

STEP 5

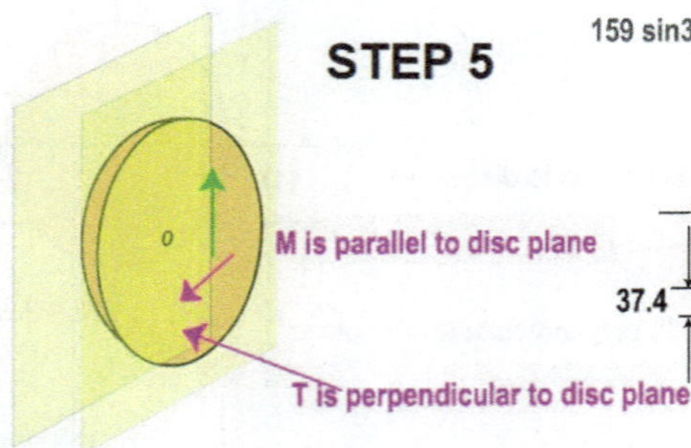

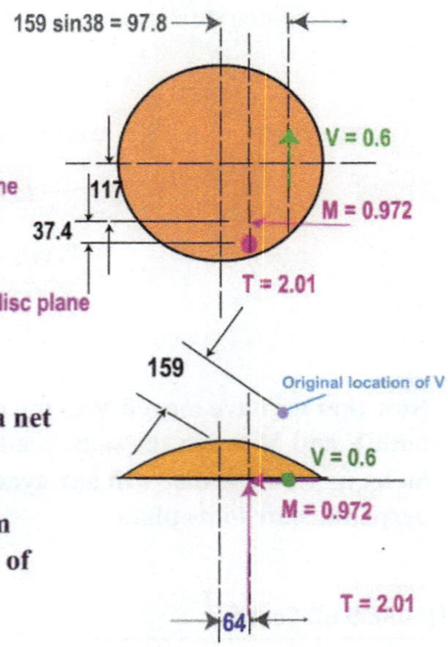

We have V and M on the same plane causing a net moment about the centre of disc as

M x 117 (CW) - V X (97.8 + 64) (CCW)

= 0.972 X 117 - 0.6 X 161.8 = 16.586 kNmm

This moment is as caused by the resultant 'U' of V and M i.e. $U^2 = V^2 + M^2 = 0.6^2 + 0.972^2$

and U = 1.14 kN

FIGURE 5.6E Cont'd

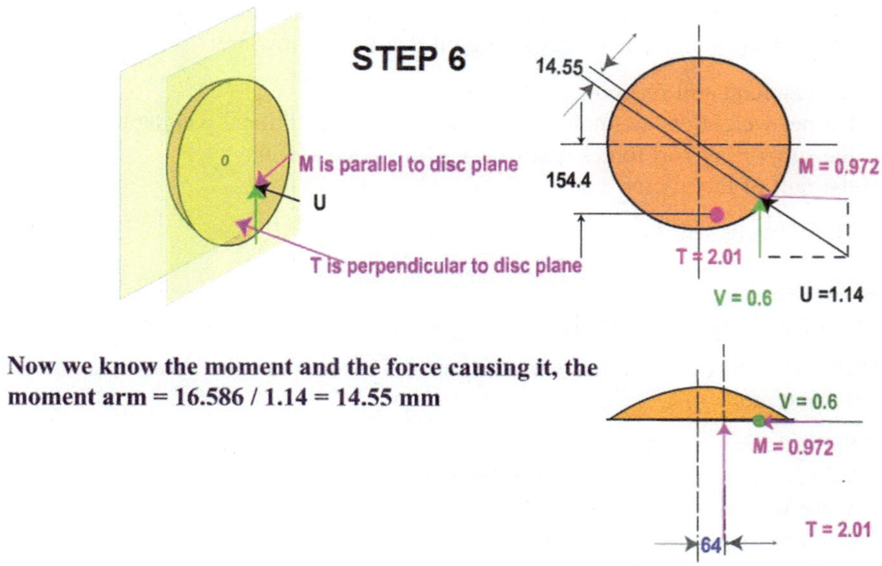

FIGURE 5.6F Cont'd

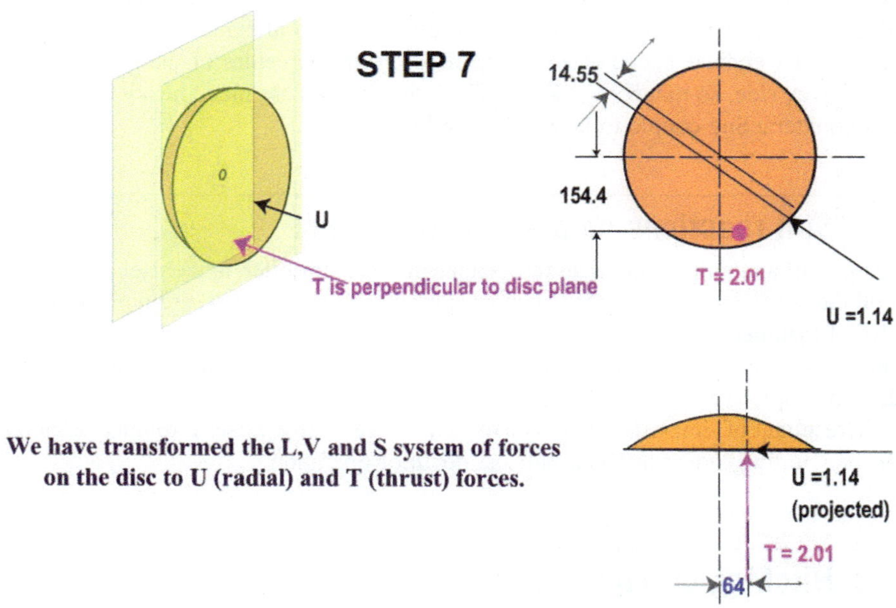

FIGURE 5.6G Cont'd

a gang of discs can be considered as a sum of all the reactions acting on the individual discs of the gang. The forces acting on the disc harrow are:

- the resultant soil reaction on each gang;
- the net weight of the implement along with any additional weight added;
- the supporting soil forces (parasitic) of the wheels, if any;
- the pull of the tractor.

These forces will be in equilibrium at any instant.

5.6 Leveling of disc harrow gangs for uniform penetration

The forces acting on the disc gang (Figure 5.7) relative to the main frame of the implement are:

1. S, the thrust parallel to the gang axis, arising out of the soil reaction acting below the gang axis at about one-third the depth of harrowing;
2. V, the vertical force acting on the gang, as viewed from the rear, which is the soil reaction opposing penetration;
3. S', the equal and opposite reaction to S at the bearings along the gang axis from the frame;
4. W, the net weight of the frame along with additional weights added in the tray.

Since S and S' are offset from each other by a distance of "f," they cause a couple on the frame, making the convex side of the gang lift up from the soil. This problem of gang level affects the levelness of the tilled soil. To alleviate this problem, an additional weight is added at an offset of "e" from the centre of the gang toward the convex side, so that $S \times f = W \times e$. The bumpers also are made heavy to provide this counteracting couple to keep the gang level.

5.7 Soil reactions on a symmetrical tool

Primary tillage tools, such as an indigenous plough, subsoiler, and chisel plough, as well the secondary tillage tools, such as cultivator tines and hoes, are typical examples of symmetrical tools. Here, there are no lateral forces perpendicular to travel and the soil reactions exist only in the vertical plane parallel to travel. This implies that no couple is existing perpendicular to travel. R_V and R are one and the same and are a resultant of simply L and V. The resultant R generally passes through a point on the penetrating shovel at about half the depth of operation.

5.8 Hitching of implements

To fully comprehend the complete set of forces acting on any tillage tool, the hitching (connecting) of a tool to the pulling source should be studied in detail. The type

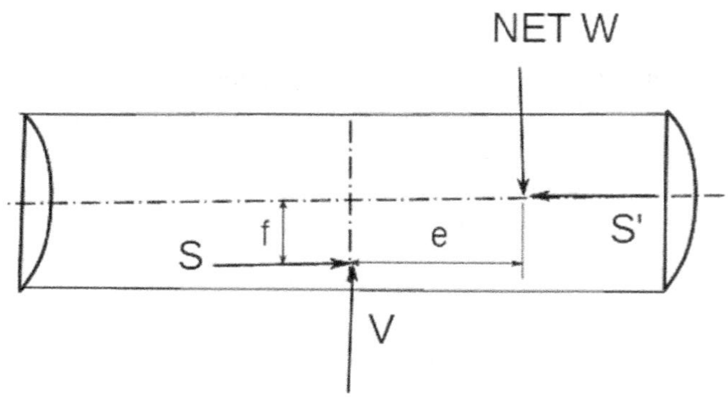

FIGURE 5.7 Forces on a Disc Harrow Gang

No Permission Required

of hitching that one uses to connect the soil working tool to the tractor or any other pulling source decides the system of forces on the tool.

5.8.1 Hitching of trailed implements

Hitching is the process of connecting the tractor (or any other power source) to the implement so that the implement can be pulled. For a trailed implement, the centre of pull is on the tractor at the centre between the rear wheels just a bit on the front above the rear axle. But, for practical purposes, it can be taken as the hitch point on the centre of the hitch bracket or drawbar. The line of pull will then emanate from this point and pass through the hitch point on the implement, where the pulling tow bar or chain is fastened to the implement. The line of pull will pass through an imaginary point known as the centre of resistance of the implement. This is the point of intersection of all forces and soil reactions acting on the tool and the point of application of the pull acting along the line of pull to maintain equilibrium. When the hitch point on the implement is changed, the soil forces decrease or increase as to locate the imaginary centre of resistance on the changed line of pull toward maintaining equilibrium. Proper hitching is then to find the hitch point that will yield reduced soil forces but yield a stable tool operation.

5.8.1.1 Vertical hitching of trailed implements

For convenience of analysis, we usually consider the vertical forces and horizontal forces separately. Vertical hitching is the process of locating the ideal hitch point in terms of the vertical forces involved (Clyde, 1934, 1935; Kepner, Bainer, and Barger, 1978). Here, we consider only the vertical forces on trailed implements.

Figure 5.8 shows the vertical force disposition of a trailed implement in a plane parallel to travel. It is known that a trailed implement receives no vertical support from the tractor and is wholly supported by a set of three wheels on the ground. These wheels are, namely, the land wheel rolling on the unploughed land, the front

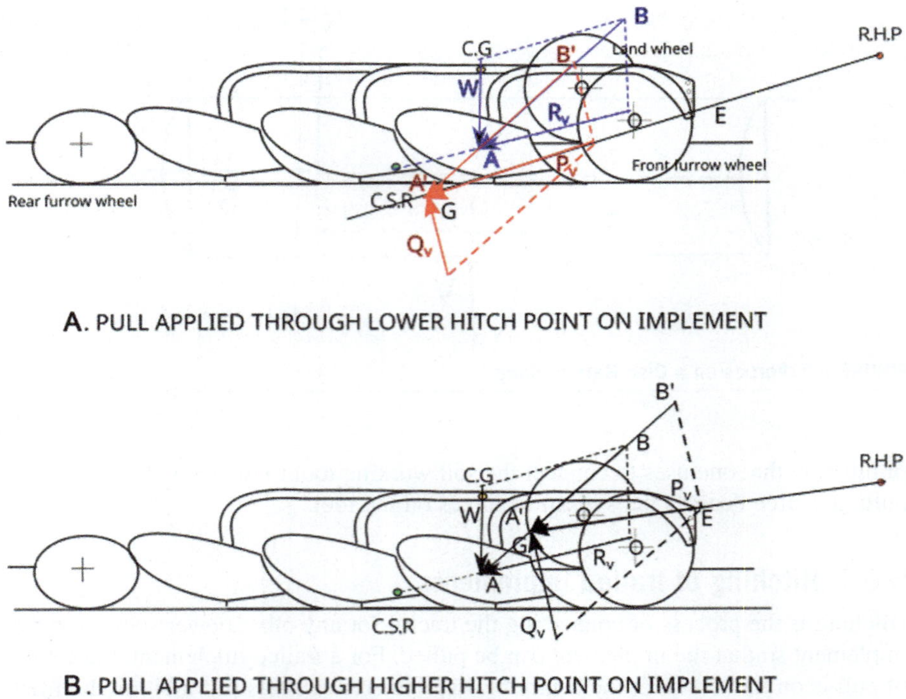

FIGURE 5.8 Force Relations in a Vertical Plane, for a Trailed Implement

No Permission Required

furrow wheel running in a furrow previously made, and the rear furrow wheel at the rear rolling in the last furrow made presently.

As we know, W, R_v, P_v, and Q_v are to be in equilibrium (Figure 5.8A), where:

- W is the weight of the trailed implement passing through its centre of gravity;
- P_v is the component of the total applied pull, P, in the vertical plane parallel to travel;
- R_v is the resultant of the horizontal soil reaction, L, and the vertical soil reaction, V, which are components of the total useful soil reaction, R, in the vertical plane parallel to travel;
- Q_v is the parasitic soil reactions in the vertical plane parallel to travel.

The weight, W, and location of the centre of gravity of the implement can be found easily, implying that the vector, W, is fully known in magnitude and location with its direction being vertically downward. The soil reactions, *viz.*, L and V, are to be measured for an individual tool or to be assumed based on available experimental data. L, the horizontal soil reaction, can also be calculated approximately from available data on soil resistance values for typical soils and types of implements. V, the vertical soil reaction, can be calculated approximately with available experimental data as a percent value of the calculated horizontal soil reaction, L. For example, the V/L ratio is typically 20 percent in loamy soils. The resultant of L and V is then R_v and is now known in magnitude and direction. W passes through the CG vertically downward, and R_V passes through the centre of the soil resistance at an angle of $\tan^{-1}(V/L)$ deg.

Based on experiments, the centre of the soil resistance is located for the mould board plough at approximately a distance equal to about one-fourth the width of the cut on the face of the bottom to the right of the shin and near the joint between the shin and mould board. The centre of the soil resistance for the whole implement will be approximately halfway between the centres of the soil resistances of a two-bottom plough or it is that of the middle bottom of a three-bottom plough. (This fact is applicable for both vertical as well horizontal hitching.) Having the magnitude and direction of W and R_v and the points through which they are passing, their resultant AB can now be found graphically by making them intersect at point A in Figure 5.8A.

The line of pull is to be drawn next as it should pass through the real hitch point (RHP) on the tractor and through the implement hitch point E. This line of pull and the resultant AB (or its extension) intersect at G, the centre of resistance of the implement. We know that P_v, the vertical component of the total pull, is the resultant of all the vertical forces and reactions, namely, R_v, the useful soil reaction; Q_v, the parasitic soil reaction; and W, the weight of the implement. Since the calculated AB is the resultant of W and R_v, P_v will be the resultant of AB and Q_v, summing all the three together.

Now, to solve for the forces, we know the following:

- P_v is known in direction as passing through E and RHP, but not in magnitude yet.
- AB is known in magnitude and direction.
- Q_v is unknown in magnitude, but its direction can be approximately deduced because it is purely due to the rolling friction of the three supporting wheels of the implement. Q_v will then be upward and inclined backward from vertical at an angle corresponding to the angle of the rolling friction of the three wheels (15° to 20°) supporting the whole implement vertically (Figure 5.8). It is actually the resultant of all the vertical forces of support from the soil acting upward and the frictional force of rolling acting horizontally, resisting travel.

Now, vector AB is graphically displaced along its line of action to A'B' until it meets the line of pull at point G. (A' and G are now the same.) Q_v, known only in direction, is now drawn through G, and the force parallelogram of GB' and Q_v is completed graphically with the resultant P_v. The graphical vector diagram will now yield the magnitude and direction of all the involved forces and reactions. Since G is the point at which all forces and reactions are in equilibrium, it is called the centre of load or resistance of the implement. It may be noted that the centre of the soil resistance of the implement is the point (at which soil resistances L and V meet) through which R_V passes, whereas the centre of resistance or load for the implement is G, the intersection of pull, AB, and Q_V, the parasitic force.

The discussed example is a desirable hitch arrangement for a trailed mould board plough on which the selected hitch point, "E," on the implement makes all the three wheels support the implement equally and Q_V is located well behind the front furrow wheel. The other example in Figure 5.8B is for the same plough, the hitch point of the implement is made higher on the tool frame. Since "*E*" is so high, graphical calculations yield Q_V being almost under the front wheels, with all the weight transferred from the rear wheel to the front wheels, making the rear wheels unstable. Lowering the hitch point, E, on the implement will reverse the situation in which more weight will be on the rear furrow wheel. Though the change in magnitude of the pull is not much by these changes, it causes undue instability in the operation of the implement since the furrow wheels that compensate the side draft of the implement cannot function properly. The improper choice of hitching alters the pull required marginally by 3% as well the draft by 4%, which contributes to the power required to pull the implement. Most importantly, it makes the operation very difficult. The above analysis was shown with a mould board plough as examples.

As for vertical hitching of a disc plough, the rules are the same as for mould board ploughs. But, unlike the mould board plough, the side thrust of the soil reaction is totally taken up by the furrow wheels, and the furrow wheel should stay in the furrow for proper operation. So, if needed, the hitch point on a trailed disc plough may be lowered to bring about more weight on the rear furrow wheel to keep it positively inside the last furrow made.

This principle of analysis is common for any type of trailed implement. In the case of a single-axle implement, as the set of wheels on a single axis supports the whole piece of equipment, the location of Q_V is fixed on a line passing slightly behind the axle centre line (Figure 5.9). This helps to overcome the wheel-bearing friction to cause rotation of the single set of wheels. Its direction is, as before, inclined backward at an angle of rolling friction. In the previous case of a trailed implement, the line of action (location) and direction of P_V were known and the direction of Q_V was known but not its location. However, here, Q_V is known in direction as well its line of action (location); P_V is known only to pass through the tractor's hitch (location) but not in direction (angle at which it passes). But, this information is adequate for completing the parallelogram of forces of A'B', Q_V, and P_V (Figure 5.9), so as to find their magnitudes and directions. The procedure of determining the force, AB is e the same as that explained earlier for trailed implements.

The following problem, shown in Box 5.4, is solved graphically to fully comprehend the idea behind solving for the forces on a simple indigenous (trailed) implement. The graphical procedure explained already applies to the example below, too.

The following examples in Box 5.5 and Box 5.6 show the calculation of pull and parasitic forces of a trailed mould board plough and a trailed single-axle implement, respectively.

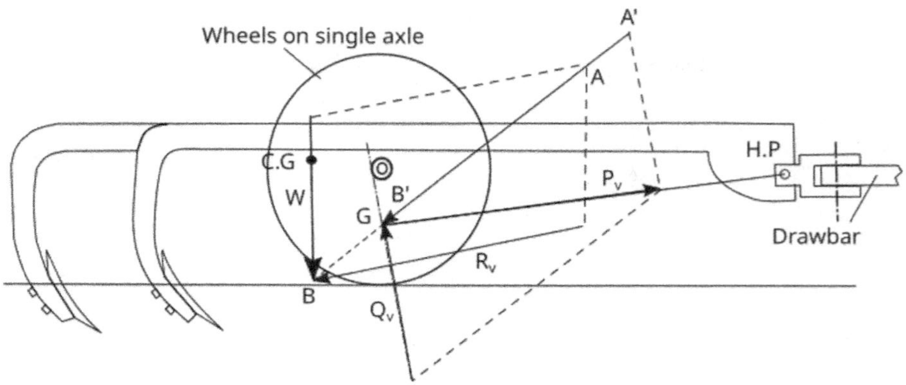

FIGURE 5.9 Vertical Force Relations for a Single-Axled Trailed Implement

No Permission Required

5.8.1.2 Vertical hitching of trailed disc harrows

A typical trailed disc harrow that does not have any support wheels is devoid of parasitic soil reactions. The line of pull, $P_{V,}$ of the implement is decided by the choice of hitch point on the implement (Clyde, 1935; Kepner, Bainer, and Barger, 1978; McKibben, 1926). It emanates from the hitch point on the tractor's hitch bracket and passes through the selected hitch point on the harrow. The intersection of the known line of pull and the weight, W, acting through the centre of gravity of the harrow gives G, where W, $P_{V,}$ and R_V are in equilibrium. R_V is the resultant vertical soil reaction of all the gangs of the disc harrow. Figure 5.10A shows the force diagram of the vertical hitching of a tandem disc harrow.

Of these forces:

- W is known in magnitude, direction, and location (CG);
- P_V is known in direction and location only.

Total R_V is known in direction and location as the R_V of each gang is known based on the typical L/V ratios of each disc harrow gang. Here, the L/V ratio is assumed as 1.2 for both the front and rear gangs; thereby, R_V, the sum of R_{Vf} and R_{Vr} will point to the rear of the harrow at an angle of 50° to the vertical. The vectorial sum R_V is shown in the figure. Since R_{Vf} and R_{Vr} are parallel but unequal, their vectorial sum (by the principle of moments) is closer to the larger R_{Vf} of the front gang.

The force polygon can now be completed by first moving W and R_V along their lines of action until they cut at the centre of resistance, G, and then connect the hitch point and G. This will yield the magnitudes of P_V and R_V.

If the hitch point on the implement is raised (Figure 5.10B), then the line of pull changes and the magnitudes change correspondingly. Although the resultant R_V (vectorial sum of the R_V of the front and rear gangs) has increased marginally by 0.6 kN, the resultant vertical reaction on the front gang is substantially more than that at the back. This physically means that the front gang is pressed into the soil

Box 5.4 Example 5.4

Example 5.4. The pull is measured on a country plough as shown in Figure 5.2. The angle of pull was measured as 20° to horizontal. The plough weighs 300 N. The centre of gravity of the plough is located as shown. Assume that the horizontal soil reaction per unit furrow area of the plough is 6 Ncm^{-2} and the L/V ratio is 2.5. The centre of the soil resistance where L and V meet is shown as CSR. Graphically, find out the location of the plough's centre of resistance and the magnitude of parasitic soil reaction due to the plough body sliding on the furrow bottom. Also, find the measured pull exerted by the bullocks on the plough. The coefficient of the sliding friction is 0.3. Assume the dimensions of the furrow section of the plough as triangular 15 cm × 20 cm.

(N.B.: You may photocopy the page having Figure 5.2 and take a printout of the same to try solving it on the image.)

Calculate the horizontal soil reaction, L, assuming the furrow section of an indigenous plough as triangular with a height of 20 cm and width of 15 cm.

L = horizontal soil reaction per unit furrow area × area of furrow c/s = 6 × 0.5 × 15 × 20 = 900 N. *(Please note that the horizontal soil reaction per unit furrow area is not same as the unit draft. Unit draft is derived from the horizontal component of pull, not the total soil reaction. But, if the horizontal soil reaction per unit area is not readily available, the unit draft can be used in its place allowing marginal loss in accuracy of calculations.)*

Then V = 900 / 2.5 = 360 N.

Then, R_v, the resultant of L and V, = (900^2+360^2)= 970 N and its angle with horizontal = $\tan^{-1}(360/900)$ = 21.8°.

Then, select a suitable scale for the force vectors. Say, 1 cm = 200 N. Draw vector R_v of length 4.85 cm (i.e., 970N) at 21.8° to horizontal and at the point CSR directed opposite to the travel direction (since this is a soil reaction opposing plough travel). The soil reaction will be downward since the country plough has a downward suction, V. Extend this line of action for use later. Draw vector W (the plough weight) to 1.5 cm (i.e., 300 N) through the centre of gravity of the plough CG such that it meets the line of action of R_v at point A. Then, shift the vector R_v along its line of action so that the vectors meet at A. Then, complete the force parallelogram of R_v and W, which will yield the resultant vector AB.

We know that the pull along the rope towing the plough is the resultant of R_v, W, and Q_v, the parasitic soil reaction due to the plough bottom sliding on the soil beneath. Of these three, we have already (vectorially) added R_v and W as vector AB. Then, pull, P, will be the resultant of AB and Q_v.

Here, Q_v is the resultant of the total upward vertical reaction from the soil (opposing the net vertical forces acting down) and the sliding frictional resistance of the bottom on the soil = μ × upward vertical soil forces. Then, μ = 0.3 and, hence, Q_v will be inclined backward at $\tan^{-1}(0.3)$=16.7° to vertical. But, its location and magnitude are unknown.

We know P in direction, that is, along the rope inclined at 20° to horizontal (the rope applying pull from the yoke at CP to the point of hitch on the plough HP).

We know vector AB both in magnitude and direction.

Extend the line of action of pull, P, and that of AB until they intersect at the point known as the centre of tool resistance, G. Since P is the resultant of AB and Q_v meeting at G, Q_v also must pass through the same point G. Shift AB to G as GB' and complete the parallelogram, forming GB' and Q_v. Then, measure the length of the resultant P and parasitic force, Q_v, to convert them to their magnitudes.

Box 5.4 Example 5.4—cont'd

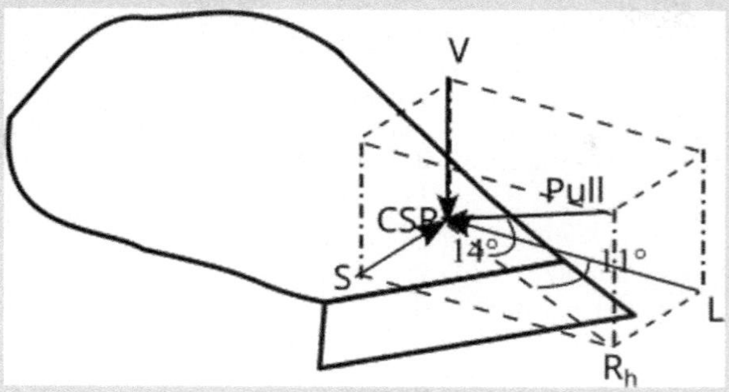

It may be noted that the country plough is a symmetrical soil tool, which does not have any lateral forces or side draft. So, here P and P_v are the same since there are no lateral forces with which to contend. All the forces are acting only in the vertical plane parallel to the travel direction. This will be true for all symmetrical soil working tools such as the cultivator.

An accompanying computer utility is available, which the reader is encouraged to use to solve this kind of vectorial analysis graphically. Extensive video tutorials are also available on how to use the utility.

more by the line of pull, raising up the rear gang out of the soil. If a lower hitch point is chosen on the harrow, the opposite will happen in which the front gang will rise out of soil. Both these cases are undesirable and hence the right choice of the implement hitch point is desirable.

L/V ratios were assumed to be equal in the above analysis to make it simple. In the actual operation, these ratios vary between the front and rear gangs. However, the analysis teaches us the general principle of force disposition while hitching a trailed disc harrow to the tractor. The example problem in Box 5.7 shows the analytical calculation of the soil reactions and the pull of a trailed disc harrow.

5.9 Horizontal hitching of soil working tools

Similar to vertical hitching, the purpose of horizontal hitching is to look at the disposition of forces in the horizontal plane, so as to make the implement operate properly.

Box 5.5 Example 5.5

Example 5.5. A three-bottom, 40 cm, trailed mould board plough is operated at a ploughing depth of 15 cm. The illustration shows the real hitch point on the tractor, the hitch point to which the plough is hitched, the centre of gravity, and the centre of soil resistance for the whole plough. The implement weighs 3 kN. The total vertical parasitic force as offered by the support wheels is inclined backward from the vertical by 12°. Graphically find the pull and the total parasitic force. Assume an L/V ratio of the plough as 2.3 and the horizontal soil reaction per unit furrow area as 30 kPa.

No Permission Required

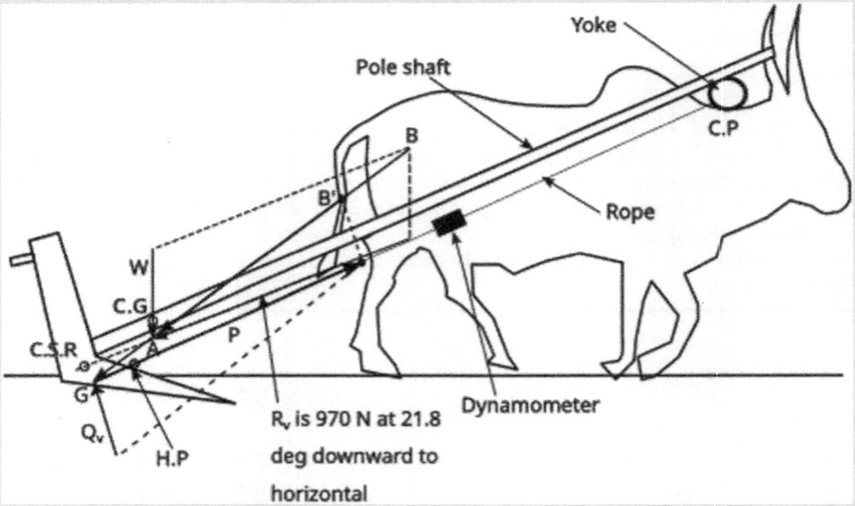

Horizontal soil reaction per unit furrow area = 30 kPa.
Effective furrow cross-sectional area = $3 \times 40 \times 15\ cm^2 = 0.18\ m^2$.
The horizontal soil reaction, $L = 30\ kPa \times 0.18\ m^2 = 5.4\ kN$.
$L/V = 2.3$.
$V = 5.4/2.3 = 2.35\ kN$.
R_v, the total vertical soil reaction located at the CRS, $= \sqrt{L^2 + V^2} = \sqrt{5.4^2 + 2.35^2} = 5.89kN$
$\tan^{-1}(V/L) = 23.5°$.

We then have W = 3 kN acting vertically down through the CG and $R_v = 5.89$ kN acting at the CRS of the plough at 23.5° downward from the horizontal.

Copy the figure of the problem and print it on paper. Choose an appropriate scale and then draw W and R_v at their locations and directions. Form the vector polygon and find the resultant as AB.

The line of pull is through the two given hitch points, one the RVP and the other on the implement. Connect these two points and extend. Move the vector AB on its line of action until it intersects the line of pull at G, the centre of resistance of the implement. Q_v, the parasitic force, should also pass through G since P, the pull, is the resultant of vector AB (which is the resultant of W and R_v) and Q_v. So, draw a line representing Q_v at G with its direction at 12° backward.

Then, knowing the magnitude of vector AB, complete the vector polygon comprising AB, P, and Q_v. Then, measure the lengths of P and Q_v. Covert to magnitudes to compute their values. Refer to Fig. 5.8 for guidance.

Solution: $P_v = 6.36$ kN; $Q_v = 4.11$ kN. (Try using the software given with this book to solve the same problem graphically and verify.)

Box 5.6 Example 5.6

Example 5.6. A 5 m wide sweep cultivator weighing 4 kN is trailed by a tractor. The horizontal soil reaction of the tool has been recorded as 1.5 kN per metre width of the tool. The vertical soil reaction is negligible. The tool's depth is controlled by a set of runners and the soil-runner frictional coefficient is 0.3. The line of pull from the tractor's hitch point through the unknown centre of load on the implement is at 20° to the horizontal. Analytically find the magnitude of pull applied as well the parasitic soil reaction on the runner, both in the vertical plane.

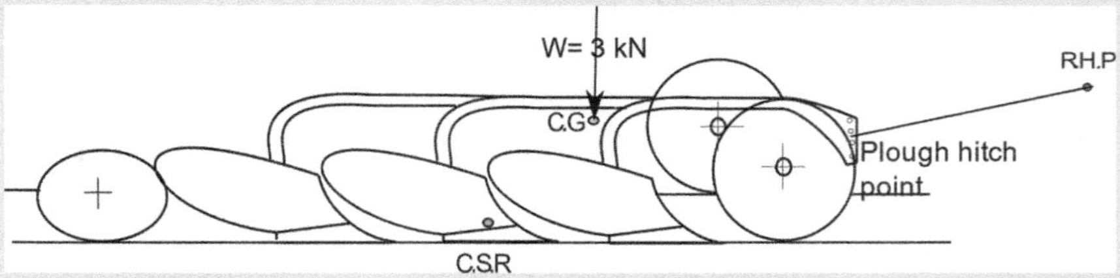

$R_v = L = 1.5$ kN/m x 5m = 7.5 kN since V is negligible.

W = 4 kN.

Resultant of R_v and W = AB = $(R_v{}^2 + W^2)^{0.5} = (7.5^2 + 4^2)^{0.5} = 8.5$ kN.

Angle between AB and $R_v = \tan^{-1}(W/R_v) = \tan^{-1}(4/7.5) = 28°$.

The line of pull is inclined at 20° to the horizontal. Q_v is located at the centre of the runner length and inclined backward at the angle of friction between the runner and soil. Since the frictional coefficient μ is 0.3, the angle of Q_v to the vertical is $\tan^{-1}(0.3) = 17°$ and, when it is relocated along its line of action, it will intersect the line of pull at G, the centre of resistance. Then, relocating AB to A'B' along its direction until it intersects the line of pull at point G, a parallelogram of forces can be formed from the known A'B', unknown Q_v, and unknown P_v. Since we are trying to solve the problem analytically, let us first deduce the different included angles of the triangle formed by Q_v, P_v, and the side A'B' in the force diagram.

The angle between P_v and horizontal is 20° and the angle between Q_v and horizontal is (90-17) ° = 73°, making the angle between Q_v and P_v = 20+73 = 93°.

The angle between A'B' and R_v (horizontal) is already found = 28°, making the angle between A'B' and P_v = (28-20) ° = 8°

$\frac{Q_v}{\sin 8} = \frac{A'B'}{\sin 93} = \frac{P_v}{\sin 79}$ where AB = 8.5 kN

$\frac{Q_v}{0.14} = \frac{8.5}{0.99} = \frac{P_v}{0.98}$

$Q_v = 1.2$ kN; $P_v = 8.4$ kN.

The location of G, the centre of resistance, is slightly at the back of the runner's midpoint and up above the runner, which can also be deduced. Try it as an exercise.

5.9.1 Horizontal forces on a mould board plough bottom

The forces acting on a typical mould board plough in the horizontal plane (Figure 5.11A) show how R_h, the horizontal component of the useful soil reaction (the resultant of L and S), is oriented with reference to travel direction. S, the lateral soil reaction, is generally about 20 percent to 25 percent of L in clay loam soils and up to 40% in sandy soils (Kepner, Bainer, and Barger, 1978; Randolf and Reed, 1938). If the soil is uniform in strength all along travel and if a pull can be applied directly opposite to R_h, then the plough can theoretically move forward without the necessity of a landside. However, this may not be practically feasible because the plough furrow has to align on the inner edge of the tractor's right rear wheel. If the horizontal pull, P_h, is applied in the direction of travel ($P_x = P_h$), a parasitic soil reaction, Q_h, is inevitable because of S. The useful lateral soil reaction, S, is now counteracted by an equal and opposite reaction of the furrow wall through the landside and a sliding frictional force of the landside rubbing on the furrow wall. Q_h is the resultant of the reaction to S acting perpendicularly to the furrow wall and the friction force, μS, acting along the landside, opposing the forward movement of the plough (μ is the soil-metal friction coefficient). This implies that Q_h is inclined backward at the friction angle ($\tan^{-1}\mu$) (Figure 5.10A). Since Q_h is the parasitic reaction provided by the landside, it is positioned at the lengthwise centre of the landside. Now, we have the following forces at hand, namely:

- R_h is known in magnitude, direction, and location since L, S, and the location of the point through which it passes (the centre of the soil resistance on the plough bottom) are known.
- Q_h is known in direction and location but not in magnitude yet. Direction is at a frictional angle aligned backward and located at the lengthwise centre of the landside.
- P_h, the component of pull, is known in direction along travel; P_x and P_h, being the same as the pull, are straight.

The force vector, R_h, can be drawn to scale, and then vector Q_h can be drawn in the known direction and location. The vectors can be made to intersect at H, which is the horizontal centre of the resistance of the plough bottom. P_h, which will be passing through H in the direction of travel, can be drawn next. Since these three forces are to be in equilibrium, the parallelogram of the three forces, R_h, Q_h, and P_h, is completed next to get the magnitudes of Q_h and P_h. The direction of P_x can be the ideal direction of pull, which can be practically feasible if certain conditions are met.

If for some reason the horizontal component of pull, P_h, is applied toward the left of the plough as seen from the rear (Figure 5.11B), a similar vectorial solution of the three forces will yield an increased Q_h due to the component of pull in the lateral direction also pushing the plough more against the furrow wall along with S.

Figure 5.11C (similar to Figure 5.11A) shows how the solution changes when a longer landside is used while pulling the bottom along the direction of travel. Q_h, the parasitic soil reaction, moves backward correspondingly and relocates H, the centre

Box 5.7 Example 5.7

Example 5.7. A trailed disc harrow weighing 2.5 kN is hitched to the tractor with the line of pull inclined at 15° to the horizontal. The locations of the centres of soil resistance for the rear and front gangs are shown in the figure below. The centre of gravity is at a distance of 50 cm from the centre of soil resistance of the front gang. The L/V ratios of the front and rear gangs are 1.2 and 0.9, respectively. The ratio of horizontal soil reactions between the front (L_f) and rear gangs (L_r) is 1.4. Analytically calculate the pull and the horizontal and vertical soil reactions of the front and rear gangs.

No Permission Required

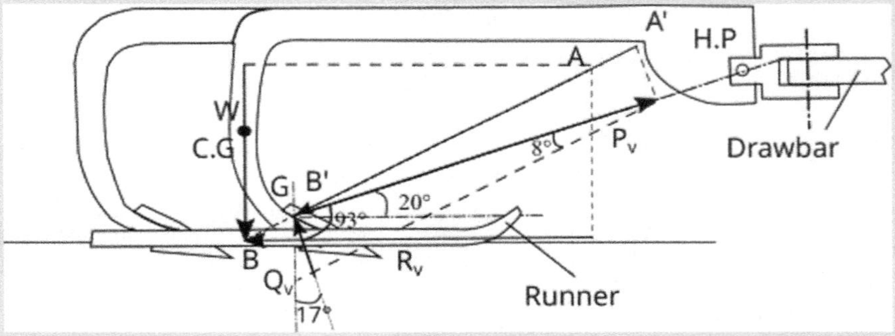

$\frac{L_f}{V_f} = 1.2; \frac{L_r}{V_r} = 0.9; \frac{L_f}{L_r} = 1.4$

$V_f = \frac{L_f}{1.2}; V_r = \frac{L_r}{0.9}; L_r = \frac{L_f}{1.4}$

We can calculate R_v as a resultant of the sum of all horizontal soil reactions and the sum of all vertical soil reactions.

The sum of all horizontal reactions, $R_{vH,} = L_f + L_r = L_f + \frac{L_f}{1.4} = 1.71L_f$.

The sum of all vertical reactions, $R_{vV,} = V_f + V_r = \frac{L_f}{1.2} + \frac{L_f}{(0.9 \times 1.4)} = 1.62L_f$.

$R_v = \sqrt{R_{vH}{}^2 + R_{vV}{}^2} = \sqrt{(1.71L_f)^2 + (1.62L_f)^2} = L_f\sqrt{(1.71)^2 + (1.62)^2} = 2.35L_f$.

The angle between R_v and $R_{vH} = cos^{-1}\left(\frac{1.71}{2.35}\right) = 43.3°$.

In the vector triangle of the three forces, namely, the known W and the unknown P and R_v, the angle between W and P is (90-15) = 75°, that between P and R_v is (43.3+15) =58.3°, and that between W and R is 180-58.3-75 = 46.7°.

$\therefore, \frac{P}{\sin(46.7)} = \frac{R_v}{\sin(75)} = \frac{W}{\sin(58.3)}$

$\frac{P}{0.72} = \frac{R_v}{0.97} = \frac{2.5}{0.85}$

P = 2.11 kN; R_v = 2.85 kN

Since R_v = 2.85 = 2.35 L_f;

∴, L_f = 1.21 kN,

V_f = 1.21/1.2 = 1 kN,

L_r = L_f /1.4 = 1.21/1.4 = 0.86 kN,

V_r = L_r /0.9 = 0.96 kN,

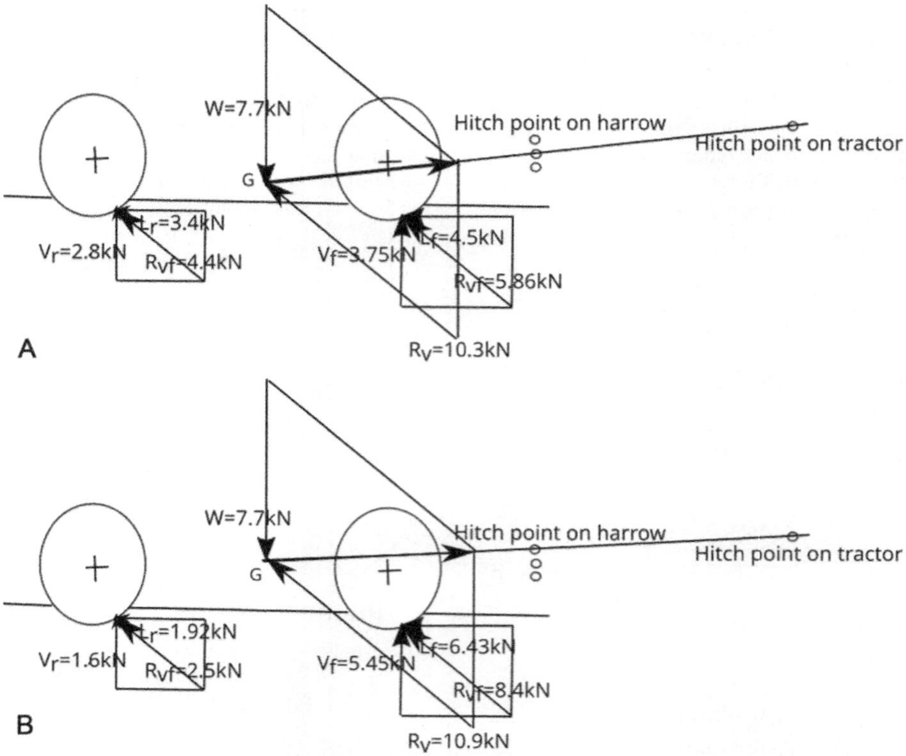

FIGURE 5.10 Vertical Force Relations for a Trailed Disc Harrow

No Permission Required

of the soil resistance, also backward and closer to the landside. This is so, since the direction and location of R_h does not change in all the three aforesaid cases. The above explanations are elaborated further through a worked-out example shown in Box 5.8.

5.9.2 Horizontal forces on a disc plough bottom

Unlike a mould board plough bottom, a disc plough does not have provisions built on it to compensate the useful lateral soil reaction, S. So, it is difficult to contend with these forces on a single-disc bottom. It has to be operated at least with a rear furrow wheel and more support from other wheels to make an analysis. Q_h, the parasitic reaction, is given hence by these wheels. However, the general principle studied for the mould board plough bottom is quite applicable for disc plough bottoms, too. The centre of resistance of the disc ploughs is closer to the furrow wall than on mould board ploughs. It is generally located to the left and below the centre of the disc

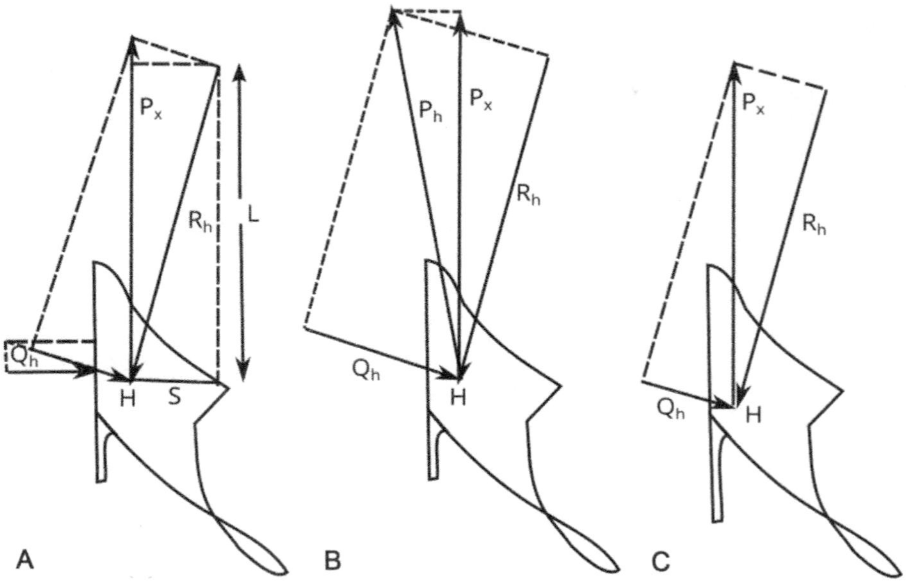

FIGURE 5.11 Horizontal Forces on a Mould Board Plough Bottom

No Permission Required

as seen from the front. The exact location is influenced by the disc and tilt angles as well the depth of the operation.

5.9.3 Horizontal forces on a symmetric soil working tool such as a chisel plough or cultivator shovel

As already explained, lateral horizontal forces perpendicular to travel do not exist in these plough bottoms. Hence, the only force with which you must contend is the horizontal soil reaction, L.

5.9.4 Horizontal hitching of trailed ploughs

An implement that has provisions to counteract the lateral soil reaction (side thrust of useful soil reaction) by landsides and/or furrow wheels will run straight if pulled along the line of travel. So, the best horizontal hitching is that in which the tread width of the tractor is adjusted exactly to match the implement width and the centre of pull is directly in line with the overall centre of the load or resistance, H, of the implement along the direction of travel (Figure 5.12A). This implies that the rightmost edge of the first plough bottom has to be in line with the inner edge of the tractor's right rear wheel as seen from the back. This is essential to align the plough passes by riding the right rear wheel in the last furrow made in the previous

Box 5.8 Example 5.8

Example 5.8. A mould board plough bottom is trailed with its pull straight ahead along the direction of travel. The resultant (R_h) of the horizontal and lateral soil reactions is at 12° from the travel direction. Analytically find the percent increase in draft, pull, and parasitic force if the horizontal component of the pull is changed to an angle of 15° toward the landside from the travel direction. The coefficient of soil-metal friction is 0.33.

No Permission Required

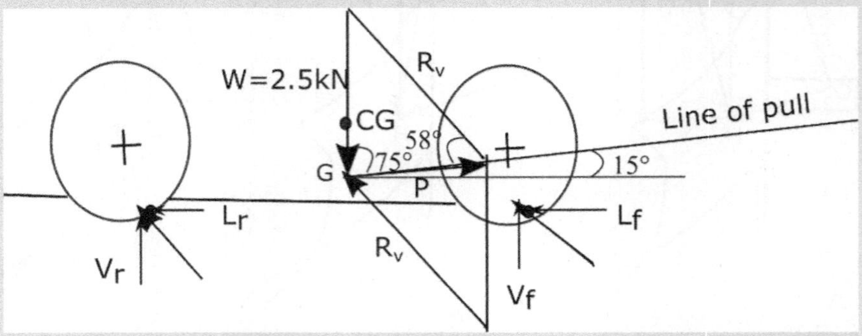

When pulled straight ahead:

The horizontal soil reaction L = $R_h \cos(12°) = 0.98\ R_h$ (not needed for the solution but just for observation).

The lateral soil reaction, S, = $R_h \sin(12°) = 0.2\ R_h$ (not needed for the solution but just for observation).

Q_h, the parasitic force, on landside is the resultant of the reaction to the lateral soil reaction from the furrow wall and the frictional force parallel to the landside furrow wall interface and opposite to the travel direction.

$Q_h = (S^2 + (0.33S)^2)^{0.5} = 1.05\ S$ (not needed for the solution but just for observation).

The angle subtended between Q_h and the direction of travel = $90 - \tan^{-1}(0.33) = 72°$.

In the parallelogram of forces of R_h and Q_h solving for P_x, considering the triangle of these three forces and applying the sine law,

$$\frac{R_h}{\sin 72} = \frac{Q_h}{\sin 12} = \frac{P_x}{\sin(180-72-12)}$$

giving $P_x = 1.046\ R_h$

where P_h = Px since the pull is in the travel direction.

$Q_h = 0.22\ R_h$.

When pulled at a 15° angle to travel,

because of the angled pull, the plough bottom is pressed further against the furrow wall by the lateral component of the new horizontal component of pull, P_h. This adds the total thrust on the furrow wall.

The horizontal component of pull, P_h, is the resultant of Q_h and R_h.

The angle subtended between Q_h and the direction of travel = $90 - \tan^{-1}(0.33) = 72°$ (as before).

This does not change since it is by virtue of the soil-metal friction angle.

Then, the angle subtended between Q_h and $P_h = 72 - 15 = 57°$.

The angle subtended between R_h and $P_h = 12 + 15 = 27°$.

Let us now consider the vector triangle formed by P_h, Q_h, and R_h. Applying the sine law,

$$\frac{R_h}{\sin 57} = \frac{Q_h}{\sin 27} = \frac{P_h}{\sin(180-57-27)}$$

$P_h = 1.19\ R_h$,

Box 5.8 Example 5.8—cont'd

$Q_h = 0.54\ R_{h,}$

$P_x = P_h \cos(15°) = (1.19\ R_h) \times 0.965 = 1.15\ R_{h,}$

Increase in pull = (1.19-1.046)/1.046 = 13.7%,

Increase in Q_h = (0.54-0.2)/0.2 = 170%,

Increase in Draft, P_x = (1.154 – 1.046)/1.046 = 10.3%,

From this example, the reader should note that the horizontal soil reaction and draft are entirely two different parameters. The horizontal soil reaction, L, is the component of the soil reaction and draft is the horizontal component of the pull, the pull being the resultant of the parasitic force and resultant soil reaction. The same kind of difference exists between the side draft and lateral soil reaction, S. The same logic should be extended to the forces in vertical hitching, too.

ploughing pass. It may also be noted that the centre of pull of a tractor is at the centre between the two rear wheels and a little ahead of the rear axle of the tractor. The horizontal centre of resistance, "H," of the whole implement is the midpoint of the line joining the centres of resistance of all the individual bottoms. But, this arrangement may not be possible for a very narrow implement pulled by a tractor having a wider tread width (Figure 5.12B) However, a mould board plough with a landside can work well, as shown, even when the line of pull is at an angle from the line of travel, as discussed in the chapter on mould board plough bottoms (Figure 5.11B). It may be noted once again that the horizontal centre of resistance, "H," of the two bottoms is the midpoint of the line joining their centres of resistances. It may be noted that RHP (Figure 5.12) is the real hitch point, in which the implement is physically connected to the tractor drawbar. So, in the case of a narrow two-bottom mould board plough, by offsetting the real hitch point (RHP) on the tractor to the right adequately, the line of pull along H and RHP is made to pass at an offset to the right of the centre of pull of the tractor.

This will make the first bottom align with the right wheel. However, the offset in the line of pull from the centre of pull will introduce a horizontal couple on the tractor's rear that may cause steering trouble if overdone. As for a disc plough, all the side thrust on the implement is borne by the furrow wheels. So, unlike the case of mould board ploughs, the pull is applied through a free link between D and RHP (Figure 5.13). So, the line of pull is fixed by the positions of points D and RHP. The position of the horizontal centre of resistance and the location of the resultant side force, $Q_{h,}$ are decided by the point of intersection of P_h, the horizontal component of the pull and R_h (resultant of L and S) rather than by the plough and soil characteristics, as in the mould board plough. As may be noted, the front furrow wheel swivels and is made to automatically align with the furrow wall by the links to make this possible. Nevertheless, an excessive offset of pull can cause a couple on the tractor.

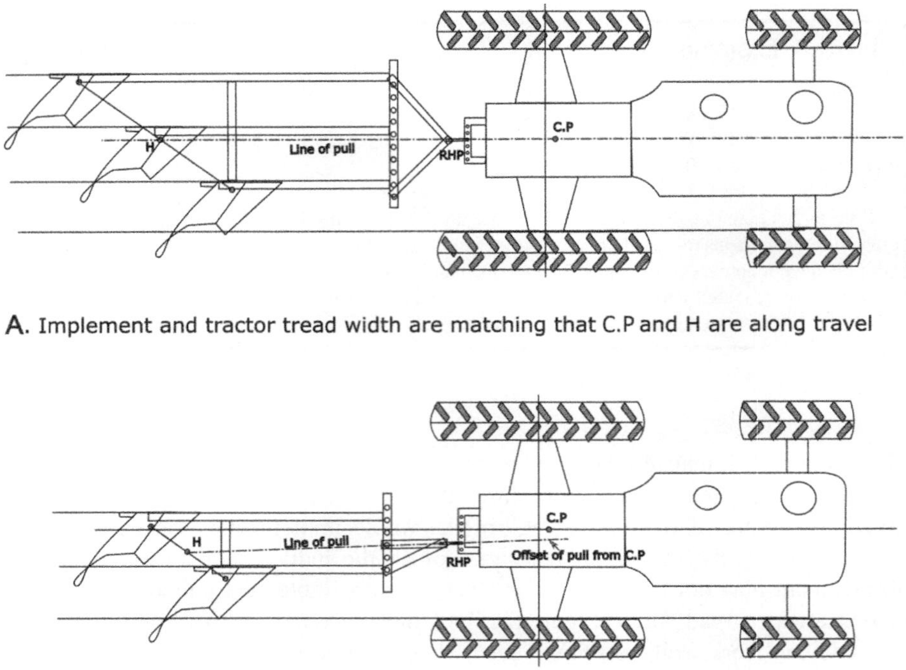

A. Implement and tractor tread width are matching that C.P and H are along travel

B. Tread width of tractor is more than implement width

FIGURE 5.12 Horizontal Hitching of Trailed Mould Board Plough

No Permission Required

5.9.5 Horizontal hitching of trailed disc harrows

A single-action disc harrow is a perfectly symmetrical tool in which the side thrusts, S, acting on the left and right gangs cancel each other. When the resultant soil reactions on each gang, R_h, are extended, they intersect at point H, which is the horizontal centre of the soil resistance of the harrow. The parallelogram of forces, when constructed to add the two resultants at H, give the net resultant soil reaction against which the pull P_h is to be applied. As can be seen from the values of L, the longitudinal soil reaction, and S, the lateral soil reaction, $P_h = 2L$ since the lateral reactions, S, of the two gangs cancel out each other. This implies that a trailed single-action disc harrow will travel in a straight line, following the centre line of the tractor. The same applies to a double-action disc harrow, too, except that the lateral soil reactions, S, of the rear set of gangs, instead of facing each other, oppose each other but, once again, cancelling out each other. Thus, P_h of the double-action or tandem-disc harrow = 4L, corresponding to the four gangs, and will be along the centre line of the implement.

However, the horizontal hitching of a trailed offset disc harrow is much different. When a trailed offset disc harrow is pulled by a free link, one end of which is pivoted

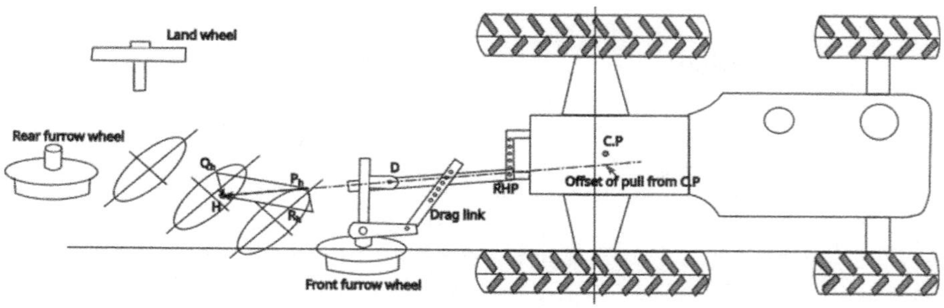

FIGURE 5.13 Horizontal Force Relations and Hitching for a Trailed Disc Plough

No Permission Required

to a real hitch point, F_1, located on the centre line of the tractor, and the other end is pivoted to the harrow, the harrow will seek an orientation of their gangs at certain angles so that the net side draft, acting on the implement, is compensated by the lateral component of the pull applied on the implement. In other words, the gangs orient themselves automatically to a set of front and rear gang angles so as to bring the system into equilibrium. The resultant soil reactions of the front gang, R_f, and the rear gang, R_r ,when made to intersect at point H and the parallelogram of forces is formed, the pull, P_h, to be applied is found. The pull has components: P_x along the travel direction and P_y in the lateral direction (Figure 5.14A). Since the algebraic sum of lateral soil reactions on both the gangs and the lateral component of the pull, P_y, are in equilibrium in the lateral direction, the implement runs straight behind the centre line of the tractor without any offset. But, when the hitch point on the tractor, to which the trailing link is connected, is shifted to F_2, the implement reorients itself to another angular position, changing its gang angles due to the soil reactions, bringing about an equilibrium with the pull applied. But, during this process, the implement not only changes its gang angles but moves laterally to an offset position from the tractor's centre line of travel (Figure 5.14B). The location of the horizontal centre of resistance, H, is determined by the intersection of R_f and R_r. This implies that every time the hitch point is shifted from the centre line of the tractor, the disc harrow rotates and moves laterally to bring the forces into equilibrium. The tractor driver can hence control the offset of the implement on the go by shifting the hitch point from the driver's seat. This is possible by swinging the hitch bar on the hitch bracket of the tractor, which can be controlled by a rope system from the driver's seat. The driver can hence change the implement's offset to easily maneuver between trees to do the harrowing operation in an orchard. The same manipulation is also possible by having an arrangement to change the gang angles relative to the mainframe. Figure 5.14C shows one particular case in which the tractor hitch point is directly in line with the centre of resistance, H, (the point of intersection of the resultant reactions of the gangs). Here, the

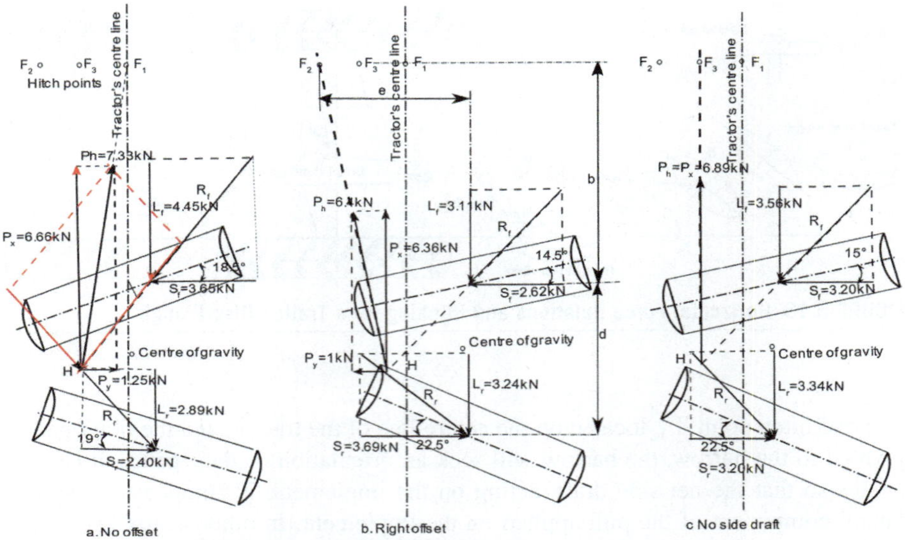

FIGURE 5.14 Horizontal Hitching of a Trailed Offset Disc Harrow

No Permission Required

implement operates without any side draft, implying that the lateral reactions of the gangs are equal and opposite and that they cancel out each other.

Taking moments about the hitch point, F_2, (Figure 5.13B), we get

$$eL_f + eL_r + bS_f - (b + d)S_r = 0$$

where e is the lateral distance of the hitch point F_2 to the centre of width of cut,

d and b are the longitudinal distances along the travel direction between the centres of two gangs and from the centre of the front gang to the hitch point, F_2, respectively.

L_f, L_r, S_f, and S_r are the longitudinal and lateral soil reactions of the front and rear gangs.

$$\therefore e = (b(S_r - S_f) + dS_r)/(L_f + L_r)$$

For the condition of no side draft $S_r = S_f = S$,

$$e = (dS_r)/(L_f + L_r)$$

The worked-out example (Box 5.9) on the horizontal hitching of a trailed disc harrow illustrates the method of solving for the pull.

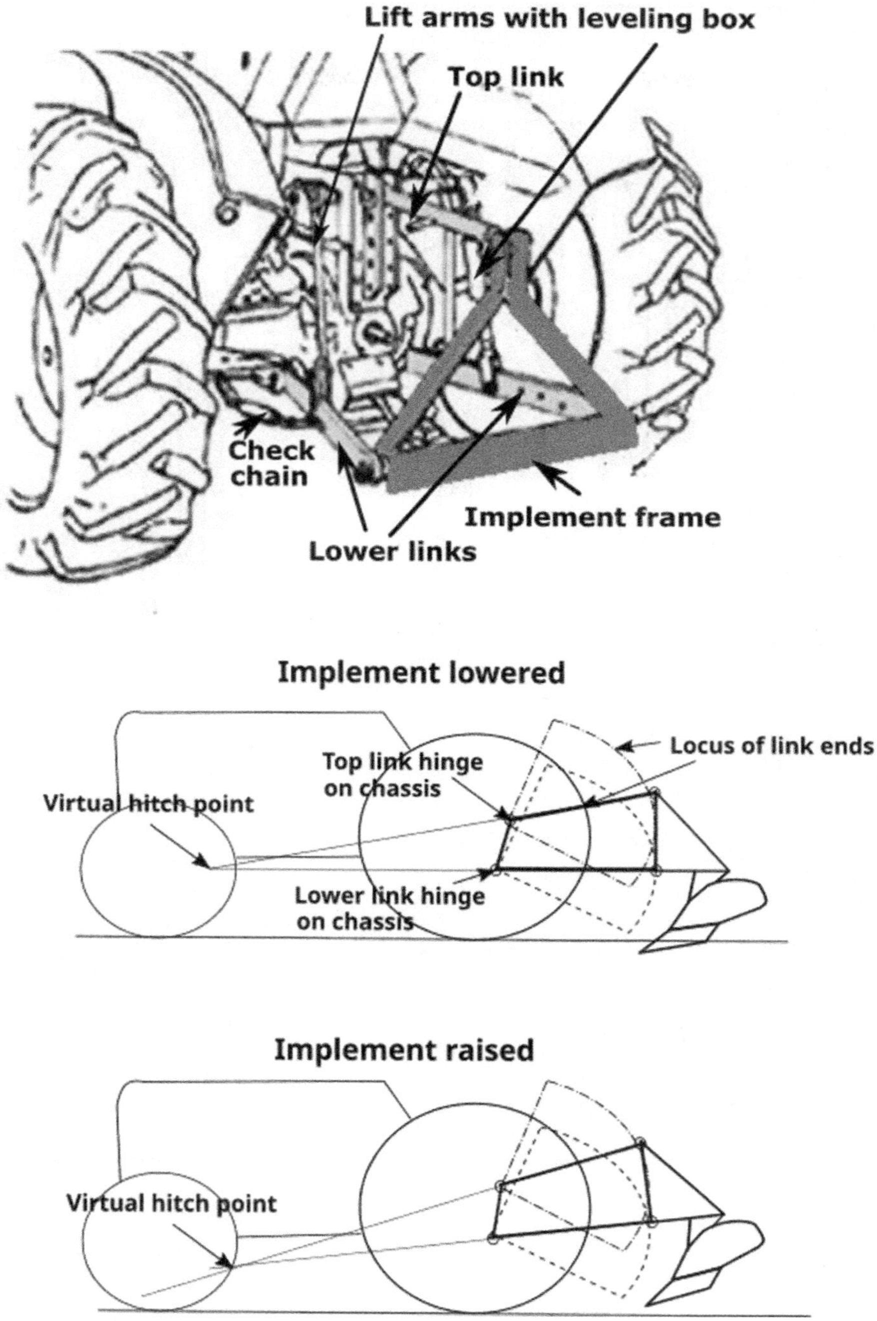

FIGURE 5.15 Three-Point Linkage of a Tractor

No Permission Required

Box 5.9 Example 5.9

Example 5.9. The orientation of a trailed offset disc harrow is shown below. The L/S ratio for both the front and rear gangs is 1.2. Graphically solve for the pull exerted by the tractor when it is trailed by a hitch point on the tractor's centre line, as shown.

No Permission Required

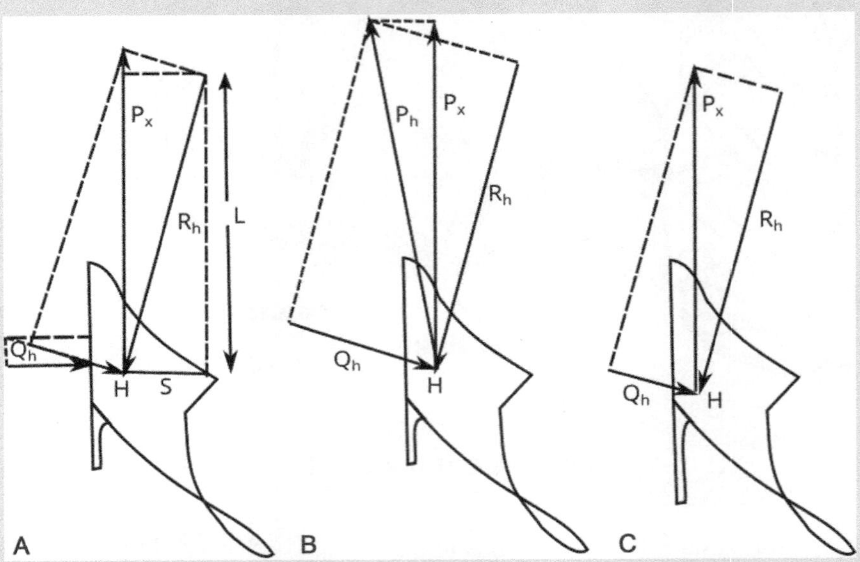

Let us first find the lateral reactions on the front and rear gangs using the given L/S ratios.

$S_f = 3\,/1.2 = 2.5$ kN

$S_r = 2\,/1.2 = 1.7$ kN

Then, select a suitable scale and draw the lateral reactions, S_f and S_r, at the centres of the soil resistances on the front and rear gangs. (Note that S_f and S_r are opposite in directions.) Add the L and S vectors on each gang by forming the polygons to find the R_{Hf} and R_{Hr}, namely, the resultant soil reactions on the two gangs (on the horizontal plane). Then, move the R_{Hf} and R_{Hr} vectors in their directions to make them intersect each other. At the point of intersection, draw a vector polygon, adding R_{Hf} and R_{Hr} to find the total resultant soil rection, R. The pull is applied opposite to R in the same direction intersecting the tractor's centre line at the hitch point. Convert the drawn vector to magnitude to compute the pull as 4.8 kN.

No Permission Required

Box 5.9 Example 5.9—cont'd

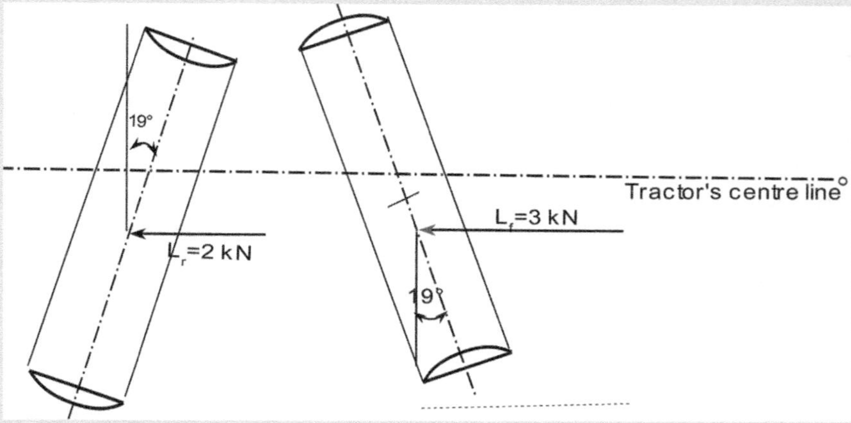

5.10 Hitching of mounted implements

Mounted implements use a three-point linkage (Figure 5.15) provided on the rear of the tractor that holds the implement completely on three points of attachment through a set of three linkages, namely, two lower links and one top link. The plough is lifted by the hydraulic power of the tractor. The plough or implement has a cross shaft on its front that carries the hitch pins at their extreme ends where they connect to the two lower links. The other ends of the two lower links are pinned to the tractor chassis. These freely hinged lower links are lifted up by lift arms connected to the tractor's hydraulic lift system. The length of the cross shaft is about the overall width of the implement, giving support for the implement on both sides. One end of the detachable, length-adjustable top link is pinned to a bracket that is provided on the implement frame at a height above the cross shaft and at the centre of the cross shaft's width. The other end is pinned to the tractor's chassis, as shown. The top link is constructed as a turnbuckle and is length-adjustable. Some ploughs have an adjustment for the cross shaft to be moved sidewise to position the width of the cut with respect to the tractor's right wheel rear. The lower link bars are adjustable for the height of their ends with reference to one another by way of a levelling box on the right lift arm. This provides a means for levelling the plough laterally. The suction of the plough is controlled by adjusting the length of the top link. When the hydraulic lift arms are actuated, the lift arms lift up the lower links and thereby the implement is supported on the three links. Figure 5.14 shows the locus of the links while lifting/lowering the implement.

It may be noted that this is a four-bar linkage system in two dimensions, as seen in the side view, that lifts the plough vertically without allowing too much pitching. Lines extended through the three links converge at a point, as shown, which is known as the virtual hitch point. The virtual hitch point is nothing but the instantaneous centre of rotation of the implement while lifting and dropping. It changes its position according to the depth of ploughing. At any depth, this centre will be the origin of pull on the implement and is hence the centre of pull when the three-point link system is free to rotate without restriction. The three-point linkage system may be laterally rigid or may swing laterally (within limits) about a virtual horizontal hitch point that is the intersection of lines extending from the lower links in the horizontal plane. These hitches may be operated with the hitch members acting as free links or with restrained links on which the implement is supported vertically through the lift mechanism on the tractor.

5.10.1 Free-link system

When the three-point hitch is operated as a free-link system (Clyde, 1940), the convergence of the links in a vertical plane gives the virtual hitch point, which is the centre of pull for the implement (Figure 5.16). The location of the VHP, the virtual hitch point, is changed by changing the lengths of the links (mainly the top link), and it also shifts up and down as the implement is raised or lowered. When the plough enters the soil, the VHP is lowered and promotes the rapid entry of the plough into the soil. The three-point hitching, when operated as free links, will be in equilibrium only when P_V, the pull from the tractor, emanates from the hitch point, VHP, and goes through the centre of resistance or load, "*G*," of the implement (Figure 5.16). Here, in the free-link operation of three-point linkage, the hydraulic system will not support the weight of the implement while in operation and hence requires a land or depth wheel to support the weight of the implement and keep

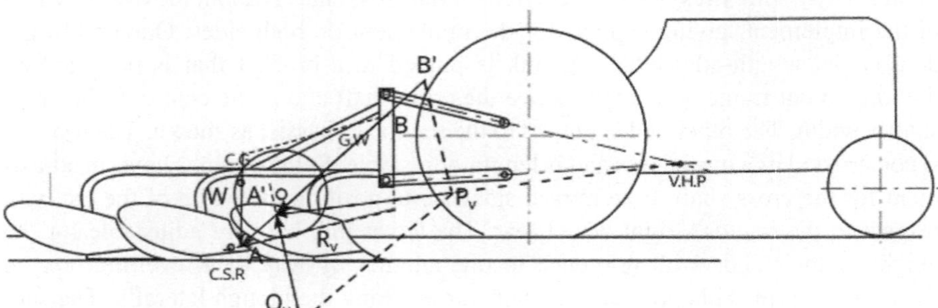

FIGURE 5.16 Vertical Force Relations for Three-Point Hitch as a Free-Link System

No Permission Required

the depth of operation in control. Sometimes, this support is rendered by a rear furrow wheel and/or a land or depth wheel rolling on the un-ploughed land. Hence, it is more like a trailed implement, pulled by the tractor through the three-point linkage. The three-point linkage is otherwise used to lift the implement up while not in operation.

The resolution of forces, as illustrated further in the example (Box 5.10), is very similar to a trailed implement with a single axle, except for the virtual hitch point being the centre of pull with its location variable at each depth of operation. Very similarly, the known forces, W and R_V, are to be added as vector AB. Since it is a free-link system, the tractor does not support the implement weight and the weight is wholly supported by the land wheel. So, the parasitic soil reaction, Q_V, is located on a line passing slightly behind the axle's centre line of the depth (land) wheel (Figure 5.16) with the aim of overcoming the wheel-bearing friction to cause the rotation of the wheel. This is very similar to the trailed single-axle implement (Figure 5.9). Its direction is, as before, inclined backward at the angle of rolling friction. Knowing Q_V in location and direction, drawing the parallelogram of the forces of Q_V, P_V, and the shifted AB to the location of Q_V, all the unknown magnitudes can be found including P_V emanating from the VHP. Although the three-point hitch system provides an up-and-down motion, it pitches the implement forward a little for easy penetration and levels off when the operating depth is reached.

5.10.2 Restrained-link system

Tractors are now provided with an automatic depth or draft control system, in which the implement is carried integrally, supported in operation by the tractor through a three-point hitch. This implies that the implement becomes an integral extension of the tractor at the rear and the linkages are restrained by the hydraulic system to be in a fixed position to control the depth. The depth control lever on the tractor allows the operator to fix the depth of operation and hence the implement is restrained to move down further. In addition to this, tractors also are provided with an automatic draft control system, in which the depth of the implement is automatically adjusted by the hydraulic system to maintain a preselected constant draft. The system automatically adjusts the depth of operation by sensing the forces on the implement (mainly by sensing the force on the top link). Here also, the three-point links are restrained completely by the hydraulic system and support the whole weight of the implement. In restrained-link operation, the plough's sole will not be supported by the furrow bottom and no depth control wheels are needed for the implement. A mounted disc plough under this system will have a rear furrow wheel to counter only the horizontal side draft, but a mounted mould board plough does not need any vertical support wheels. The hitch links also are used to lift or lower the implement.

Since the tool has no support from the soil, there is no parasitic force, Q_V, in the vertical plane. The component of pull, P_V, is simply the resultant of W and R_V (Figure 5.17). The line of pull is hence the direction of this resultant and does not pass through the virtual hitch point of the three-point hitch system. However, the

Box 5.10 Example 5.10

Example 5.10. Assuming $P_V = 6.5$ kN in the given figure of a free-linked mounted plough, find out the force in the top link and the total force in the two bottom links. The tractor's base width is 2.5 m. Solve by analytical method.

No Permission Required

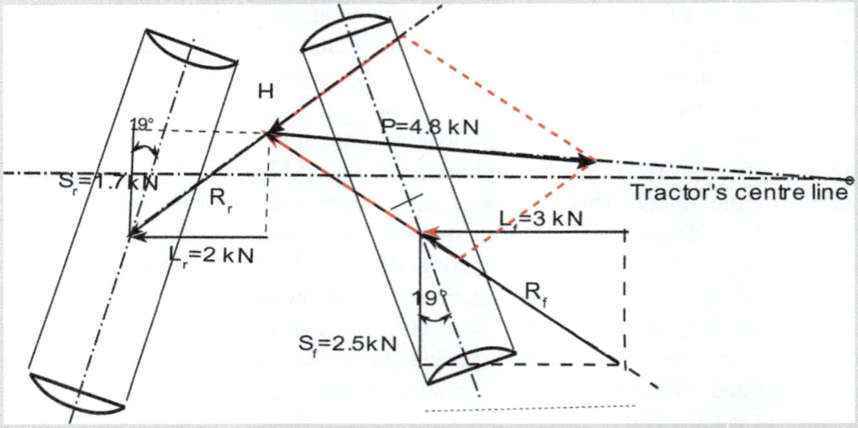

By measuring the base width in the figure and equating to 2.5 m, the scale of illustration could be ascertained. Based on the scale, the following dimensions are measured and found by conversion:

Height of bottom link pin above the pull vector = 18 cm.

Height of top link pin above the pull vector = 70 cm.

Let us shift P_v through a height of 18 cm to the bottom link pin on the implement without disturbing the force systems. This will result in:

P_v = 6.5 kN acting on the bottom link whose horizontal component = 6.5 cos (5°) = 6.47 kN causes tensions on the pair of bottom links and vertical component = 6.5 sin (5°) = 0.56 kN acts downward, causing the plough to penetrate into the soil;

a CW moment equal to horizontal component of P_v × 18 = 6.5 × 0.18 = 1.16 kNm.

The CW moment of 1.16 kNm will try to rotate the implement about the bottom link pins and cause a horizontal force, P_{TP}, (say) acting on the top link pin toward the direction of travel.

P_{TP} = the moment/the height of the top link pin above the bottom link pins = 1.16 kN m/0.52 m = 2.23 kN

P_{TP} can be resolved into a component parallel to the top link and a component perpendicular to it.

The angle of the top link relative to P_{TP} = 5° + 13° = 18°.

Compressive force on the top link = 2.23 cos (18°) = 2.12 kN.

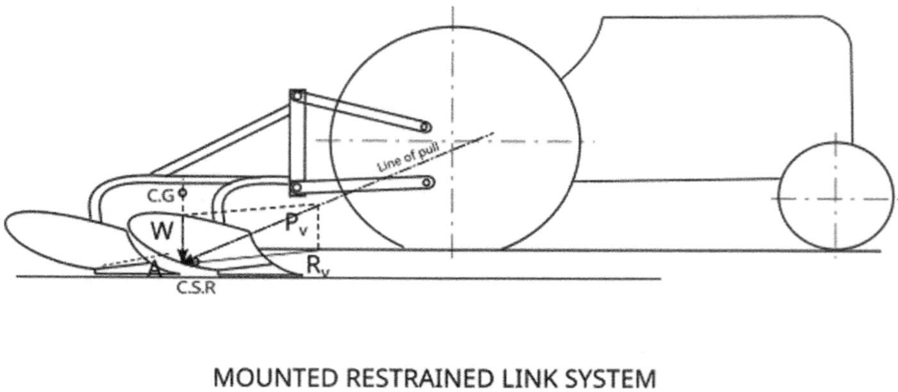

FIGURE 5.17 Vertical Force Relations for Three-Point Hitch as a Restrained-Link System

No Permission Required

line of pull cannot go below the virtual hitch point since the pull may then adversely lift the implement from the soil. A hydraulic system in older model tractors can only lift up the implement and is not capable of pushing it down to make it penetrate.

The major advantage of restrained links apart from the automatic depth/draft control of the implement is that the line of pull is much steeper than with a free-link system and provides for more weight transfer of the tractor to the rear wheel, rendering better traction to the rear wheels. The concept of weight transfer is explained later in this chapter.

As for the vertical hitching of mounted disc harrows, it is exactly like other mounted ploughs since the harrow is completely carried by the tractor. The harrow can be levelled perfectly well so that the front and rear gangs operate at the same depths. However, there will be a difference in soil reactions between the front and rear gangs because the rear gang tills the soil already tilled by the front gang.

5.10.3 Parallel-link hitching

A three-point hitch system causes the implement to penetrate the soil with an enhanced pitch at entry, which is necessary for a mould board and other suction-assisted ploughs. A parallel-link hitching system, as shown in Figure 5.18, causes no change in pitch as the depth of operation or height of lift varies. Because of this, it is an excellent choice for implements that skim the soil surface and penetrate only by a limited suction angle. This provides the best cultivation action in implements such as row crop cultivators with sweep shovels. This system is usually used as a free-link system with suitable depth wheels to support and control the depth of operation. Hydraulic rams are used just to lift the implement and not to restrain it. The virtual hitch point of the parallel links is at infinity since their lines of extension cannot intersect. Thus, P_V, the component of applied pull in the vertical plane, will be parallel with the parallel links. The direction of the vertical parasitic

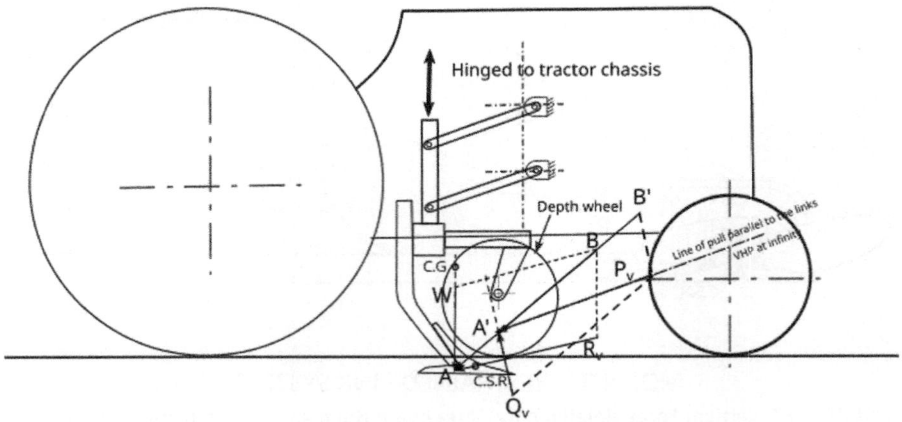

FIGURE 5.18 Vertical Forces of a Parallel-Link, Free-Link Hitch System

No Permission Required

reaction, Q_v, is at a slight offset to the centre of the supporting depth wheel, as in the case of trailed single-axle piece of equipment. Knowing Q_v in direction and location, the intersection of Q_v and AB (the resultant of W and R_v) yields "G," the centre of resistance of the implement. Drawing P_v through "G" and completing the parallelogram of forces yields the magnitude of P_v and Q_v,

5.11 Weight transfer due to the pulling effort of tractors

Figure 5.19 shows a typical tractor that applies a pull on a trailed or mounted implement. When not pulling an implement, the weight of the tractor acts down through its centre of gravity and will have reactions R_f at the front wheels and R_r at the rear wheels from the ground. These reactions can be determined easily since:

$W = R_f + R_r$ and $R_f \times x_1 = W \times (x_1 - x_2)$, where x_1 is the wheelbase, which is the distance between the front and rear wheel centres, and x_2 is the distance of the tractor's CG from the front wheel centre.

$$R_f \times x_1 = W \times (x_1 - x_2)$$

$$\text{i.e., } R_f = W \times (x_1 - x_2)/x1$$

$$\text{and } R_r = W - R_f$$

When the tractor is applying pull on the implement, let us say the pull passes through a point at a height "y" right above the rear wheel contact point with ground, which is right beneath the central axis of the rear wheel. Though the rear wheel reaction, R_r, will move slightly ahead from the ground contact point when the tractor is in motion due to rolling resistance, this can well be neglected. The pull component in

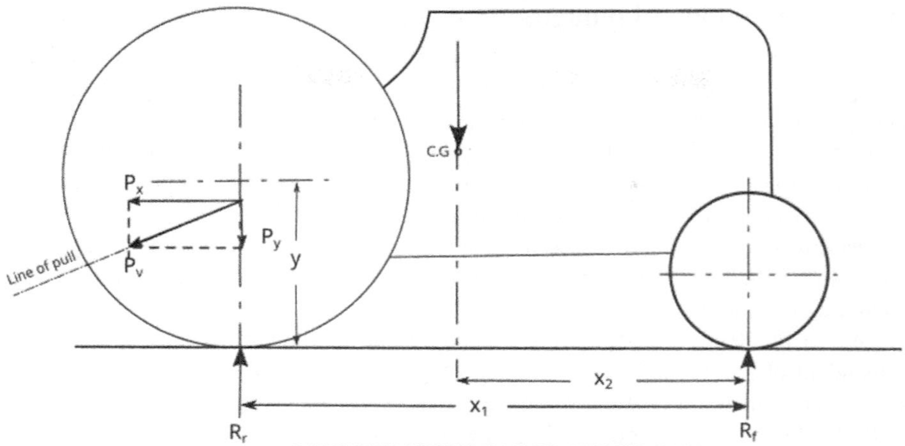

FIGURE 5.19 Weight Transfer due to Pull on an Implement

No Permission Required

the vertical plane can further be resolved into horizontal P_x and vertical P_z forces. Now let the reactions in the front and rear wheels be R_f' and Rr'.

Then, W= R_f'+ R_r' and $R_f' \times x_1 + P_x \times y = W \times (x_1 - x_2)$.

From which, $R_f' \times x_1 = W \times (x_1 - x_2) - P_x \times y$;

i.e., $R_f' = (W \times (x_1 - x_2) - P_x \times y)/ x_1$.

Comparing R_f and R_f', it can be seen that the front wheel reaction while in operation has been reduced by $(P_x \times y)/ x_1$. R_r', the rear wheel reaction, increases by a sum of this reduction of the front wheel reaction and the vertical download component of the pull. This implies that the weight of the tractor on the front wheel is transferred to the rear wheel and enhanced by virtue of the pull applied on the implement. This enhances the tractive effort of the rear wheel.

5.12 Influence of soil and machine parameters on the soil reactions of ploughs

We have already learnt that P, the pull, is the net force applied on a farm implement by a power source, such as a tractor, power tiller, or animal. The horizontal component of pull parallel to the line of travel is the "draft," and "unit draft" is the draft per unit area of the furrow cross-section. The unit draft of ploughs varies widely, influenced by implement type, soil type, and its condition, particularly the moisture, ploughing speed, type and shape configurations of the bottom, sharpness of the cutting and penetrating parts, depth of ploughing, width of plough furrow, type of attachments, and adjustment of the plough and attachments.

5.12.1 Mould board ploughs

Soil type and its condition is the most important factor influencing the unit draft of a plough. The unit draft of mould board ploughs is between 14 kPa and 20 kPa for sandy soils and 100 kPa and 140 kPa for very heavy clay soils (Kepner, Bainer, and Barger, 1978; Nichols, Reed, and Reaves, 1958). Soil moisture is very important in terms of both draft requirements and the quality of ploughing. Reduced moisture in soil obviously requires more draft, while wearing off the cutting edges more. An increase in soil moisture from 9% to 12% reduces the unit draft in a fine sandy loam by 15% to 35%. A higher degree of compaction and the presence of more weeds also increase the draft required.

As for the depth of operation, the force for cutting the bottom of a furrow is independent of the depth, but the other forces for handling the soil increase with the depth as the furrow slice volume increases. The horizontal soil reaction, L, increases with the depth of operation but not in linear proportion to the furrow cross-section. So, the unit draft does not remain constant with the depth of operation (Bainer, Kepner, and Barger, 1936).

Generally, plough bottoms of a larger working width would require more draft. With weight and parasitic forces proportionally larger for wider ploughs, the increase in draft is proportional to the area of the furrow cross-section. This makes the unit draft constant for wider ploughs operating at the same depth.

The shape of the mould board markedly influences the draft requirement of mould board bottoms. The unit draft of a stubble bottom is more than general-purpose bottoms since their abrupt curvature spends more energy in pulverizing the soil. Generally, bottoms that invert and pulverize the soil well have a greater draft requirement.

Rolling coulters and jointers contribute to a greater draft requirement. About one-sixth of the total draft is required for using rolling coulters.

The use of furrow wheels to take up the side thrust rather than the landsides of mould boards provide a reduced draft by 5% since rolling frictional forces are much less than sliding frictional forces.

The speed of ploughing influences the draft of a plough substantially (Bainer, Kepner, and Barger, 1936). As for a trailed mould board plough operated in a sandy loam soil at 3.5 kmh^{-1}, 50% of the total draft is spent on cutting the furrow, 30% is spent on transporting, pulverizing, and inverting the soil, and the rest is spent on overcoming the parasitic reactions (McKibben and Reed, 1952). When the speed is increased, the part that is spent for transporting, pulverizing, and inverting the soil also increases. The force required to cut the furrow slice does not change with the speed of operation. However, all the forces contributing to the acceleration of the soil on the mould board change theoretically with the square of the speed. This implies that the general form of the relationship between the draft, D, and the speed, S, of a mould board plough is in the form:

$$D = D_c + \alpha\,(S)^2$$

where D_c is the component of the draft due to cutting alone (independent of speed) and α, the regression coefficient of the draft, varies with the soil acceleration (Rowe and Barnes, 1961).

The equation indicates that higher speeds demand a substantial increase in draft. However, higher speeds lead to a better pulverization of the soil due to the force of the throw of the soil onto the ground. But, excessive speeds may throw the soil off the previously cut furrows and may result in a poor ploughing performance.

5.12.2 Disc ploughs

The unit draft of disc ploughs generally ranges from 25 kPa in sandy soils with 5% moisture to 150 kPa in clay soils with 15% moisture. The unit drafts for dry, hard soil are considerably higher than the above values (Gordon, 1941; McKibben, 1926; Summer, Chen, and Bierl, 1978). Similar to a mould board plough, the unit draft of a disc plough increases with the speed of operation. The side draft also increases with speed, but the vertical upward force decreases with speed. At optimal tilt angles, the speed is observed to improve penetration. But, a vertical disc blade (tilt angle = 0) has difficulty in penetrating at higher speeds. The unit draft is at a minimum at a disc angle range of 42° to 45°, which increases at greater disc angles due to a greater throw of the soil (Clyde, 1939). At smaller disc angles, it again increases because of the greater contact and the resulting parasitic friction between the furrow wall and the rear side of the disc. The type of scraper used on the disc also affects the draft because it impedes the free flow of soil over the disc. The net draft is, of course, a function of the width and depth of the operation since they contribute to the cross-sectional area of the furrow cut.

Table 5.3 shows typical soil reaction values for 65 cm discs operated at a 15 cm depth in a clay, loam soil (Bainer, Kepner, and Barger, 1936).

This table will help us understand the L/V and L/S ratios of typical discs and also how these ratios vary with the disc angle both in disc harrows (no tilt angle) and disc ploughs.

5.12.3 Disc harrows

The draft of disc harrows generally varies from 0.5 kN to 2.0 kN per m width of a single-action disc harrow and 1.2 kN to 2.5 kN per m width of a double-action, tandem-disc harrow, respectively.

5.12.4 Sweep blades

The draft of a sweep blade used on cultivators is mainly influenced by the pitch angle of the blade with reference to the ground surface. The sweep blades are usually set at a shallow forward pitch angle of 5° to 7°. Typically, when the sweeps are attached to spring tines, the pitch would increase by the backward flexing of the tine due to the horizontal soil reaction on it. The increased pitch would

Table 5.3 Typical soil reactions measured on disc blades

Disc angle°	0° tilt angle, 3.5 kmh^{-1} speed			15° tilt angle, 5.5 kmh^{-1} speed		
	L	V	S	L	V	S
35	1.11	0.67	1.11	2.68	0.80	1.11
40	1.34	0.45	1.16	2.23	0.54	0.89
45	1.34	0.36	1.18	2.01	0.40	0.80

cyclically increase the soil reactions, resulting in more flexing until the tine flings backward by spring pressure. It was found that the draft per unit width of a blade increases to about 31% from 1.7 kN m^{-1} at 2.5° pitch to 2.3 kN m^{-1} at 18.5°(Gullacher and Coates, 1980) when the sweep was operated at a depth of 40 mm at a speed of 8 km h^{-1}. An increased speed was found to increase the draft of these blades.

5.13 Implement settings and adjustments for tillage

In the hitching of implements to the tractor, there are basically three adjustments that are very important and are common to many tillage implements. Apart from these, there are some specific adjustments required depending on the implement. The first three are for adjusting the implement's vertical suction, implement levelling, and front furrow width.

5.13.1 Adjustments for mounted implements, particularly for ploughs

5.13.1.1 Vertical suction adjustment

One ideal example of a suction-operated implement is the mould board plough. Vertical suction is the vertical clearance at the heel of the plough intended to aid the penetration of the plough into the soil. The vertical suction in mould board ploughs is adjusted by pitching the implement forward so that the desired clearance is obtained (Figure 5.20A). In mounted ploughs, this is achieved by adjusting the top link of the tractor (Bell, 2003; Lovegrove, 1968). The top link is a turnbuckle link and is length adjustable. When its length is shortened, the implement supported on the three-point hitch system rotates about the ends of the lower links (Figure 5.20A) and pitches it forward to the required level. Lengthening the top link reverses the rotation.

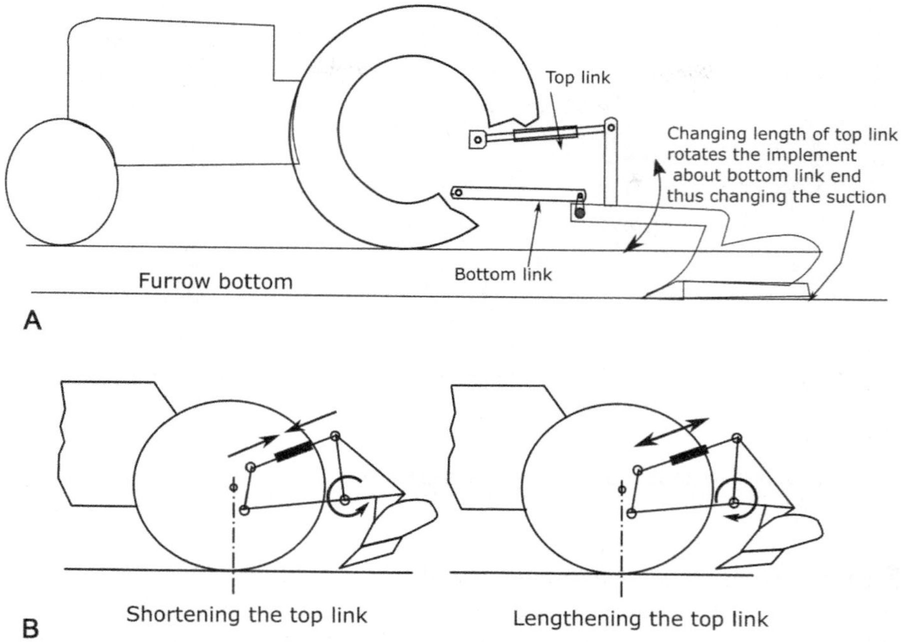

FIGURE 5.20 Vertical Suction Adjustment in Mounted Implement

No Permission Required

5.13.1.2 Implement levelling

Each plough bottom digs a furrow and throws the soil into the adjacent furrow already made by the bottom preceding it or the last furrow made by the last bottom in the previous ploughing pass. This implies that the last furrow of the previous pass is not filled by soil until the next pass is undertaken. This unfilled furrow of the previous pass acts as a guide for the tractor to align the present pass being taken with that of the previous pass. This is achieved by the tractor operator guiding the right rear wheels (both front and rear) into this unfilled furrow while ploughing. This aligns the present ploughing exactly with that of the previous pass. The furrow slice dug by the first bottom of this pass now fills the last furrow of the previous pass. This procedure makes the tractor travel with an angular orientation to the right (Figure 5.21A) because the right wheels are in the furrows and the left wheels are on level soil. But the plough should not be inclined along with the tractor. It has to be levelled horizontally so that all plough bottoms dig uniformly. The levelling box (Figure 5.15 and 5.21B) is an assembly on a link pin connected to the right-side lift arm of the tractor and the right lower link. The crank on this box, when used, will alter this link's length, thus lifting the right lower link independently from the left lower link. Since the implement is supported on the lower links, the implement can be levelled parallelly to ground irrespective of the tractor's angular orientation.

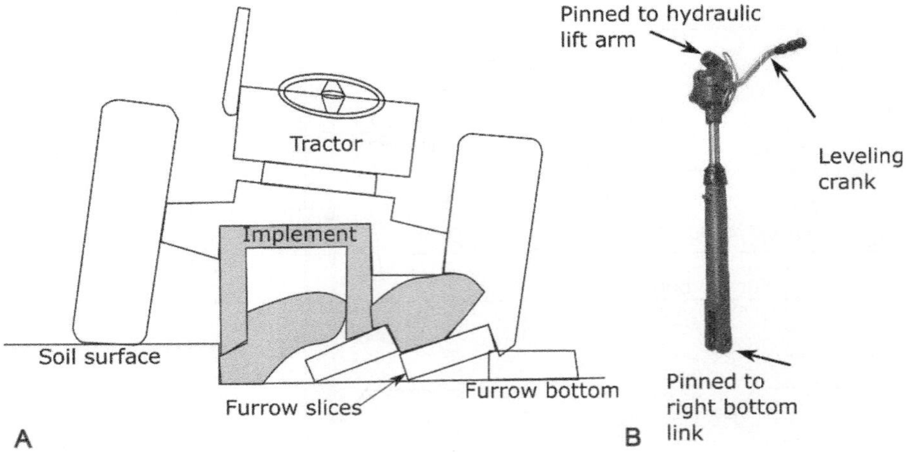

FIGURE 5.21 Implement Levelling in Mounted Implement

No Permission Required

5.13.1.3 Front furrow width adjustment

The implement has a cross shaft on its front that carries the hitch pins at their extreme ends. These pins are hitched to the two lower links. The length of the cross shaft is about the overall width of theplough , giving support for the implement on both sides. It was explained in the previous paragraph that the tractor's right-side wheels ride in the already made furrow to align the present pass to the previous. So, it is imperative that the inner edge of the right rear wheel exactly aligns with the first bottom's share wing (Lovegrove, 1968) so that the first bottom digs a furrow exactly next to the furrow in which the right wheels are riding. There are two adjustments to be made for achieving this.

Coarse adjustment

The wheel tread of the tractor, which is the lateral distance between the two rear wheels, should be first adjusted so that the inner edge of the rear wheels aligns with the first bottom's furrow width.

All tractors have provisions to do this alteration. The wheel discs and rims of the rear wheels have arrangements to alter the distance between the wheel centre and the tractor chassis's centre line. The wheel disc has a concave shape (Figure 5.22) with its central surface fastened to the tractor's wheel hub. The disc can be fastened on the hub with the concave side facing toward ("disc in") the hub (Figure 5.22A and B) or facing away from (Figure 5.22C and D) ("disc out") it. The disc's outer edge is fastened to the inner surface of the wheel rim through spacer bushes welded to the inner rim (Figure 5.22). The wheel disc can be fastened to either side of the spacer bushes. "Rim in" is when it is fastened on the outer side of the spacer bushes (Figure 5.22A and C) and "rim out" is when it is fastened on the inner side (Figure 5.22B and D). The rubber tire is fitted to the wheel rim.

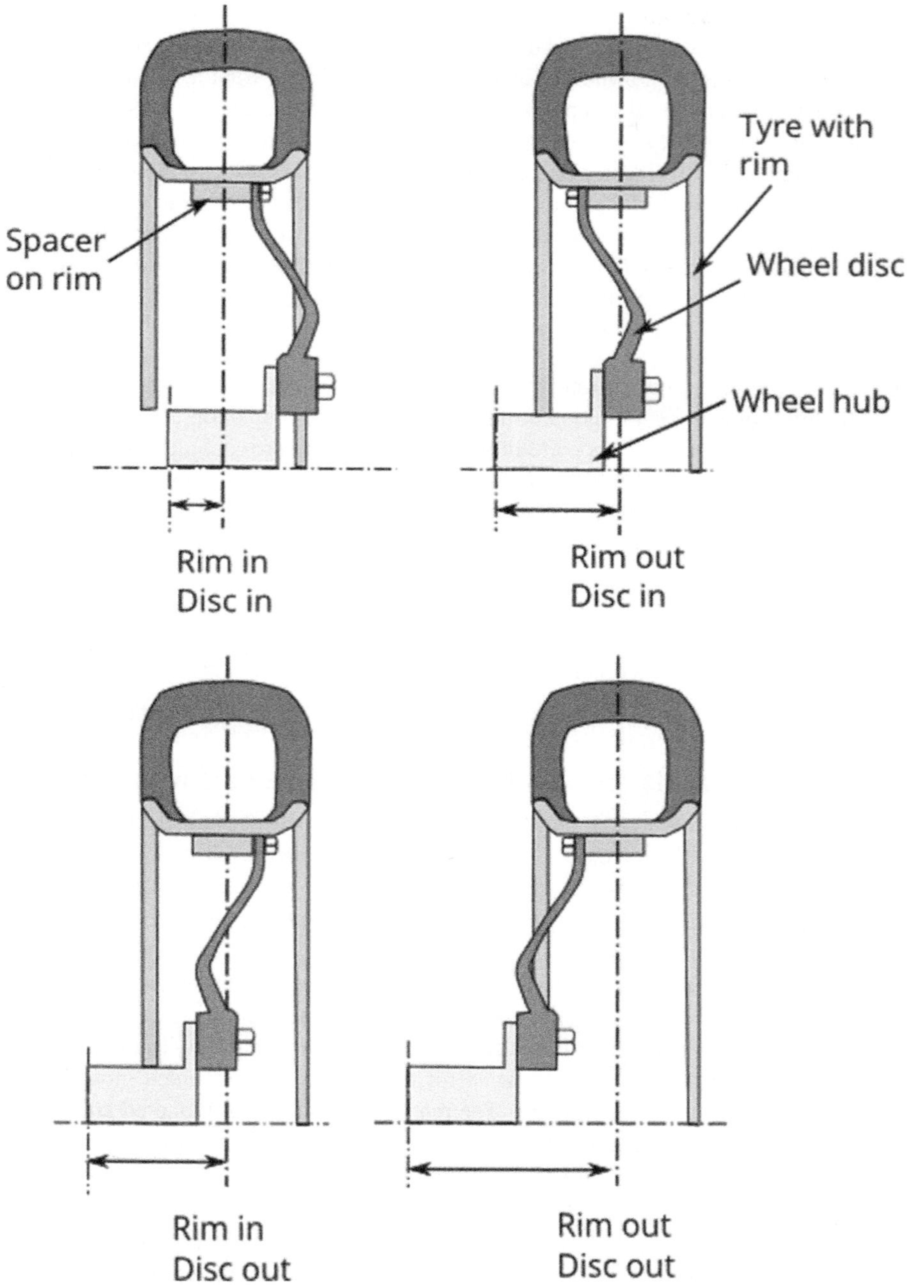

FIGURE 5.22 Tractor's Wheel Tread Adjustment

No Permission Required

So, there are four possibilities of arranging the disc and the concave with respect to the wheel rim and tire. As illustrated in Figure 5.22, each arrangement (A to D) increases the offset of the wheel center from the wheel hub (tractor's center line) in steps. The same choice of arrangement can be made on the left rear wheel also, so that the tread width between the rear tires can be set to the requirement of adjusting the front furrow width. Tractor operator manuals provide the data on the stepped offsets that are available for each model of tractor. The front wheel tread width is also easily adjusted through a telescopic arrangement on the front axle. The front tread width is matched equally to the rear. The tread width setting is only a coarse adjustment that will bring about the front furrow width adjustment to a certain level.

Fine adjustment

The cross shaft is a cranked rod (Figure 5.23) that is on the front of the plough and has the hitch pins to accommodate the lower links. The cranks on either side have eccentricity in diagonally opposite directions. When the cross shaft is rotated, the implement is rotated in the horizontal plane and thus provides for shifting the share wing slightly toward (Figure 5.23B) or away from (Figure 5.23A) the line of rear wheel inner edge. The cross shaft and the ball joints at the lower link ends make this rotation possible.

Nowadays, some ploughs have an easier provision for the cross shaft to be moved sidewise with respect to the implement to position the first furrow with respect to the tractor's right rear wheel (Figure 2.5). The cross shaft provides a precise front furrow width setting.

5.13.2 Adjustments for trailed and semi-mounted ploughs

The trailed implements are completely supported on three wheels, namely, the land and front furrow wheels on the front and the rear furrow wheel on the rear of the implement. The semi-mounted implement is supported by the tractor at the front and by a rear furrow wheel at the rear. The rear furrow wheel can be raised up or down with respect to the implement frame so that the plough can be pitched forward or backward to adjust the vertical suction of the bottoms. The levelling of the trailed implement with respect to the tractor is achieved by adjusting the heights of the three supporting wheels accordingly. The levelling box is used in the semi-mounted plough similarly to the mounted ploughs. Front furrow width adjustments for trailed and semi-mounted ploughs are same as for that of mounted ploughs. The rear furrow wheels of mounted and semi-mounted disc ploughs may move away laterally from the last furrow of the ploughing pass when the cross shaft rotation is made to correct the front furrow width (Figure 5.13). There are some provisions available for the wheel to be repositioned. It may also be noted that the rear furrow wheel of a disc plough (Figure 5.13) is inclined away from the furrow wall so that it can roll as well as take up the side thrust. Here, the whole of the side draft is supported by the rear furrow wheel, unlike the mould board plough for which landsides compensate that force.

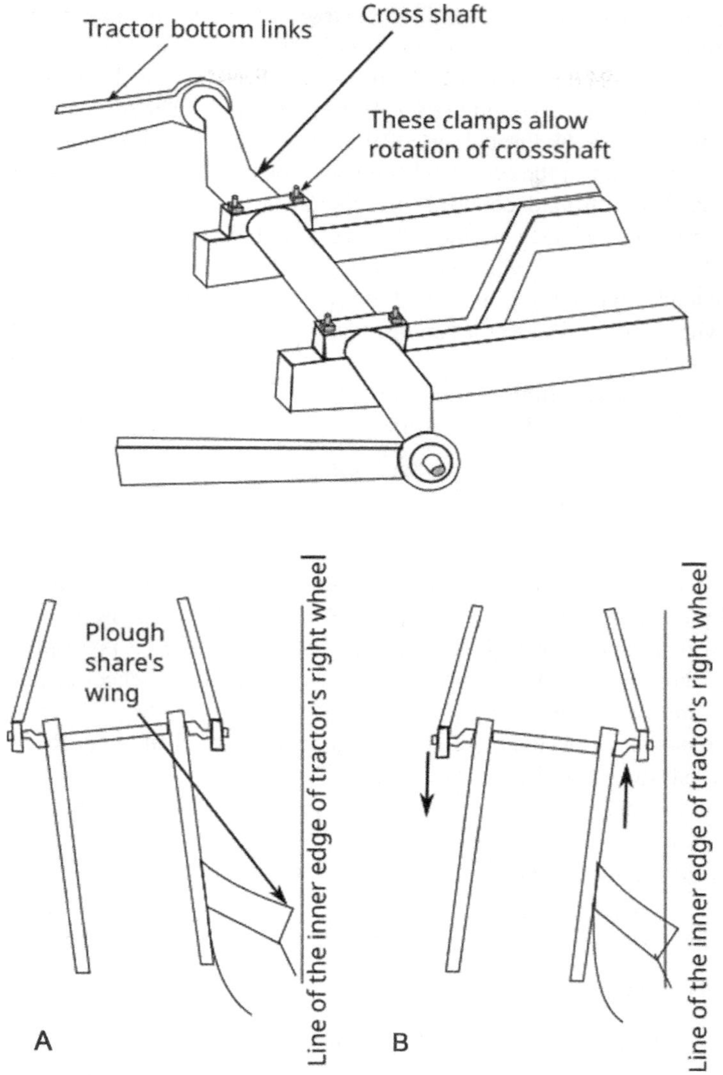

FIGURE 5.23 Front Furrow Width Adjustment

No Permission Required

5.14 Locating the center of gravity of a tractor

Let us take an example of a tractor weighing 24 kN with a wheelbase of 2080 mm. Before solving this problem, let us first learn the procedure for using the weighing method to locate the centre of gravity. After which, try solving for the CG using the given observed values in the laboratory.

Firstly, read and record the weight of the tractor from its operation manual. Using a jib crane shown in the illustration, the front axle can be chained or roped appropriately on both sides and made ready for lifting the tractor's front. The rear wheel may be locked in place with cogs at the back of the rear wheels. The rope or chain may be linked to the crane hook through an axial load cell or a dynamometer of at least 15 kN capacity. Then, use the crane to lift up the front of the tractor to make the front wheels just leave ground contact. The clearance between the front wheels and ground should be the least possible. The rope's or chain's verticality may be assured using a plumb bob; check whether the rope is in line with the front wheels' center. Then, the load cell reading is measured; it is equivalent to the front wheels' reaction, R_f. The wheelbase of the tractor can be taken from the manual or can be accurately measured as x_1.

Then, from the illustration in Figure 5.24A, take moments about the rear wheel ground contact point in illustration (a):

$$Wx_2 = R_f x_1$$

$$\therefore x_2 = \frac{R_f x_1}{W}$$

Then, the crane is used to lift up the tractor's front, adequately assuring the verticality of the lifting chain. The crane is to be moved a bit to assure this. The height through which the front wheels' bottom-most points are lifted is precisely measured as H. The horizontal distance (projected) between the front wheel center and rear wheel ground contact point is recorded as x_1'.

Simultaneously, the load cell reading is also recorded, now as R_f'. It will be observed that the $R_f' < R_f$.

Ensure that this difference is adequate. If not, lift the tractor a bit more and record the observations.

Then, the angle through which the tractor is raised is

$$\theta = tan^{-1}\frac{H}{x_1'}$$

and, also, moments are

$$Wx_2' = R_f' x_1'$$

$$x_2' = \frac{R_f' x_1'}{W}$$

In the illustration in Figure 5.24B, note that

$x_3 = \frac{x_2'}{cos\,\theta}$ and in Figure 5.19

but $x_2 = x_3 + x_4$;

$\therefore$, $x_4 = x_2 - \frac{x_2'}{cos\,\theta}$ (we have already calculated x_2)

but $tan\,\theta = \frac{x_4}{y}$;

$$\therefore, y = \frac{x_4}{tan\,\theta}$$

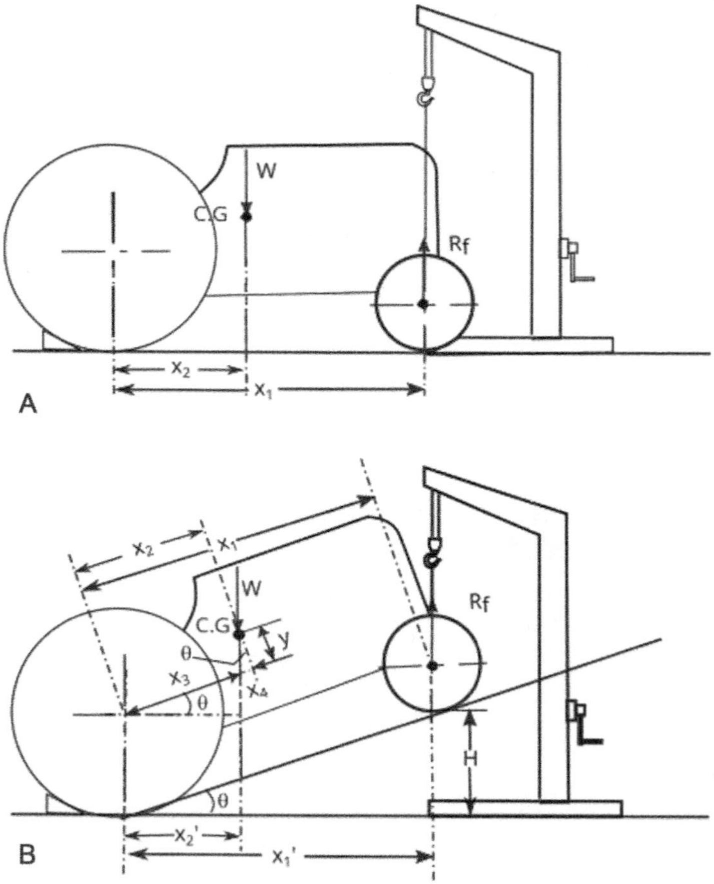

FIGURE 5.24 Locating the Centre of Gravity of a Tractor

No Permission Required

The CG was located at coordinates (x_2 , y) relative to the rear axle as the origin. Then, for the calculations with the given experimental observations (say),

$$W = 24 \text{ kN}$$

$$x_1 = 208 \text{ cm}$$

$$x_1' = 210 \text{ cm}$$

$$H = 71 \text{ cm}$$

$$R_f = 10.1 \text{ kN}$$

$$x_2 = \frac{R_f x_1}{W} = 87.7 \text{ cm}$$

$$R_f{}' = 9.2\ \text{kN}$$

$$x_2{}' = \frac{R_f{}' x_1{}'}{W} = 80.1\ \text{cm}$$

$$\theta = tan^{-1}\frac{H}{x_1{}'} = 18.7^\circ$$

$$x_3 = \frac{x_2{}'}{cos\ \theta} = 84.6\ \text{cm}$$

$$x_4 = x_2 - x_3 = 87.7 - 84.6 = 3.1\ \text{cm}$$

$$y = \frac{x_4}{tan\ \theta} = 9.15\ \text{cm.}$$

The coordinates of the CG are (87.7, 9.15) with respect to the rear axle in cm.

5.15 More accurate soil-tool interaction models for assessment of soil reactions

Sohne's model of a simple inclined tool was discussed in Section 5.1 to understand what we mean by soil reactions and how they arise in the analysis of forces on soil working tools. It helped us to understand how these reactions manifest themselves into the pull applied on the tool and thereby the power requirement for its operation. The model explained was a very simplistic model, which, many times, can over- or under-predict the reactions and, hence, cannot be used for an accurate assessment toward the design of a soil working tool. Over the years, many prediction models were evolved to predict the soil reactions more precisely. The universal earth-moving equation (Reece, 1965) was developed based on the passive earth pressure theory used in the subject of basic soil mechanics and foundation engineering. Although it improved the assessments many folds, it was mainly applicable for wide soil working tools, where the ratio of the depth of operation and tool width tended to zero. A bulldozer blade, an earth-moving blade, is an example of wide soil working tool. The Reece model and many refinements made on it were all basically two-dimensional models. Simultaneously, three-dimensional models that predict the soil reactions of narrow soil tools (Clyde, 1944; Perumpral, Grisso, and Desai, 1983) such as chisel ploughs were developed. Let us present the McKyes and Ali model (McKyes and Ali, 1977), which is one such effective model that can be universally applied both on narrow as well as wide soil working tools.

When a soil working tool with an approach angle or rake angle, α, interacts with the soil, it causes a failure plane emanating on the cutting edge and extending as a failure plane rising to the soil surface. The soil failure plane is at a shear failure angle, β. The failure plane angle depends on:

1. the shear strength of the soil, explained by Coulomb's equation (Section 5.1) involving the cohesion, c, and angle of internal friction, ϕ, of the soil;

2. the adhesion equation applicable to the soil-tool interface of the tool's frontal surface, which was also explained earlier (Section 5.1) and involves the adhesion, a, and the soil metal friction angle, δ;
3. the component due to the soil density, implying the weight of the soil in the failure wedge;
4. the surcharge pressure, q, on the soil that is vertically acting downward on the tool, which is any additional pressure due to a uniform load on the soil above the failure wedge; although this can be calculated, it is assumed as zero in most tillage situations.

The failed soil is in the form of a wedge of width equal to the tool width, w, and a maximum depth equal to the depth of the operation of the tool, d. In the earlier wide tool models, Reece (1965) considered only this central soil wedge (Figure 5.25) for the prediction because the soil failure on the tool sides is insignificant in wide tools. But, in narrow soil tools, the soil failure significantly extends to the tool sides, too. The McKyes and Ali (1977) model rightly assumes the shape of the soil failing on the sides as circular crescents, as shown (Figure 5.25), which significantly contribute to the soil reactions along with the central soil wedge.

The resultant soil reaction due to the tool causing soil failure was explained by Reece (1965) by the universal earth-moving equation of the form:

$$P = \gamma d^2 N_\gamma + cdN_d + adN_a + qdN_q$$

where $N_\gamma, N_C, N_a, and\ N_q$ are dimensionless factors concerning the soil density, γ; the soil cohesion, c; the soil metal adhesion, a; and the surcharge pressure, q.

McKyes and Ali (1977) derived these factors for the three-dimensional model they proposed as below. The derivation of these factors is much involved and beyond the scope of this text but can be learnt from their referred paper. The angle of soil failure plane, β, is the only unknown parameter, influenced by all the other known parameters either predefined or measurable for the soil and soil-metal interaction. These factors are:

$$N_\gamma = \frac{0.5(cot\ \alpha + cot\ \beta)\left(1 + \frac{2d}{3w}\sqrt{cot^2\ \beta + 2\ cot\ \alpha.cot\ \beta}\right)}{\sin(\alpha + \beta + \delta + \varphi)} \sin(\alpha + \delta)$$

$$N_C = \frac{cos\ \varphi}{sin\ \beta} \frac{\left(1 + \frac{d}{w}\sqrt{cot^2\ \beta + 2\ cot\ \alpha.cot\ \beta}\right)}{\sin(\alpha + \beta + \delta + \varphi)}$$

$$N_a = -\left(\frac{\cos(\alpha + \beta + \varphi)}{sin\ \alpha\ \sin(\alpha + \beta + \delta + \varphi)}\right)$$

$$N_q = \frac{(cot\ \alpha + cot\ \beta)\left(1 + \frac{d}{w}\sqrt{cot^2\ \beta + 2\ cot\ \alpha.cot\ \beta}\right)}{\sin(\alpha + \beta + \delta + \varphi)} \sin(\alpha + \delta)$$

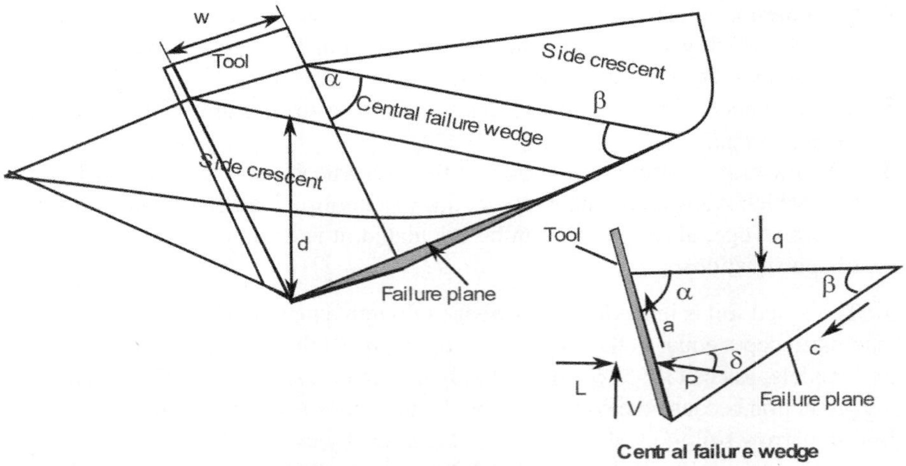

FIGURE 5.25 McKyes and Ali Model of Soil-Tool Interaction

No Permission Required

N_γ can now be minimized using a simple computer code for a range of β values, and the critical β value causing the minimum N_γ is substituted into the other dimensionless factors and finally substituted into the earth-moving equation to find the resultant soil reaction, P, of the tool. The horizontal, L, and vertical, V, soil reactions are then calculated after duly applying the adhesion also to the tool surface. These are of the forms:

$$L = Psin(\alpha + \delta) + cwdcot(\alpha)$$

$$V = Pcos(\alpha + \delta) - cwd$$

The accuracy of prediction using these narrow tool models has improved so much that they can be used as a tool for the design of a soil working tool. The model calculations (Box 5.11) may be tried by the reader or by using the computer code available with this book.

Box 5.11 Example 5.11

Example 5.11. Compute the horizontal and vertical soil reactions of a chisel plough of 5 cm width working at a depth of 40 cm and at a rake angle of 40°. The soil parameters are soil density = 16 kN m^3, angle of internal friction = 30°, and soil-metal friction angle = 23°. The soil may be assumed as cohesionless and adhesionless.

(A computer code with GUI is available that can compute the soil reactions using the above model. Students are requested to try it out for different values of soil and tool parameters.)

Exercise problems

1. A single-bottom mould board plough is hitched to a three-point, free-link system of a tractor, illustrated below. The tractor's base width is 2.5 m. By graphical methods, determine the location of the virtual hitch points, when the plough is operating at a depth of 15 cm and when the plough point is just entering the soil surface.

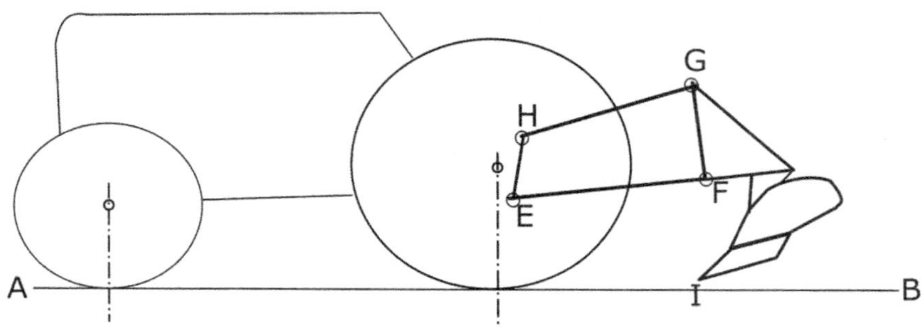

2. In the Example problem 10, calculate the change in the front and rear wheel reactions of the tractor weighing 22 kN pulling a mounted plough. $P_V = 6.5$ kN directed downward at 5° to horizontal. The wheelbase of the tractor is 2.5 m. The center of gravity of the tractor is 1.2 m ahead of the rear axle.
3. The total draft of a 30 cm, three-bottom plough working at a depth of 15 cm and at a speed of 5 kmh^{-1} is 15 kN. Find the unit draft and horsepower requirement. If the field efficiency is 75 percent, calculate the actual field capacity.
4. Find out the wheel tread setting for a tractor pulling a trailed, three-bottom, 40 cm mould board plough without side draft. The tractor has 32 cm wide rear tyres. Show the pertinent dimensions on a diagram.
5. The magnitudes and orientations of the thrust and radial forces acting on a single 50 cm disc blade of a disc harrow gang are shown in the figure. The gang has seven such discs and is supported by two radial thrust bearings assembled 110 cm apart. Assume that the weight of the gang is completely supported by the soil. Find the radial and thrust loads on each bearing assembly. Chilled cast iron bearings have a 5 cm diameter cylindrical bearing surface. The thrust surface in each bearing is on an average diameter of 4 cm. Assume the coefficient of friction of the bearing surfaces as 0.15. Verify whether the rotating torque generated in the gang is adequate to overcome the bearing friction.
6. Find the magnitude and direction of the vertical component of pull, P_V, applied by the tractor on the mould board plough weighing 2000 kN. The plough is hitched to a free three-point link system. The magnitudes of R_V, the resultant of

the soil reactions in the vertical plane, and W, the weight of the implement, are given. The direction and location of Q_V, the parasitic force, is also given. Take a photocopy of the figure, assume a suitable scale for the given magnitudes, and solve the problem graphically.

No Permission Required

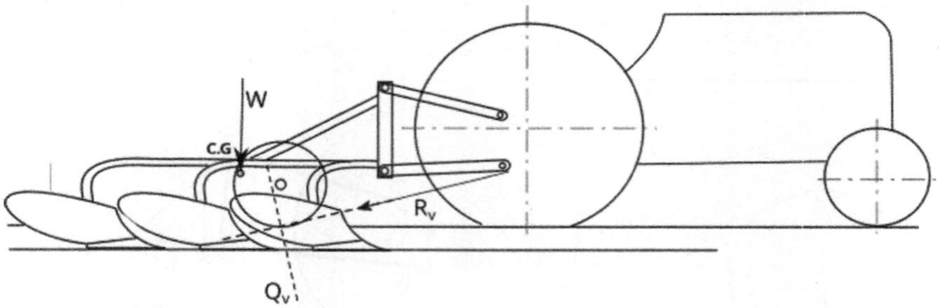

7. Graphically find the magnitude of the pull to be applied to the sweep cultivator weighing 1000 kN hitched to a parallel0link system of the tractor. The magnitudes of R_V, the resultant of the soil reactions in the vertical plane, and W, the weight of the implement, are given in the figure. The direction of Q_V, the parasitic force of the depth wheel, is also marked. Take a photocopy of the figure, assume a suitable scale for the given magnitudes, and solve the problem vectorially.

No Permission Required

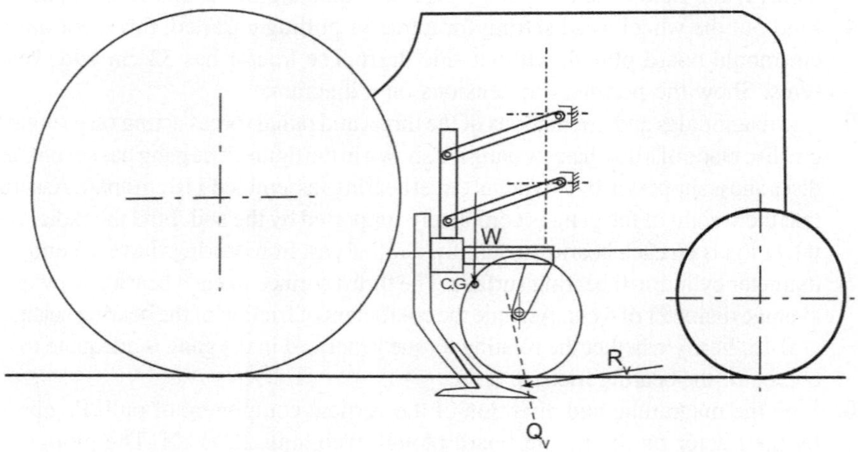

8. Find the magnitude and direction of the parasitic force on the trailed mould board plough weighing 5 kN. The magnitudes of R_V, the resultant of the soil reactions in the vertical plane, W, the weight of the implement, and P_V, the vertical component of pull applied through the real hitch point of the tractor and implement, are given in the figure. The magnitude and directions of P_V, R_V, and W are given. Take a photocopy of the figure, assume a suitable scale for the given magnitudes, and solve the problem graphically.

No Permission Required

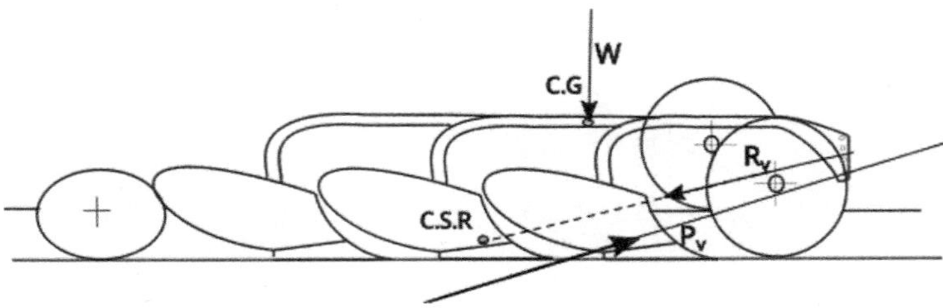

9. A trailed disc harrow weighing 3.0 kN is hitched to the tractor by a drag link. The L/V ratios of the front and rear gangs are both 1.0. The locations of the centres of the soil resistance for the rear and front gangs and the center of gravity of the harrow are shown in the figure below. Assuming that the drag link is connected to the middle hitch hole of the harrow's drag bar, find the pull exerted as well the horizontal and vertical soil reactions of the front and rear gangs.

No Permission Required

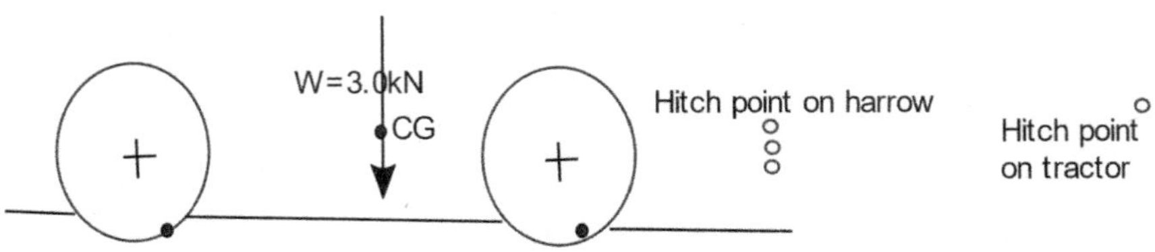

10. The distance between the two gang centers of an offset disc harrow measured along the direction of travel is 63 cm. The distance between the front gang's centre and the hitch point is 109 cm. The centers of resistances of each gang coincide with the gang center as viewed in the plan. The ratio between the horizontal soil reactions of the front and rear gangs is 1.07. The L/S ratios of the front and rear gangs are 1.13 and 1.04, respectively. Find the lateral offset of the gang centers from the hitch point.

11. Use Excel to compute the horizontal and vertical soil reactions of a chisel plough of 5 cm width working at a depth of 40 cm and at a rake angle of 40°. Assume the soil density as 15 kNm^3, angle of internal friction as 30°, and soil-metal friction angle as 23°. The soil may be assumed as cohesionless and adhesionless. Use the McKyes and Ali model for the computations.

No Permission Required

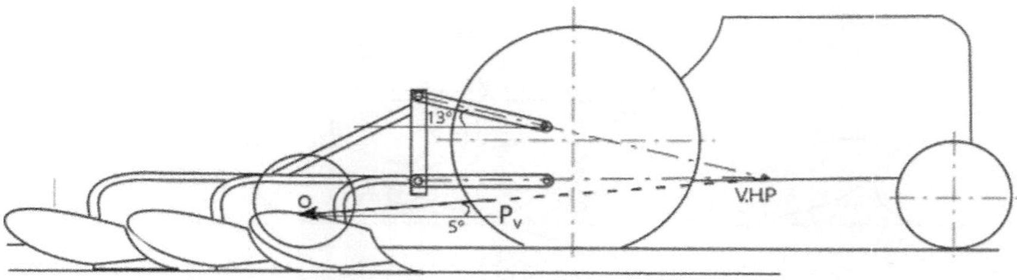

Bottom link parallel to soil surface

References

Bainer, R., Kepner, R. A., & Barger, E. L. (1936). Principles of farm machinery. Measurement of forces on soil tillage tools. *Agricultural Engineering, 17*, 5–9.

Bell, B. (2003). *The tractor ploughing manual.* UK: Old pond Publ. UK.

Clyde, A. W. (1934). Mechanics of plow and tractor hitches. *Agricultural Engineering, 151*, 388–390.

Clyde, A. W. (1935). Vertical hitching of farm implements. *Agricultural Engineering, 16*, 358–360.

Clyde, A. W. (1936a). Measurement of forces on soil tillage tools. *Agricultural Engineering, 17*, 5–9.

Clyde, A. W., & Measurement of Forces on Soil Tillage Tools. (1937). Load studies on tillage tools. *Agricultural Engineering, 17*(1936), 117–121.

Clyde, A. W. (1939). Improvement of disk tools. *Agricultural Engineering, 20*, 215–221.

Clyde, A. W. (1937). Load studies on tillage tools. *Agricultural Engineering, 18*, 117–121.

Clyde, A. W. (1940). Mounted plows and their effects on the tractor. *Agricultural Engineering, 21*, 167–170.

Clyde, A. W. (1944). The cutting of soil by narrow blades. Pennsylvania Ayr. *Expt. Sta. Bull., 405*(2).

Durairaj, C. D., & Balasubramanian, M. (1997). Influence of tool angles and speed on the soil reactions of a bent leg plough in two soils. *Soil and Tillage Research, 44*(3–4), 137–150. https://doi.org/10.1016/S0167-1987(97)00048-2

Gill, W. R., Vander, & Berg, G. E. (1968). *Soil dynamics in tillage and traction. Agricultural Handbook No. 316 USDA-ARS.*

Godwin, R. J., Reynolds, A. J., O'Dogherty, M. J., & Al-Ghazal, A. A. (1993). A Triaxial dynamometer for force and moment measurements on tillage implements. *Journal of Agricultural Engineering Research, 55*(3), 189–205. https://doi.org/10.1006/jaer.1993.1043, 00218634.

Gordon, E. D. (1941). Physical reactions of soil on plow disks. *Agricultural Engineering, 22*, 205–208.

Gullacher, D. E., & Coates, W. E. (1980). *Effects of cultivator sweep pitch on tillage forces.* ASAE. Paper No. 80-1567.

Kepner, R. A., Bainer, R., & Barger, E. L. (1978). *Principles of farm machinery.* Avi Publishing Company.

L Nichols, M., Reed, I. F., & Reaves, C. A. (1958). Soil reaction: To plow share design. *Agricultural Engineering, 39*, 336–339.

Lovegrove, H. T. (1968). *Crop production equipment.* Hutchinson press.

McKibben, E. G. (1926). A study of the dynamics of the disk harrow. *Agricultural Engineering, 7*, 92–96.

McKibben, E. G., & Reed, I. F. (1952). 1 1952/01 SAE Technical Papers 10.4271/520142 26883627 SAE International undefined the influence of speed on the performance characteristics of implements. http://papers.sae.org/.

McKyes, E., & Ali, O. S. (1977). The cutting of soil by narrow blades. *Journal of Terramechanics, 14*(2), 43–58. https://doi.org/10.1016/0022-4898(77)90001-5, 00224898.

Perumpral, J. V., Grisso, R. D., & Desai, C. S. (1983). Soil-tool model based on limit equilibrium analysis. *Transactions of the American Society of Agricultural Engineers, 26*(4), 991–995, 00012351.

Randolf, J. W., & Reed, I. F. (1938). Testing of tillage tools: II. Effect of several factors on the reactions of fourteen inch moldboard plow. *Agricultural Engineering, 19*, 29–33.

Reece, A. R. (1965). *"The fundamental equation of earth moving mechanics". Symposium on Earth Moving Machinery. 179.* Institute of Mechanical Engineers.

Rowe, R. J., & Barnes, K. K. (1961). Influence of speed on elements of draft of a tillage tool. *Trans. of the ASAE., 4*(1), 55–57.

Sohne, W. (1956). *Some principles of soil mechanics as applied to agricultural engineering. Grundlagen der Landteckink* (pp. 11–27). NIAE Translation 53. 7.

Summer, M. S., Chen, S. H., & Bierl, J. F. (1978). *Disk blade performance.* ASAE Paper No. 83-1537.

Bernachi, H., Haman, J., & Kanafojski, C. (1972). *Agricultural machines, theory and construction, 1.* Published for USDA and NSF by Foreign Cooperation Center of the Central Institute for Scientific, Technical and Economic Information.

Klenin, N. I., Popov, I. F., & Sakun, V. A. (1970). *Agricultural machines . USDA and NSF by amerind.* Publishing Co. Pvt. Ltd, USDA and NSF by Amerind Publishing Co. Pvt. Ltd.

Smith, H. P. (1965). *Farm machinery and equipment.* Tata Mc-Graw Hill Pub. Co. Ltd.

CHAPTER

Crop-planting machinery

6

Chapter outline

Seeding or planting is to apply seeds in well-tilled soil and create a conducive environment for them to germinate and establish as a crop. Row crop planting is to sow seeds in rows with spacing to allow inter-crop machines to traverse between rows and perform inter-cultivation. If the rows are narrower or if the seeds are sown without any row or seed spacing, then it is known as solid planting.

Machines of the Farm. https://doi.org/10.1016/B978-0-443-32906-7.00006-2

Seeds may be sown in the following methods, namely:

1. broadcasting or random scattering seeds over the surface of a field and then incorporating them in tilled soil by a harrow or cultivator;
2. drilling of seeds by dropping seeds continuously in the dug furrow and covering the seeds to give definite rows; but when the seeds grow, the plants are in a continuous row and have to be thinned to the desired plant population per unit area;
3. precision planting of single seeds in rows at accurate seed-to-seed spacing;
4. hill dropping, or the process of sowing seeds in hills or groups, at required hill spacing on the rows;
5. check row planting, the sowing of seed hills at accurate hill as well row spacing, with the hills aligned in mutually perpendicular directions; this allows inter-cultivation along the rows as well across the rows. This system is not used anymore because of newer machines and technology available now.

Seed germination and plant stand depend on many factors such as the quality or vitality of the seeds, soil tilth, the depth of seed placement, availability of moisture and nutrition in the soil, and the firm contact of seed to soil, to name a few. Among these, the depth of placement and the soil-to-seed compaction are the parameters concerning the sowing machine. Generally, to counter the effect of a lower percent of germination, drilled crops are sown at a higher seed rate and later thinned down to the desired plant population. But this is not feasible where the cost of seed is high.

6.1 Broadcasting machines

The machine broadcasting of seeds is generally done with centrifugal spreaders, providing a fast and inexpensive method of sowing small amounts of seeds and grasses. Broadcasting is usually adopted in rain-fed areas, where inexpensive seeds, which depend on rain for germination, are utilized. The machine (Figure 6.1) is also used for the application of fertilizer in open fields before sowing. Seeds are metered from the seed hopper by a bulk-flow seed-metering device to drop seeds on one or two horizontal discs, radially ribbed as a fan. These discs rotate at high speeds, of 300 rpm to 500 rpm. The seeds falling into the eye of the radial fan are ejected from the disc rim by centrifugal force. The width of coverage of such a centrifugal broadcaster depends on the diameter of the spreader disc as well as the speed of rotation. It also depends on the physical characteristics of the seed such as size, shape, and specific gravity. If broadcast seed is to be covered, this is generally done as a separate operation with an inter-cultivator or a spike-toothed harrow.

6.2 Seed drills and crop planters

The row planting of crops involves:

- Making furrows in tilled soil at a required row spacing and to a required depth;

FIG. 6.1 Centrifugal Broadcasting Machine (Exacta EL, Kverneland Group, Norway)

FIG. 6.2 Seed Drill (Simulta 2500KH, Junkkari Oy, Finland)

- Metering the seeds into the furrows either with seed-to-seed spacing or by drilling the seeds continuously;
- Closing the furrows to cover the seeds;
- Applying a compaction on the closed furrow for the seeds to make proper contact with the soil.

A typical sowing machine (Bell, 2005; Culpin, 2013; Kepner, et al, 1978; Lovegrove, 1968; Smith, 1965) has a seed hopper to hold the seeds to be sown. The seed drills (Figure 6.2 and Figure 6.3), which sow the seeds continuously into the row, are provided with a common seed hopper, which may or may not have compartments, serving the seed-feeding devices intended for each of the rows. These machines are available as seed-cum-fertilizer drilling machines that can deposit both seed and fertilizer into adjacent rows. However, a crop-planting machine (Figure 6.4), which provides seed-to-seed spacing on the rows, is provided with independent seed hoppers for each row. The seed-metering or -feeding devices are fitted at the

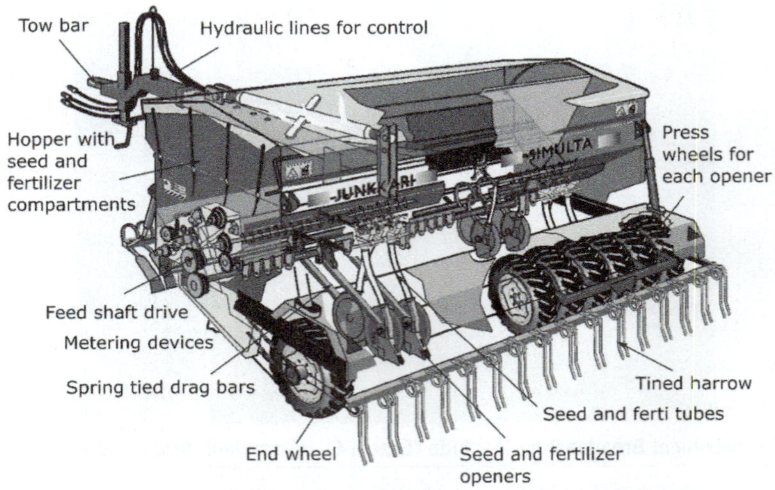

FIG. 6.3 Construction of a Typical Seed Drill (Courtesy of Junkkari Oy, Finland)

FIG. 6.4 Precision Crop Planter (Optima V, Kverneland Group, Norway)

bottom openings of the seed hopper and serve to meter out a controlled quantity of seeds from the hopper to the furrow. These devices are of different types for the seed drill and the row crop planter. The mechanical drive to these feeding devices is typically derived from either a common ground wheel or end wheel of the machine, as in a seed drill, or from the press wheels of the planting assembly of the respective row in a row crop planter. The ground wheel transmits a rotary motion relative to the forward distance travelled by the machine. This makes it possible for the machine to sense the distance travelled and meter the quantity of seed accordingly. These machines have a transmission system between the ground wheel and seed-metering device, either as a variable-speed drive or fixed-ratio drive of gears or chain or toothed-belt drives. This allows the operator to choose the transmission ratio suitable for the seed-metering devices to meter the required seed quantity per unit run of the machine. The type of transmission drive used depends on the seed-metering mechanism being used in the machine. The seeds that are metered out of the bottom of the metering devices fall into a funnel, and a seed tube, which is made of flexible plastic, and the tubes lead the seeds to the seed boot of the furrow openers to place them right into the opened furrows. The shape and configuration of the furrow openers also are many and depend on the type of soil, the crop being sown, and the required uniformity of seed placement depth. These furrow openers are either fixed rigidly to the machine frame or they are fitted through drag and pressure bar linkages, so that each furrow opener can follow the field terrain precisely and independently. The latter is very important to maintain the depth of seed placement uniformity. The drag bars simply tow the furrow openers and the spring-controlled pressure bars press down the openers into the soil, controlling the furrow depth. This system allows the operator to adjust the pressure bars to cut the furrow to a desired depth in a particular type and condition of soil. However, generally in crop planters, the furrow openers are fitted to the individual frames of the row they serve. Figure 6.5 shows a compact, four-bar mechanism that is used in modern planters and drills and maintains a straight-line, up-down motion for the furrow opener or the pressure wheel, so that it can follow the field undulation precisely. The pressure of the spring restraint on the movement is adjustable according to the field soil condition. In automated crop planters, a hydraulic/pneumatic cylinder is used instead of the spring to adjust this pressure from the driver's console. Right

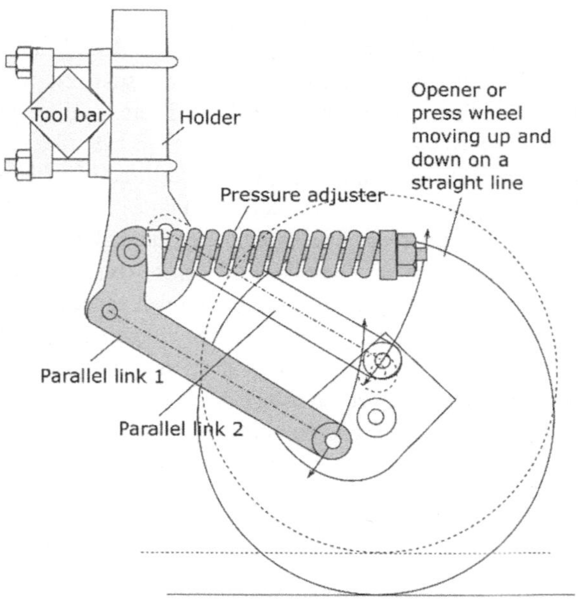

FIG. 6.5 Parallel Link Mechanism of Furrow Opener/Press Wheel

No Permission Required

behind or integral to the furrow opener's rear, furrow closers are provided to cover up the soil and level it. Here, again, furrow closers are of many types, starting from simple drag chains to disc-type closers. Many times, pressure wheels are used instead of simple closers, which are constructed to push the soil into the furrow to close it and then apply a downward pressure to compact the closed furrow. This compaction helps in proper seed-to-soil contact and enhances the germination performance of the sown seeds. The press wheel has its own system of drag and pressure linkages similar to that of furrow openers, and the compaction pressure can be adjusted according to the type and condition of the soil. A pair of row markers is provided on the machine to align seeding rows in subsequent passes. These are simple disc tools extending to the side of the machine and make a furrow mark in the soil when a seeding run is made. In a subsequent pass, the operator aligns the tractor's wheel on the previously made mark and does the seeding run. The mark is placed such that this process aligns and places the present seeding rows precisely next to the previously sown rows.

6.2.1 Seed-metering devices for drilling

6.2.1.1 Bulk-flow metering devices

Seed-metering devices for drilling and bulk-flow metering devices are simple devices (Figure 6.6) in which a mechanical agitator is made to revolve above an orifice with an adjustable opening. The agitator is driven by the ground wheel of the seeding machine. The agitator does not allow the bridging of the seeds, which are flowing from the orifice. The amount of seed drop depends on the size of the opening (Beverloo, et al., 1961; Giannini, et al., 1967; Moysey, et al., 1988), and the seed rate (weight of seed sown per unit area) is adjusted accordingly. As can be visualized, the seeds flow out in bulk and are led to the furrow in a continuous flow. There is no seed-to-seed spacing on the row and, hence, seeds are drilled. After the crop emerges, the continuous row of plants is thinned down, if needed, to bring about the desired plant population.

The flow rate of the seeds or granular material, such as fertilizer, flowing from an orifice at the bottom of a hopper was proven by experimentation that it is independent of the height (head) of the granular layer above the orifice (Beverloo, et al., 1961). It is also independent of the dimensions of the holding hopper, especially when the hopper bottom's dimension is more than 2.5 times the orifice diameter. It was also proven that the mass of granular material over an orifice of diameter, D, has a free-fall behaviour due to gravity up to a height of D over the orifice. We know that the free-fall velocity of such a mass with zero initial velocity is proportional to the square of the height of the fall, $D^{1/2}$. And, the mass rate of flow is proportional to the product of the orifice area, D^{2}, and velocity, $D^{1/2}$. Therefore, the mass flow rate will be proportional to $D^{5/2}$.

Based on this fact, Beverloo's law expresses the flow rate as

$$W = C\rho\sqrt{g}\left(D - kd_p\right)^{5/2}.$$

where W is the average mass discharge rate, g s^{-1} through the orifice, D is the orifice diameter in cm, d_p is the spherical size of the seed, ρ is the bulk density in

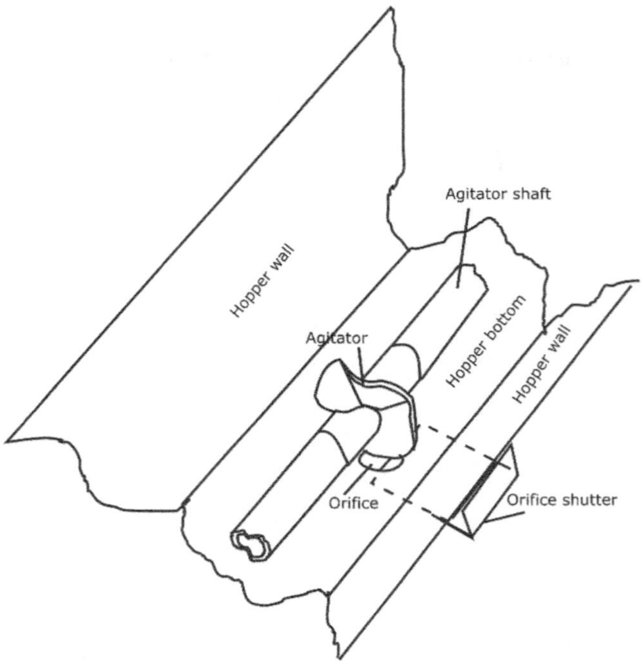

FIG. 6.6 Bulk-Flow Metering Device

No Permission Required

g cm^{-3}, g is the acceleration due to gravity in cm s^{-2}, and C and k are empirical discharge and shape coefficients, respectively. C depends on the internal friction coefficient of the seed material and was found to range between 0.55 to 0.65. For most seeds, such as wheat and rapeseed, C is 0.58. The correction coefficient for particle size, k, was found to be 1.4 for many granules including seeds. Beverloo's law, known also as the 5/2 power law, is precise for cases when $D > d_p$, when there is no flow impediment due to the bridging of the granules above the orifice. The law is, hence, valid in the zone of free-fall head above the orifice, limited by the formation of granular arching or bridging. Mechanical agitation prevents the formation of such bridging. The example given in Box 6.1 illustrates the use of this law in understanding the seed flow rate of bulk-flow metering devices.

6.2.1.2 Internal double-run device

An internal double-run device has a cast housing with a seed-metering disc rotating inside, dividing the housing into two lateral halves (Figure 6.7). The disc has helical indents cast on their inner periphery on both sides, which make it an internal double-run device. The housing is open on the top, with its top openings leading to both sides of the disc for the seeds to flow in from the seed hopper. The metering disc has a square hole at the centre and is driven by a matching square-sectioned shaft.

Box 6.1 Example 6.1

Example 6.1. A seed drill with a bulk-flow metering device is sowing wheat with its circular orifice adjusted to 12 mm diameter. It sows at a travel speed of 5 km h^{-1} with row spacing set at 150 mm. An agitator is used above each orifice of the drill to prevent any bridging of the seeds. The bulk density of wheat seeds is 750 kg m^{-3}, and their spherical size 0.5 cm. Calculate the seed rate it offers.

According to Beverloo's law,

$W = C\rho\sqrt{g}(D - kd_p)^{5/2}$,

where W is the average mass discharge rate, g s^{-1}

D the orifice diameter = 1.2 cm,

d_p the spherical size of the seed = 0.5 cm

ρ the bulk density = 750 kg cm^{-3} = 0.75 g cm^{-3}

g the acceleration due to gravity = 980 cm s^{-2}

C = 0.58 and k = 1.4

$W = 0.58 \times 0.75\sqrt{980}(1.2 - 1.4 \times 0.5)^{5/2} = 2.40$ g s^{-1}

The flow rate of the seeds in each row is 2.4 g s^{-1} = 0.0024 kg s^{-1}

The area covered by one row per unit time = 1 row × 0.15 m spacing × (5000/3600) m s^{-1} = 0.21 m^2 s^{-1}

∴Seed rate rendered by the drill = flow rate / area covered per unit time = 0.0024 kg s^{-1} / 0.21 m^2 s^{-1} = 0.0114 kg m^{-2} = 114 kg ha^{-1}.

When the disc is revolved, the seeds are drawn inward by the disc and fed down to the bottom of the housing. The seeds that are forced down fall through openings provided in turn with flap valves. The seeds are thus fed out from both sides of the disc and the amount of seed metered depends mainly on the speed of rotation of the disc. The flap valves are used to give finer control to the amount of seed metered. The drive shaft, known as the feed shaft, gets its drive from the ground wheel of the seeding machine through a variable-speed gear box. The transmission ratio can be varied to drive the disc at desired speeds to acquire the desired seed rate. Similar to a bulk-flow device, this device also feeds the seeds in a continuous flow and is, hence, a seed-drilling device. One such device is used for each row of sowing, and is driven by a single feed shaft and transmission system. The mass flow rate is dependent on the speed at which the disc revolves. Usually, a continuously variable transmission is

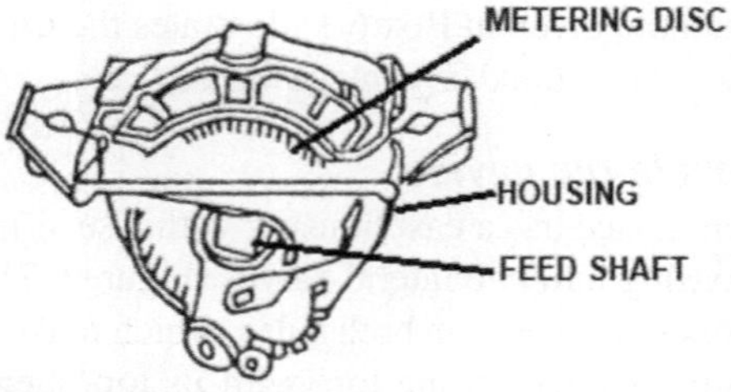

FIG. 6.7 Internal Double-Run Metering Device

No Permission Required

provided between the ground wheel and the feed shaft; this drives all the metering discs on individual seeding rows together. This allows for adjusting the speed ratio continuously to achieve the desired seed rate.

6.2.1.3 Fluted roller device

A fluted roller device is also a bulk-flow metering device but performs much better in controlling the amount of seed metered. The device (Figure 6.8) has a cast housing in which a fluted roller, with flutes large enough to accommodate a bulk of large seeds such as pulses, is rotated. It fits snugly in the housing and is driven by a square-sectioned feed shaft. The roller is also pinned to the feed shaft. The fluted roller has a cut-off device arranged in series with it on the feed shaft. The cut-off is of the same diameter as the fluted roller and is keyed to the housing so that it can move axially in and out of the housing. It has a bushing, in which the feed shaft rotates freely. However, the cut-off is locked from moving axially on the feed shaft by a collar and pin on one end and the fluted roller on the other end. So, when the feed shaft is shifted axially, the cut-off moves along with the fluted roller but allows the fluted roller to rotate. When the fluted roller with the cut-off is moved in or out of the housing, the external surface of the fluted roller that remains inside the housing changes.

This changes the volume of flutes exposed to the seed mass inside the housing. When the roller rotates, the seeds that are filled in the exposed flute volume are ejected from the bottom of the housing. The amount of seed metered depends on

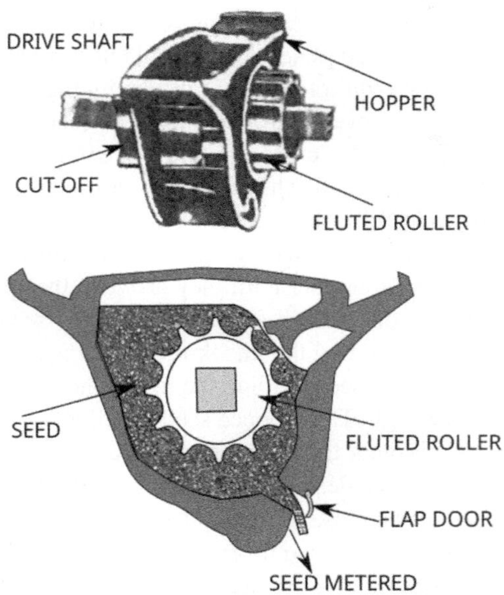

FIG. 6.8 Fluted Roller Device

No Permission Required

the speed of rotation of the roller as well the axial volume of the flutes on the roller that is exposed to the seed mass. A flap valve is also provided to acquire a finer control on the amount of seed metered. A collar, following the profile of the roller and rotating inside the side opening of the housing, prevents seeds from leaking from the roller through the side of the housing. The feed shaft is driven by the ground wheel of the seeder through a chain drive or a notched belt drive. It has a common collar, which is shifted by a yoke and indexing lever, which axially moves the fluted roller and cut-off assemblies of all such metering devices of the machine, relative to their respective housings. The index lever moves on a sector calibrated and marked with the seed rate settings required for common seeds. Similar to the bulk flow, this device also feeds the seeds in a continuous flow and is, hence, a seed-drilling device. It has better control on the seed rate and can accommodate relatively larger seeds than the internal double-run device. The example problem in Box 6.2 shows how the exposed volume of the flute influences the seed rate of this device.

6.2.2 Seed-metering devices for crop planting

Seed-metering devices used for crop-planting machines have cells on a moving disc or endless belt. These cells are sized to accommodate single seeds or hills of seeds. The cell plates or belts are arranged to rotate in vertical, horizontal, or inclined planes, so that they can pick up the seeds from the hopper and release them appropriately into the furrows.

6.2.2.1 Horizontal plate planter

A typical horizontal plate planter (Figure 6.9) has such a cell plate and rotates at the bottom of a seed hopper. The seed plate is replaceable in order to change the number and size of seed cells to the requirement. It is driven by the ground wheel or press wheel of each row through a bevel gear assembly. The seed hopper has a crowned bottom covering the entire seed plate, leaving an opening along the side wall of the hopper for seeds to drop through. The edge drop seed plates (inset of Figure 6.9) allow the seeds to drop and fill in the space between the cell wall and the hopper wall. As the seed plate rotates, the seeds accommodated in the cells are moved along the inner periphery of the seed hopper until they arrive at the spot of drop opening. At that location, a spring-loaded, knock-off pawl pushes the seeds down, preventing clogging of the cells. Just preceding the knock-off mechanism, a simple brush-type cut-off pushes away the excess seeds, riding the top of the seeds in the cells so that excess seeds do not fall through. The dropped seed or hill of seeds falls into a funnel-shaped receptacle, flows through the seed tube, and falls to the furrow bottom.

The size of the seed cells decides whether it accommodates single seeds or a seed hill. If the seed plate is located close to the ground and the seed tube is avoided, the seed or hill dropping down has a near-uniform seed-to-seed or hill-to-hill spacing on the furrow bottom. The seed tube, if provided, generally breaks the fall of the seeds and disturbs the seed-to-seed spacing inherently provided by these planters. But, basically, since these are not bulk-flow devices and can reasonably provide crop

Box 6.2 Example 6.2

Example 6.2. A 20-row seed drill is used for planting wheat at a 200 mm row spacing and at a seed rate of 100 kg ha^{-1}. It sows at a travel speed of 5 km h^{-1}. It has a fluted roller seed-metering device with eight flutes distributed over its peripheral diameter of 30 mm, with the flute being 10 mm deep and in a circular-segment shape. The speed ratio between the ground wheel of 75 cm diameter and the fluted feed shaft is 4:1. Find the length of the fluted roller that is exposed to the seed mass in the hopper to dispense the required seed rate. The bulk density of wheat seeds is 680 kg m^{-3}.

Roller's peripheral diameter = 30 mm

Fluted roller circumference = $\pi(30)$ = 94.2 mm

Width of each flute = 94.2 / 8 = 11.5 mm = approximately 10 mm wide flute with 1.5 mm spacing between flutes

Depth of flute = 10 mm

Area of the flute section = $\pi(10)^2/8$ = since the flute is a circular segment groove = 39.25 mm^2

Volume of the flute holding seeds = (39.25L) mm^3, where L is the length of flute exposed to seed feed

Seed rate for wheat = 100 kg ha^{-1}

Effective width of sowing = Row spacing × No of drilled rows = 0.2 × 20 = 4 m

Weight of seeds sown by the drill per m = (100 kg / 10000 m^2) × (4 m) = 0.04 kg m^{-1}

Weight of seeds sown per row per m = 0.04 / 20 = 0.002 kg m^{-1}

$$\text{Volume of seeds sown per row per m} = \frac{\text{Weight of seeds sown per row per m}}{\text{Seed bulk Density}}$$

$$= \frac{0.002\text{kg}m^{-1}}{680\text{kg}m^{-3}} = 2.94e^{-06}m^3m^{-1} = 2940mm^3m^{-1}$$

Ground wheel diameter = 0.75 m

Ground wheel circumference = $\pi(0.75)$m = 2.36 m

Revolution made by ground wheel to cover a meter run of the drill = 1 / 2.36 = 0.42

Corresponding revolution made by the feed shaft = 0.42 / speed ratio = 0.105

∴Volume of seeds sown per revolution of fluted roller = 2940 mm^3 /0.105 = 28000 mm^3

Then, volume of seeds sown per flute = 28000 / 8 = 3500 mm^3

$$\text{Length of flute exposed to seeds, L} = \frac{\text{Volume of seeds sown per flute wn per row per m}}{\text{Area of section of flute}}$$

= 3500 / 39.25 = 89.11 mm.

spacing, they are classified as seed-planting devices rather than drilling devices. A large collection of seed plates is supplied with the machine to meet the requirements of different crops (seed sizes) and crop spacing. Mainly, they are used for bulky seeds such as maize (corn) and ground nut (peanut) and also for medium-sized seeds such as grams and pulses.

6.2.2.2 Inclined plate planter

An inclined plate planter (Figure 6.10) has its seed plate oriented at an inclination of 60 deg to the horizontal. Here, the seed plate is an edge drop-type and picks up the seed at the bottom of the hopper and carries it to the top, where it is released by gravity. The seed plate is rotated by a suitable drive from the press or ground wheel and rotates inside the seed mass of the hopper. A baffle plate prevents the seed mass from collecting over the revolving seed plate and allows the seed to enter only through the sloping bottom of the hopper to the edge of the plate where the seed cells are moving. This avoids undue friction between the seed mass and the rotating plate that may

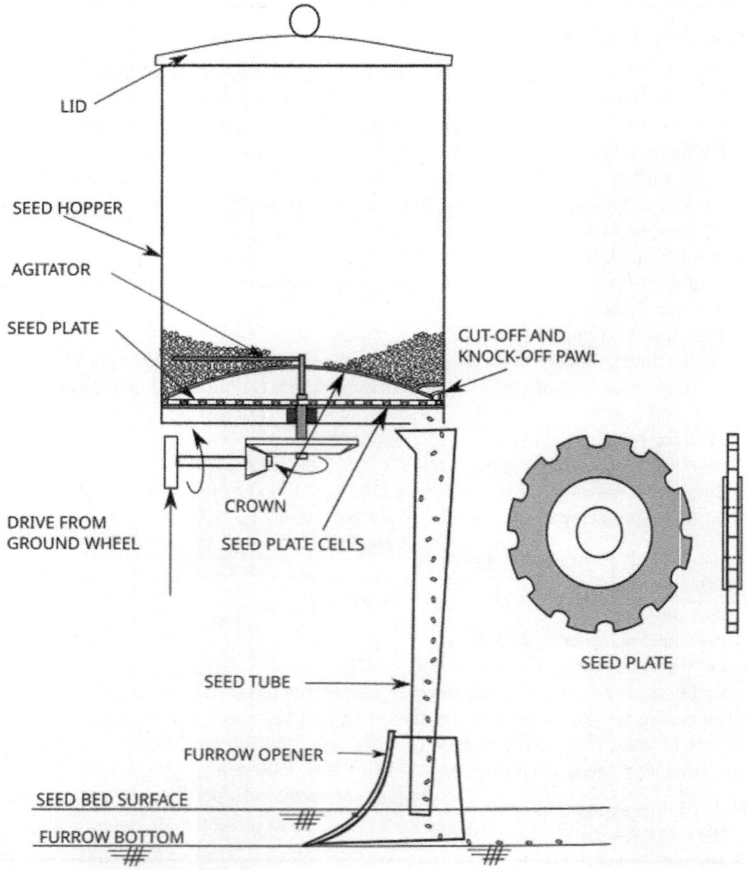

FIG. 6.9 Horizontal Plate Planter

No Permission Required

cause a great deal of damage to the seed coats. The seeds fill the plate cells and are carried up for a drop at the top. A seed funnel catches the dropped seeds and leads them to the furrow being made by the machine. These devices are again suitable for bulky seeds and medium-sized seeds as well. Since the device drops the seeds or seed hills at a time interval, these are also classified as crop planters. As in the case of horizontal cell planters, these also can offer reasonably better plant-to-plant spacing in a row. The advantage of this device is that it does not have cut-off and knock-off pawl and is of a simple design. The only disadvantage is that the seed is raised to the top of the seed plate and dropped; that height is in addition to the total drop height of the seed and may cause some underperformance in plant-spacing uniformity.

6.2.2.3 Vacuum seeder

Although all the above-mentioned planter mechanisms can meet their objectives of placing the seeds at a spacing on the row, there can be misses and multiple picks, which may require a

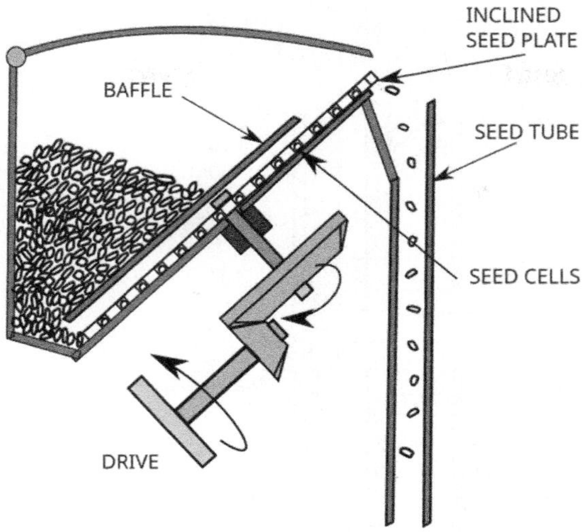

FIG. 6.10 Inclined Plate Planter

No Permission Required

filling up or thinning, as appropriate, after the crop emerges. The seed spacing rendered is also not very precise as is desired for planting high-value vegetable crops. Precision planters (Giannini, et al., 1967) are seeding machines that offer precise seed-to-seed spacing on the row. Moreover, they offer the exact depth of placement that can enhance the quality of crop stand in the field. The vacuum-seeding mechanism (Figure 6.11) is used in precision planters to acquire positive seed singulation and can avoid missed pickups to a great extent. It has a seed plate whose cells face the seed mass on one side and a vacuum chamber on the other. The seed plate has vacuum seals at appropriate places on the right half of the housing acting as a vacuum chamber. The chamber is connected to a vacuum source driven by the tractor PTO. The cells on the seed plate have through-holes that connect them to the vacuum chamber. When the seed plate is rotated by the drive shaft, the seed cells pick up the seeds from the seed chamber positively due to the vacuum pressure present on the cells. When each such cell reaches the bottom of the housing, a small, sealed chamber releases the vacuum pressure of that particular cell by connecting it to the atmosphere. The seed is, hence, released from the cell and falls downward to the furrow. Seeds are fed into a seed chamber on the left half of the housing. Because of the vacuum pressure, the seed pickup is positive; only one seed is accommodated in each cell, as the cell shape is configured accordingly.

6.2.2.4 Seed plate performance

As the seed plate rotates in the seed mass, the seeds are expected to fill the cells positively, so that there will not be any misses in the pick and drop performances of the seed plate (Kepner, et al., 1978). The percent of cell fill is the ratio of the number of cells that are filled by the seeds at a point on the periphery of seed pickup to the total number of cells passing by that point. The factors that influence the cell fill of a

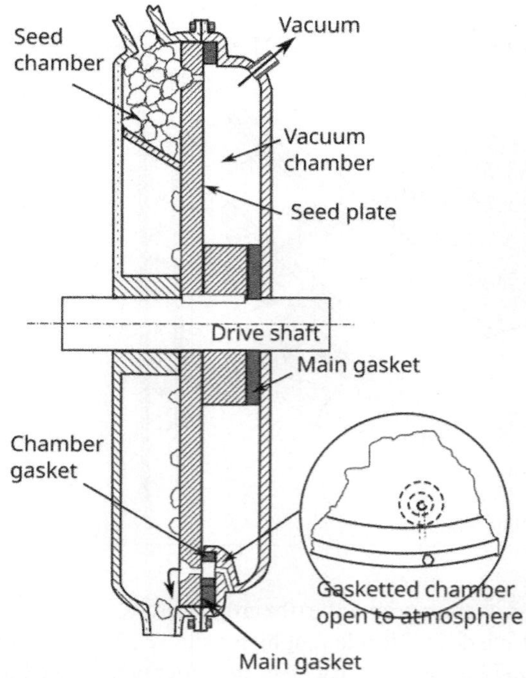

FIG. 6.11 Vacuum Seeder

No Permission Required

particular seed plate are influenced by the uniformity of seed shapes and sizes in relation to the cell size and configuration and, most importantly, the peripheral speed of the seed plate. The seeds are usually graded to acquire the required uniformity of sizes to suit the selected seed plate. The peripheral speed of the plate generally has a linear relationship with the percent of cell fill. Logically, as the peripheral speed is less, the cell fill percent will be higher and, as the speed increases, the percent of fill decreases. It was found that a peripheral speed in the range of 0.25 to 0.3 ms^{-1} provides about 95 percent to 100 percent cell fill. However, cell fill percent does not take into account the multiple seeds that are filled in a single cell owing to the lack of uniformity in seed sizes or the orientation of flat seeds such as maize while filling the cell. The optimal peripheral speed for a better cell fill is to be considered while designing the seed plate's diameter and the number of cells on the periphery of the plate. The forward speed of the planter and the transmission ratio between ground wheel and plate are also factors that influence the peripheral speed of the seed plate. The following problems shown in Box 6.3, Box 6.4, Box 6.5, Box 6.6, and Box 6.7 have to do with the metering performances of different seed-metering devices and illustrate how the speed ratio between the ground wheel and the feed shaft, the number of cells on the seed plates, the peripheral speed of the plate, and the speed of the planter influence the seed rate, plant population, cell fill efficiency of seed plates, and the seed spacing on a row.

6.3 Furrow openers

There are many types of furrow openers (Figure 6.12), and the choice of one depends on factors such as the depth of placement and the soil type and its condition, particularly the soil moisture. Seeds of some crops are very sensitive to placement depth and have to be placed at a correct and uniform depth to germinate and grow to a desired population. The runner is a bladelike device that cleaves the soil very narrowly. It works excellently when the seed bed is of friable soil. It works at depths of 50 mm to 75 mm on trash-free soil surfaces. It is generally used in maize planters along with horizontal plate devices. Skids or floats are attached to the sides of the runner to limit the depth of penetration. These floats are simple horizontal plates and do not allow the runners to penetrate beyond the float surfaces. There are special

Box 6.3 Example 6.3

Example 6.3. Let us say that a horizontal plate planter is generally required to travel at a speed of 3 kmh^{-1}. Let the transmission ratio between the ground wheel of 50 cm diameter and the seed plate be 1.5:1. For 90 percent to 100 percent cell fill, the peripheral speed of the cells is to be 0.3 ms^{-1}. Find the design diameter of the seed plate. Also, find the seed-to-seed spacing it offers on the seed bed if the seed plate has 30 cells on its periphery and the seed drop height is very low.

Linear speed of the ground wheel $= 3\ kmh^{-1} = 3000/3600 = 0.83\ ms^{-1}$

$= \pi\ DN/60$ where D, ground wheel diameter $= 0.5$ m

$\therefore$ N, the rotary speed of ground wheel $= (60 \times 0.83)/\ (\pi \times 0.5) = 31.7$ rpm

Then, the rotary speed of the seed plate = Rotary speed of ground wheel / transmission ratio

$= 31.7/1.5 = 21$ rpm

Optimal peripheral speed of seed plate should be $= 0.3\ ms^{-1} = \pi\ dn/60$

where n = rotary speed of the seed plate = 21 rpm

$\therefore$ d, the seed plate diameter $= (60 \times 0.3)/\ (\pi \times 21) = 0.27$ m = 27 cm.

The ratio of peripheral speeds of seed plate and ground wheel $= (0.3\ ms^{-1} : 0.83\ ms^{-1}) = 0.36$

The linear distance moved by seed plate for each seed drop = the spacing of cells on plate

$= \pi$ d/no of cells $= \pi \times 27/30 = 2.82$ cm = 0.0282m

$\therefore$ Distance moved by the seeder for one seed drop i,e. seed-to-seed spacing

$= 0.0282/0.36 = 0.0782\text{m} \approx 8$ cm.

{Looking at it another way)

Time interval for one seed drop = reciprocal of peripheral speed ($s\ m^{-1}$) $\times$ linear distance between the cells (m) $= (1/0.3) \times 0.0282 = 0.094$ s

Distance moved by planter for one seed drop = time for one seed drop $\times$ Linear speed of planter

$= 0.094 \times 0.83 = 0.0782\ \text{m} \approx 8$ cm.

The example teaches us that the seed-to-seed spacing is dependent on the cell spacing on the seed plate and the peripheral speed ratio of the seed plate to the ground wheel. It is not dependent on the travel speed of the seeder. But ground speed will affect the peripheral speed of the seed plate and thereby the percent of cell fill.

Box 6.4 Example 6.4

Example 6.4. A precision planter is used to plant maize at a row spacing of 75 cm. The recommended plant population for a maximum yield is 20,000 plants per hectare. Find the seed spacing on the row. Assuming a seed germination of 80 percent, find the number of plant locations to be reseeded. Then, if a hill seeder is used instead and the number of seeds per hill varies from 2 to 3 at a probability of 70 percent and 30 percent, respectively, find the seed rate required to achieve 20,000 hills per hectare. Assume the 1000 seed weight of maize is 400 gm

Seed spacing = 10000/(0.75 × d) = 20000; d = 1/(2 × 0.75) = 0.67 m = 67 cm

Reseeded plant locations = 20000 × 0.2 = 4000 plants

Requirement = 20000 hills per ha

Seeds required = (total hills × prob. of getting 2 seeds × 2 seeds) + (total hills × prob. of getting 3 seeds × 3 seeds) = (20000 × 0.7 × 2) + (20000 × 0.3 × 3) = 28000 + 18000 = 46000 seeds per ha

1000 grain weight = 400 g; ∴ Seed rate = 46000 × 0.400/1000 = 18.4 kg ha^{-1}

Box 6.5 Example 6.5

Example 6.5. A horizontal plate planter is traveling at 5 kmh^{-1}, sowing seeds at a 75 cm spacing on the row. The seed plate is of 20 cm diameter and has 16 edge cells that can pick single seeds only. Find the peripheral speed of the seed plate.

Rate of seed drops = 5000 mh^{-1} / (0.75 m × 3600 s h^{-1}) = 1.85 seeds s^{-1}

Peripheral distance moved by plate to drop one seed = seed plate circumference / no of cells = π (0.20 m) /16 = 0.039 m/seed

Peripheral speed of plate = distance moved by plate (m/seed) × rate of seed drop (seed s^{-1}) = 0.039 × 1.85 = 0.072 ms^{-1}

Box 6.6 Example 6.6

Example 6.6. An inclined plate allows the seeds to fall freely and directly into the furrow. The point of release of the seed is 30 cm above the furrow bottom. The planter is traveling at a speed of 3 km h^{-1}. Find the horizontal distance the seed moves between the point of release and the point of hitting the furrow bottom.

Time (t) for the seed to travel the drop = $\sqrt{\frac{2(height\ of\ drop)}{g}}$ = 0.6 / 9.81) = 0.247 s

(as per $s = ut + 0.5gt^2$; $u = 0$)

Horizontal distance moved by seed at the same time = vt = (3000/3600) ms^{-1} × 0.247 s) = 0.20 m = 20 cm backward

runners, which are shorter in length in medium-textured soil. Hoe-type openers are suitable for rough seed beds with more clods on the surface. They dig deep and are used for root-infested soils to some extent. They have a conical seed boot or space into which the seeds fall to be placed in the furrow. The seed boot can also be in a shoe shape. Replaceable shovels are used on these openers to cleave the soil. Single- or double-disc openers are suitable for hard soils and also can work with scrapers in clayey soils. The single-disc type is a miniature disc plough but with a shallow disc

Box 6.7 Example 6.7

Example 6.7. A horizontal plate crop planter has 20 cm diameter seed plates with 60 cells on the edge. The ground wheel diameter is 60 cm. The speed ratio between the ground wheel and the feed shaft is 2:1. The bevel gear drive from the feed shaft is from a 10-tooth bevel gear to a 20-tooth bevel on the seed plate stub shaft. Find the seed spacing on the row. If a plate peripheral speed of 0.3 ms^{-1} gives 95 percent cell fill, find the travel speed of the planter. If a seed spacing of 20 cm is desired with the same plate driven at a peripheral speed of 0.3 ms^{-1}, what are the changes required in the planter?

No. of gear teeth on feed shaft: No. of gear teeth on seed plate = rotary speed of seed plate: rotary speed of feed shaft = 10: 20 = 1:2

Speed ratio between the plate and ground wheel = $(1/2) \times (1/2) = 1/4$

Angular displacement between cells on plate = 360 / no of cells = 360 / 60 = 6 deg

Angle of ground wheel rotation per seed drop = Angle of seed plate movement per drop / transmission ratio between plate and ground wheel = $6 \times 4 = 24$ deg

Linear distance moved for 6 deg angle = $(24/360) \times \pi D$ = seed spacing on the row

where D is ground wheel diameter = 0.6 m

$= (24 \times \pi \times 0.6)/360 = 0.1256$ m = 12.56 cm.

Peripheral speed of plate = π dn/60 where d is plate diameter = 20 cm and n is rotary speed, rpm

If 0.3 ms^{-1} peripheral speed is desired for 95% cell fill,

rotary speed of the plate, n = $(60/\pi \times 0.2) \times 0.3 = 29$ rpm

∴ Rotary speed of ground wheel = plate speed × speed ratio between plate and ground wheel $= 29 \times 4 = 116$ rpm

Then, linear speed of planter = peripheral speed of ground wheel = π DN/60

where D is ground wheel diameter = 60 cm and N is rotary speed = 116 rpm

$= (\pi \times 0.6 \times 116) / 60 = 3.64\ ms^{-1} = (3.64 \times 3600)/1000 = 13.1\ kmh^{-1}$

Desired seed spacing on the row = 20 cm = 0.2 m

Angle of ground wheel movement per seed drop = $\{0.2 /(\pi D)\} \times 360$ deg = $(0.2 / \pi \times 0.6) \times 360$ = 38.2 deg

Angle of seed plate movement per seed drop = 6 deg (already calculated earlier)

Speed ratio between ground wheel and seed plate = 38.2 : 6 = 6.4:1

Travel speed for percent of cell fill = $0.3 \times 6.4 = 1.92\ ms^{-1} = 6.9\ kmh^{-1}$.

A higher travel speed is desirable for better field coverage. Even if the travel speed is reduced, the peripheral speed will drop to a very slow level but will not affect the percent of cell fill. One other way to change the seed spacing is to use a seed plate with a lesser number of cells. Try doing this calculation with 30 cells instead of 60; you will understand the option.

and is set at lower disc and tilt angles. They are mounted on bushes or bearings that they can rotate on their axes and make the cut. They penetrate more effectively in trashy and hard seed beds and make a relatively wider and deeper furrow than the double-disc type. They are always used with drag and pressure links since they do not have the suction to penetrate and have to be pressed down forcibly to penetrate. These links help to precisely adjust the depth of operation. They are usually used on seed drills rather than crop planters. Double-disc furrow openers are constructed with two flat, sharp discs converging together at the front. They are mounted on a set of bearings, and, while rotating, they cut a wedge-shaped furrow at the front. They have self-cleaning blades and work very well even in moist and sticky soils. They have ring-type skids around their periphery to control the depth of

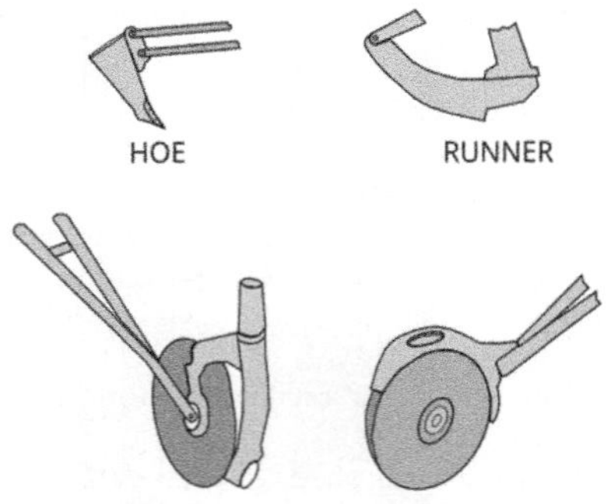

FIG. 6.12 Types of Furrow Openers

No Permission Required

operation. The seed boot is constructed in between the discs at the back. They are well suited for seeding row crops of low to medium planting depth.

6.4 Furrow closers

Common furrow-closing devices provided on seeders are drag chains, drag bars, press wheels, or double-disc types. An ideal closer covers the seeds in the furrow completely so that the soil and seed are in close contact, aiding quick germination. If the soil moved to cover the seeds is compacted down firmly around the seeds, the seed germination is enhanced. The closers should not dislocate the seeds placed in the furrow and should leave them at the proper depth. Some seeds are not very sensitive to such compaction and simple drag chains, which merely cover the seeds with loose soil and are found to be adequate. This is true for seed drills under most conditions, as with crop planters, which handle expensive seeds of commercial crops. Drag chains are just chains dragged behind the furrow opener. Drag bars are simple steel bars dragged behind the furrow openers. Concave press wheels or double wheels with a slight toe-in are used to give the required compaction in crop planters used for maize and pulses.

6.5 Row markers and their settings

As previously explained, these are simple disc tools extending to the side of the machine, making a furrow mark in the soil when a seeding run is made (Figure 6.13). Two such markers are provided, one on each end of the machine. The marker is mostly a disc tool at the end of a telescopic rod extending from a hinge on the machine. The whole arm can be lifted or lowered using hydraulic rams or simple tug ropes.

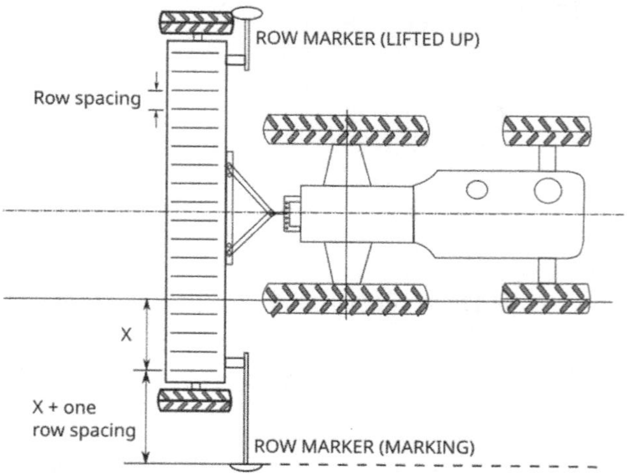

FIG. 6.13 Row Marker Settings of a Sowing Machine

No Permission Required

When the seeder is taking a run in one direction, the marker on one side is used, and the other is used when it is taking the seeding run in the opposite direction. When one is marking, the other is lifted up. If the lateral distance between the last furrow opener on the end of the machine and the centre line of the tractor wheels on that side is 'X' m, then the row marker is to be adjusted to mark a line which is (2X + one row spacing) away from the centre line of the wheel on that end. The row spacing is the spacing between the furrow openers that sow the crop rows. The length of the row marker is set by telescoping the marker mast appropriately. Once the mark is made, in the subsequent pass taken next in the same direction, the operator aligns the tractor's wheel on the mark in the present run and does the sowing. By the mentioned setting, the last furrow opener will then be running one row spacing away from the last row in the previous seeding pass, implying that the previous and present seeding passes are aligned precisely next to each other. Row markers are used on all row crop machines such as seed drills and crop planters.

6.6 Seed drill calibration

The process of setting up the drill or planter to sow a desired seed rate of a particular crop is known as calibration.

Let us say a seed drill is used to sow a crop at a recommended seed rate of 15 kg ha^{-1}. Let us assume that the seed drill has nine furrow openers, and the recommended crop row spacing is 30 cm. Let the ground wheel diameter be 1 m.

The drill is to be calibrated in the farmyard. It is jacked up suitably, such that the ground wheel or the press wheel driving the metering devices can freely rotate. The desired seeds are loaded in the seed hopper. Cloth or polythene bags are tied to each seed tube outlet to collect the metered seeds.

Then, the ground wheel is rotated manually by (say) 20 rotations, and the seeds collected from each seed tube are weighed.

Let us assume that the average weight of the seeds collected from each seed tube for 20 revolutions is 37.8 gm, giving 37.8 / 20 = 1.89 gm per revolution of ground wheel. Nine furrow openers of the drill would then meter $1.89 \times 9 = 17$ gm of seeds per revolution of the wheel. The linear distance the seeder would travel for one revolution of the ground wheel is π d $= \pi \times 1 = 3.14$ m.

Since the effective width of sowing is the number of furrow openers × row spacing $= 9 \times 0.3$ m $= 2.7$ m, the area covered per revolution of ground wheel

$$= 2.7\ \text{m} \times 3.14\ \text{m} = 8.478\ \text{m}^2 = 0.00085\ \text{ha}.$$

Then, the seed rate = 0.017 kg / 0.00085 ha = 20 kg ha^{-1}.

But, since we require a seed rate of 15 kg ha^{-1}, the drill's metering mechanism needs to be adjusted to reduce the seed metered per ground wheel revolution. The weight of the metered seeds is measured again, and the seed rate is calculated again. This activity is repeated until the required seed rate is achieved at a setting.

6.7 Hill drop planters

The seeds available on the market are not 100 percent viable and the germination percent will be about 95 percent. Moreover, as already seen in the introductory paragraphs, seed germination is influenced by many factors such as the depth of placement, seed-to-soil contact, soil conditions, etc. Hence, plant emergence after sowing will be around 60 percent to 70 percent at best. So, while using a planter that provides seed-to-seed spacing, there is no guarantee that singulated seeds will emerge and make the desired plant stand. So, it is customary to plant two to three seeds per hill with the desired hill spacing on the row. This would ensure the desired plant population after emergence. The hill drop planter that does this job is just an augmentation to the regular cell plate planter. A trip or rotary valve at the end of the seed tube collects a hill of two to three seeds and drops the hill collectively in the furrow. Since the valve is placed quite close to the furrow, hill spacing is maintained uniformly. These valves are driven from the planter's ground wheel and, hence, sense the ground travel (hill spacing). Rotary valves are more efficient in handling higher travel speeds than reciprocating valves. But the scattering of the seed hill dropped into the furrow is a problem with which we must contend at higher travel speeds. Hill drop planters are not popular now, since the trend in modern times is to perform high-speed drilling with an air drill and then perform selective chemical thinning to get the desired plant-to-plant spacing on a row.

6.8 Air seeders

Air seeders are seeding machines that have gained popularity in recent years, especially in the United States, Canada, and Australia. They are designed as huge machines, with 15 m to 20 m effective widths that render an immense area of seeding per unit of time. Unlike vacuum seeders, these machines work by blowing out the seeds through seed hoses to the furrow openers. These drills are used for sowing multiple crops, including sunflower, in very large areas in these countries. The working principle of such a seeder is shown in Figure 6.14.

An air seeder has a central unit (air cart) trailing behind the toolbar, with furrow openers of choice. This unit has all the systems for blowing out the seeds to the openers. A high-velocity blower is used; it is connected to a manifold, and from the manifold, the air is blown out through an array of outlets to which flexible hoses are connected to the furrow openers individually. Preceding these hoses, each of these outlets let the air through a passage into which a seed-metering device, such as a fluted roller unit, meters the seeds in. Typically, a venturi that sucks metered seeds into the airstream is used in this passage. This prevents the backflow of seeds into the metering unit. The flow rate of seeds from the seed hopper above is controlled precisely, as in any conventional device. The seeds that hit the airstream are carried at a high rate of speed through the individual hoses and deposited into the opened furrows behind the shovel. Precision press wheels, which have extensive control over the pressure exerted on them, cover the seeds with soil. Because of the high air velocity, the flow rate of seeds is much enhanced, permitting a higher travel speed with the air seeder (over 9 kmh^{-1}). The metering unit as well the blower speeds are controlled to output the desired seed rates at these higher speeds. The toolbar is usually very wide and is made foldable along with all the hoses and

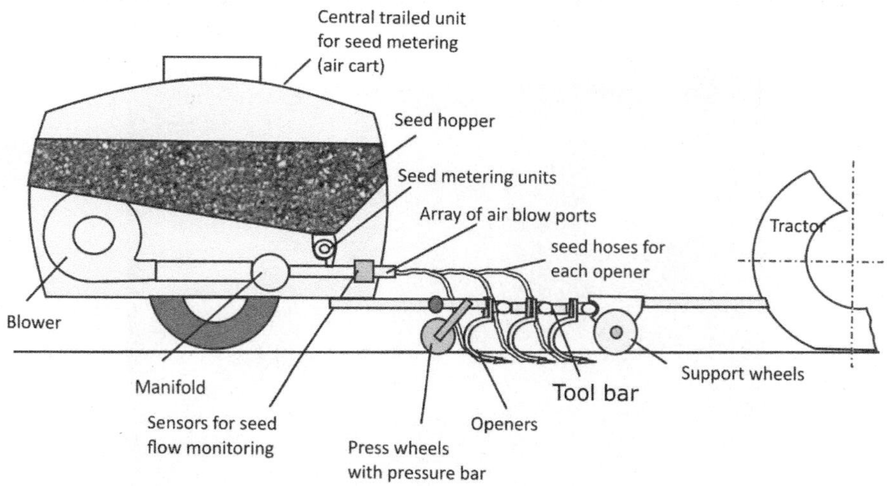

FIG. 6.14 Working of an Air Seeder

No Permission Required

openers. Depending on the need, the openers with suitable shovels and sweeps are fitted on the toolbar. The entire toolbar, with the trailed central unit behind it, is towed by a high-powered tractor (Figure 6.15). The precision press wheels are set up for a higher downward pressure in conservation tillage situations, in which the stubble on the soil surface is cut open by disc openers, the press wheel requires the compaction of the soil with trash on it, and seeds are closed in the furrows. It is an excellent tool for direct seeding in arid lands where no other tillage tools are used before sowing the seeds. There are provisions in the machine to simultaneously apply granular fertilizer in adjacent furrows of the sown furrows. Air seeders are also used along with tools for precision agriculture to meter variable seed rates at the desired field locations tagged by GPS.

6.9 Potato and tuber planters

Potatoes and root crops generally use pieces of tubers as seed material. The pieces are seen to have eyes, cut from the whole tuber, which can be used to germinate. But, in regions where the seed potato price is not substantial, whole potatoes are used as seed material. Graded whole potatoes are easier to pick and drop than tuber pieces. The planter mechanism is basically of two types, namely, a picker wheel type and a cup feed type. The former has a clawlike mechanism that is actuated by the feed drive to grasp the potato pieces or whole potato, carry them to the ground level, and drop them in the furrow made by an opener. The cup feed device (Figure 6.16) has cuplike scoops running on a belt drive that collect the whole potato and move it to the top of the conveyor. At the top of the conveyor, the potatoes drop down onto the back side of the previous cup and are taken downward, one by one. The potato is retained between the back of the cups and the conveyor housing until they reach ground level at the bottom. They fall

FIG. 6.15 A 25 m Air Drill in Operation (Case IH Precision Air 5 Series air cart and Flex Hoe 900 seeder)

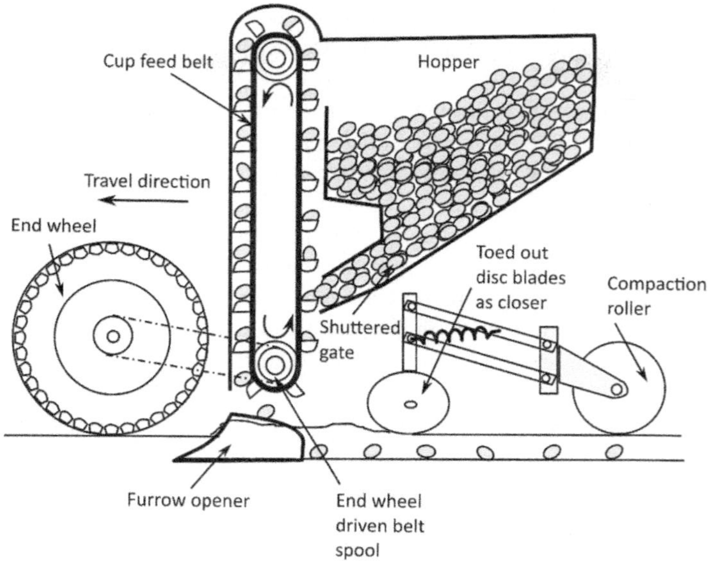

FIG. 6.16 Working of a Cup-Type Potato Planter

No Permission Required

down into the furrow, one by one, providing spacing between drops. The drive for the cup drive is taken from the end wheel of the machine through a mechanical drive. The drive ratio can alter the spacing if desired. The potatoes are fed into the cup drive from a hopper through a gated opening. These machines use wider runners or shoe-type openers that drop the seed material at a depth of about 10 cm. Double-wheel or disc furrow closers are used to close and cover the seeds at this depth. A press wheel mounted on a parallel-linked mechanism and following the furrow closer presses down the soil, thus compacting it over the seeds planted. Some designs of potato planters have vertically rotating picker wheels with arms having sharp picking pins that impale the seed piece to pick and release it above the furrow. The peripheral speed of these pickers is relatively slow, such that the picking performance is appreciable. Semi-automatic tuber planters possess horizontal or vertical rotary magazines, similar to vegetable transplanters, into which the seed pieces are fed manually by operators sitting against the device of each row. The forward speed of these planters is, hence, limited to 1 kmh^{-1}. Since the bulk density of seed material is high, bigger seed hoppers are used in these machines.

6.10 Transplanting machines

Crops such as cabbage, cauliflower, brinjal, chili?, and tomato are generally grown in a nursery, in a dense population, and are picked as young seedlings and transplanted in the field. Transplanted vegetable crops have long been proven to yield a higher productivity than directly sown plants. In recent years, vegetable growers prefer seedlings grown in plug trays possessing media-filled pockets in which vegetable seeds are sown and grown as nursery plants. A number of entrepreneurs have entered the business of raising and supplying plug tray-grown seedlings to vegetable farmers. The process of producing virile nursery plants in farms is, hence, avoided, eliminating the underlying risks and labor involvement. Transplanting machines plant these young seedling plugs in soft seed beds. Semi-automatic machines (Figure 6.17) have furrow openers and tabletop frames for placing the plug trays of seedlings. Either horizontally rotating magazines or vertical picker wheels, with pockets into which the seedlings are dropped, are driven by a ground wheel. An operator sitting against each of these mechanisms (Figure 6.18) feeds the young seedling plugs taken from the trays on a tray rack into these rotating pockets. The rotating magazine, in the shown type of machine, has receptacles in the shape of inverted bottles closed at the bottom. These are actually bill-type receptacles that have two halves that are hinged at the top and can open at the bottom like a clam shell. The operator quickly grabs the seedling plugs from the trays and puts them into these receptacles as they go around. The plugs are held in the receptacles. These receptacles, which rotate with the magazine when aligned one by one with the opened furrow, are made to open up like a clam shell and drop the plug retained in it into the furrow. The receptacles are opened automatically at these locations by suitable actuation mechanisms. The magazines or picker wheels drop the young seedlings into the furrow and toed-in press wheels cover the seedling roots and compact the soil around them. The spacing between the drop holes on these magazines (or the arms of the picker wheels) and the speed ratio between the ground wheel and these rotating magazines decide the plant spacing on the row. These machines have wide runner or

FIG. 6.17 Semi-Automatic Vegetable Transplanter

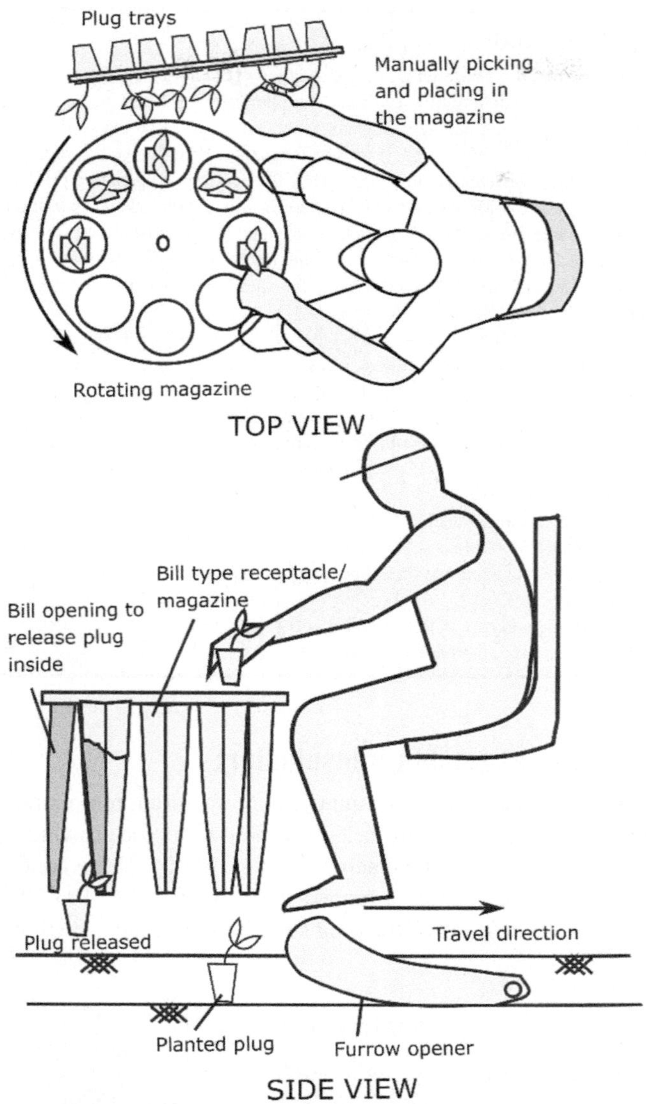

FIG. 6.18 Working of a Semi-Automatic Vegetable Transplanter

No Permission Required

hoe-type furrow openers to make furrows of a uniform depth and width that can accommodate the seedling root mass. The machine being fed by a manual means is, hence, slow, limited by the operator's ability to pick and place the plugs into the magazine holes. The example in Box 6.8 illustrates this fact as a limitation of semi-automatic transplanters.

Box 6.8 Example 6.8

Example 6.8. A two-row, semi-automatic, magazine-fed vegetable transplanter, as shown in Figure 6.16, is used to transplant chilli seedlings at a plant spacing of 50 cm on the row. The operators sitting against the picker magazines serving each row feed the young seedlings taken out of the trays into the magazine's rotating pockets. The magazines drop the young seedlings into the furrow for transplanting. The average time required by the operators to pick and place one seedling in a pocket is 2 s. This is the minimum time required for this continuous work without causing fatigue. The circular magazine, having five pockets arranged at a pitch diameter of 40 cm, is driven by a ground wheel of 70 cm diameter. What is the maximum speed the machine can travel so that the operator can do this continuous work? What is the speed ratio between the magazine and the driving ground wheel?

Plant spacing of crop = 0.5 m
Time available for dropping one seedling = 2 s
∴ Maximum speed of travel = 0.5 /2 = 0.25 m s^{-1} = 0.9 km h^{-1}
Circumference of the ground wheel = π (0.7) = 2.2 m
Revolution made by ground wheel for one seedling drop
= Plant spacing / Circumference of the ground wheel
= 0.5 / 2.2 = 0.22 rev
Rotational speed of ground wheel = 0.22 rev / 2 s = 0.11 rps = 6.6 rpm
The magazine should also rotate at a speed that each pocket faces the operator every 2 s.
Revolution made by magazine for one pocket to face operator
= 1 / number of pockets = 1/5 = 0.2 rev
Rotational speed of magazine = 0.2 rev / 2 s = 0.1 rps = 6 rpm
Speed ratio between ground wheel and magazine = 6.6 / 6.0 = 1.1

6.10.1 Automatic vegetable transplanters

Automatic vegetable transplanters designed, manufactured, and marketed by some Japanese and Chinese firms (Figure 6.19) use plug trays for growing the vegetable seedlings to be transplanted by the machine (Tsuga, 2000). These are flexible plastic trays with 128, 200, or 288 cells, arranged in grids of 8×16, 10×20, and 12×24, respectively. The shape of each of the plug cells, in which the growth media is filled

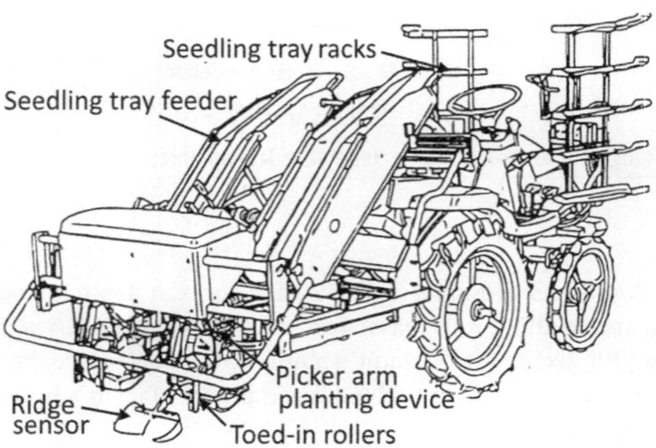

FIG. 6.19 Single-Row Automatic Vegetable Transplanter

and sown and the seedlings are grown, is an inverted hollow pyramid flattened at the bottom. The basic model of this automatic transplanter has a single- or two-row planting system mounted on a self-propelled riding- or walking-type vehicle. It is made quite compactly, with a rugged ground drive and a short turning radius to negotiate the machine even in small field plots. It consists of a tray feed mechanism, into which the trays are manually placed. The feeding conveyor positions the tray rows one by one, from which picking mechanism can pick.

A rotary seedling pickup mechanism (Tong *et al.*, 2020), which has a two-bar series system, is used; it moves an end effector on its picking arm in a particular locus (Figure 6.20) so that the arm can go over the plug tray, pick the seedling one by one from the plug tray, and drop it down into the planting mechanism on its downward path. This mechanism is an intricately designed two-bar system; it is synthesized based on the desired locus of the picking arm and is driven by a planetary gear train with special gear profiles. The picker arm, thus driven by the ground wheel, has an end effector (an actuated claw) at its tip; this penetrates into the seedling plug in an open position and grasps the plug (the plant root mass in the growth media with the plant) by closing the end effector. Further movement of the end effector on its locus extracts the seedling plug from the tray; the plug travels downward to be dropped again by opening the effector. The opening and closing of the end effectors are thus strategically performed at the right positions on the locus by appropriate means. The inset in the figure shows the actual picker drive unit; it has dual picker arms that pick twice per revolution of its main drive, making it plant faster.

The planting section of the machine has bill-type receptacles similar to semi-automatic transplanters. It is shaped like a bird's bill positioned upside down and can open into two halves because it is hinged at the top. The seedling plugs that are dropped from the picker mechanism are received into this receptacle on each row; they are then dropped gently into the opened furrow beneath. The opening

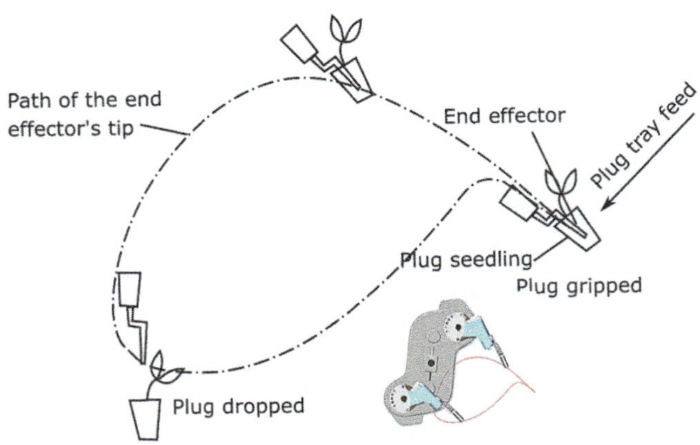

FIG. 6.20 Picker Arm Locus and the Picker Arm Drive Unit

No Permission Required

and closing of this receptacle of each row are performed by cams driven by the ground wheel and are synchronized to the plant spacing desired as well as to the picker's drop cycle. The furrow is closed over the plugs by a pair of toed-in press wheels and compacted.

A hydraulically actuated planting depth control and ridge following controls provide for uniform transplanting. Shoe- or runner-type furrow openers are used to open the furrows for plug seedlings to be planted. Petrol and diesel engine-driven versions of these machines are available. Since these machines are operated at relatively slow speeds, at about 50 plant drops per minute, the plugs are directly dropped into the furrow without the need for a zero relative velocity drop.

Yet another commercial model of tractor-drawn automatic vegetable transplanter (Figure 6.21) has plug trays that are fed from the top in a downward direction by a tray-positioning system that can feed the trays automatically downward as well as position the plug cells in the appropriate horizontal positions. A row of pneumatic ejectors ejects the plugs from the plastic trays from their back through the tray's drain hose? The row of ejectors is aligned above the rows that are to be planted. A swinging pair of arms with a receptacle tray of channels with grippers at their end receive the ejected plugs. While gripping the plugs meant for each row at their ends, the arms swing downward to a vertical position and release the plugs into the

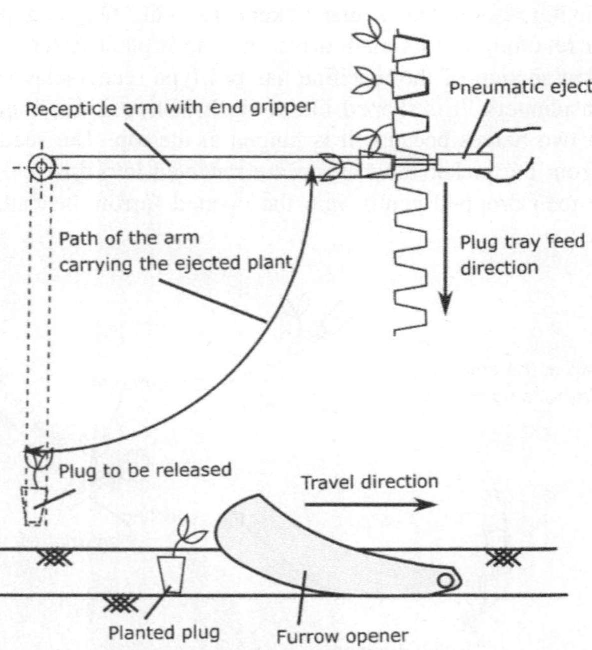

FIG. 6.21 Automatic Transplanting System With Back End Ejector

No Permission Required

corresponding furrows made by the preceding furrow openers. Pneumatic actuators are used for the grippers and arm movements. Although the response time for pneumatic actuators is quite small, it is still not fast enough to eject and drop the plants at higher travel speeds. So, the machines use two to three parallel arrays of such a system to plant each row, so that the problem of response time does not affect the travel speed of the machine. However, these are massively wide machines that are driven at a very slow speed to transplant the plugs. After the plugs are dropped into the furrows, toed-in compacting rollers close and compact the soil around the plants in the plugs. Because of the slower speed, the question of horizontal displacement and rolling of the plug while it is being dropped is avoided. The vertically fed trays are horizontally moved and positioned against the ejectors until the whole row of cells is emptied. Subsequently, the trays are moved down one row and the cycle is repeated. Automatic vegetable transplanters that pick and plant bare seedlings have also been developed (Suggs et al., 1987; Brewer, 1988). These machines grip bare-rooted plants between pairs of grippers on a wheel, which then are rotated to release the gripped plants into the opened furrow one by one.

6.11 Rice-transplanting machines

Rice-transplanting machines (Figure 6.22) pick and place rice seedlings, of predefined sizes, drawn from a mat-type nursery. The mat nursery is one in which rice seedlings are raised on a thin layer of soil and farmyard manure (FYM) or compost and placed on a polythene sheet or in a shallow plastic tray with perforations for irrigation. Raw seeds are densely sown in this media and the seedlings are ready for planting after 14 to 18 days. Either a mat grown on a plastic sheet is cut into desired sizes or seedlings grown in predefined sizes of trays are used; then, they can be fed into the seedling conveyor of the transplanting machine. These mats have their roots strongly embedded in the thin layer of media and the seedlings are firmly knitted at

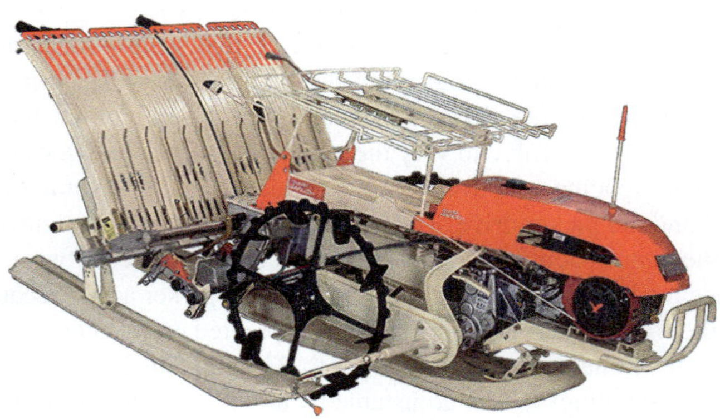

FIG. 6.22 Four-Row Walk-Behind Rice-Transplanting Machine (Sharp Garuda Farm Equipments Pvt. Ltd., India)

the root level. These mats are strong enough to be handled manually after 14 to 18 days.

Rice is generally grown in a soil puddle under submerged conditions. Young seedlings are transplanted in this soft soil puddle by these machines. The machine has two major sections, one for locomotion and the other for transplanting the seedlings. An engine drives these two sections via suitable transmission drives. The ground wheels propel the machine forward. The transplanting assembly has a frame supported by a float assembly, which allows the machine to slide and move smoothly over the wet puddle. The transplanting mechanism has a seedling conveyor, sloping downward to the back of the machine if it is a riding-type one or to the front in a walk-behind machine. It has a number of channels that accommodate the mats with the seedlings; this number corresponds to the number of rows being transplanted. The seedling mats fed into these channels slide down toward the picking assembly at the end, near the soil surface. The frame accommodating the seedling conveyor has picker arms at this end; these arms are driven by a common shaft and sets of four bar linkages for each picker on its rows. The number of picker arms corresponds to the number of rows. The locus of the picker arm is designed such that it moves the picker needles with their tips to:

- Reach into the seedling mat at the bottom edge of the seedling box or conveyor;
- Pick out a chunk of mat with seedlings;
- Move down to the soil level to plant the picked seedlings;
- Retract from the soil and return for the next cycle.

Each time the pick-and-place cycle is completed, the seedling conveyor moves sidewise laterally making it possible for the picker to pick from the next spot on the mat's row. The whole seedling conveyor carrying the mats moves laterally on the frame in steps of about 15 mm, driven by a spiral shaft receiving motion from the power transmission elements. This stepped movement is performed on the full width of the mat; then, the whole seedling conveyor cycles back to the original position laterally. Once the whole row of the seedling mat has been picked, the mat in each channel slides down by that distance and is ready for the next set of picks. The picker needle is a sharp double needle that simply penetrates into the mat and scoops the root mass along with the plants. The picker arm mechanism has a throw adjustment on the linkages that can alter the depth of reach of the picker needles, thus making it possible to vary the number of seedlings drawn out per pick. The plant-to-plant spacing on a row is varied by altering the speed ratio between the ground wheel and the picker drive. The more distance the machine moves per pick, the larger is the plant spacing. Plant spacing is usually variable, between 100 mm to 200 mm. Since the lateral spacing between is fixed the picker arms, because they are mounted on the frame, the row spacing of a transplanted crop is constant. The common row spacing on these machines is between 238 mm and 250 mm. The depth of placement of the seedlings is also adjustable by altering the height of the float relative to the frame. Machines of four-, six-, and eight-row configurations are common. Walk-behind and riding-type versions of rice-transplanting machines are available.

The example in Box 6.9 shows how the number of rice seedlings picked by hand is influenced by an adjustable reach into the mat.

Exercise problems

1. A bulk-flow metered seed drill is required to sow rapeseed at a seed rate of 12 kg ha^{-1} and at a speed of 6 km h^{-1} with the row spacing set at 25 cm. An agitator is used above each orifice of the seed drill to prevent the bridging of seeds. The bulk density of wheat seeds is 80 kg m^{-3}, and their spherical size 0.2 cm. Calculate the orifice size required to render the desired seed rate.
2. A fluted roller seed drill sows at a travel speed of 5 km h^{-1}. The metering roller has 8 mm and 12 mm deep flutes of semicircular shapes distributed over its peripheral diameter of 35 mm. The speed ratio between the ground wheel of 75 cm diameter and the fluted feed shaft is 4:1. Find the length of the fluted roller that is exposed to the seed mass in the hopper to dispense the required seed rate for each crop given in the table below. The maximum length of exposure to seeds for adjusting the seed rate is 10 cm. In that case, is it necessary for any one crop to change the speed ratio?

Box 6.9 Example 6.9

Example 6.9. A six-row rice transplanter plants mat seedlings at a row spacing of 30 cm. The picker needles are about 12 mm wide and actuated by a shaft driven from the ground wheel drive of the machine. The effective diameter of the transplanter's ground wheel is 85 cm. The axle rotates once for each picking cycle of the machine and the speed ratio between the picker axle and the ground wheel is set at 22:1. The mat tray assembly moves laterally in the direction perpendicular to travel for every pick cycle by the same width of the picker needle. The picker needle's throw or reach into the mat nursery is adjustable and set to 15 mm from the edge of the mat. Typically, plant density on a rice nursery mat is 3 plants per cm^2. How many plants are planted in each pick cycle? What would be the reach setting to pick and plant only three to four plants per hill? Also, find the plant spacing on the row at the present setting. If plant spacing on a row of 16 cm is set, what would be the speed ratio setting?

Circumference of the ground wheel = $\pi(0.85)$ = 2.67 m
Speed ratio set = 22:1
Hill spacing at this setting = 2.67 m / 22 = 0.121 m = 12 cm
If hill spacing desired is 16 cm, then speed ratio to be set = 2.67 / 0.16 = 16.7:1
Picker needle width = 12 mm = 1.2 cm
Reach of picker into the mat = 15 mm = 1.5 cm
$\therefore$ Area of mat scraped out in each pick = 1.2 × 1.5 = 1.8 cm^2
Plant density on the mat = 3 plants cm^{-2}
$\therefore$ No of plants picked per hill of pick = 3 × 1.8 = 5 to 6 plants
For planting only 3 to 4 plants per hill, the area of mat to be reached out
= 3.5 plants per hill / 3 plants cm^{-2} = 1.17 cm^2
$\therefore$Reach of the picker needle into the mat = area of mat to be scarped / picker needle width
= 1.17 / 1.2 = 0.7 cm = 8 mm

3. Hybrid corn is planted with a planter at a row spacing of 70 cm and at a forward speed of 6 km h^{-1}. It meters 10 seeds per revolution of the seed-metering disc to get a desired plant population of 20,000 plants per hectare. Find the required seeding rate of seeds possessing 92 percent germination and the seed spacing required on the row to achieve it. If the rolling radius of the planter wheels is 40 cm, calculate the speed ratio between the seed-metering shaft and the planter wheel.
4. In the previous problem, find whether the seed-metering disc of 8 cm diameter is performing with 100 percent cell fill if the maximum permitted peripheral speed of the disc is 0.25 m s^{-1}.
5. A horizontal plate planter has a transmission ratio of 2:1 between the ground wheel of 60 cm diameter and seed plate sows at a speed of 4 kmh^{-1}. The peripheral speed of the seed plate is 0.35 ms^{-1} providing 90 percent to 100 percent cell fill. Calculate the diameter of the seed plate and the seed-to-seed spacing if the seed plate has 25 cells on its periphery.
6. A planter drops the metered seeds to fall freely and directly into the furrow from a point 30 cm above the furrow bottom. If the planter is traveling at a speed of 5 km h^{-1}, find the horizontal distance the seed moves between the point of release and the point of hitting the furrow bottom.
7. An eight-row rice transplanter designed for a row spacing of 30 cm transplants mat seedlings using picker needles that are 14 mm. The rolling diameter of the transplanter's ground wheel is 90 cm. The picker mechanism is a double-throw unit that picks two bunches per each picking. The speed ratio between the picker axle and the ground wheel is set at 12:1. The mat tray assembly moves laterally in the direction perpendicular to travel for every pick by the same width of the picker needle. The picker needle's reach into the mat nursery is set to 8 mm from the edge of the mat. The plant density on the rice mat nursery is 3 plants per cm^2. Find the number of plants planted on each hill cycle. Find the picker reach setting to pick and plant 20 percent more plants per hill.
8. The automatic vegetable transplanter system shown in Figure 6.16 is transplanting tomato plug seedlings at a row spacing of 40 cm and a plant spacing of 30 cm. The minimum response time of the system for ejecting the plugs and placing them is 3 s. Find the maximum possible travel speed the system can handle for planting the plugs at the required spacings. If the swinging of the 30 cm arm takes place in 1.5 s, find the relative velocity imparted on the plug while being dropped into the furrow.
9. In the previous problem, calculate how many parallel arrays of the planting systems need to be serving each row to make the relative velocity zero such that the plug will fall vertically down. If only one parallel array is used, as in the previous problem, what would be the minimum plant spacing that the transplanter can handle to make the relative velocity zero.

References

Bell, B. (2005). *Farm machinery.* UK: Old pond Publ.

Bernachi, H., Haman, J., & Kanafojski, C. (1972). *Agricultural machines, theory and construction* (p. 1). Warsaw, Poland: Published for USDA and NSF by Foreign Cooperation Center of the Central Institute for Scientific, Technical and Economic Information.

Beverloo, W. A., Leniger, H. A., & Velde. (1961). The flow of granular material through orifices. *Journal Chemical Engineering Science, 15*, 260–296.

Brewer, H. L. (1988). EXPERIMENTAL AUTOMATIC FEEDER FOR SEEDLING TRANSPLANTER. *Applied Engineering in Agriculture, 4*(1), 24–29.

Culpin, C. (2013). *Farm machinery.* Hesperides Press, 144370301X, Hesperides Press. ISBN 144370301 :X.

Giannini, G. R., Chancellor, W. J., & Garrett, R. E. (1967). Precision planter using vacuum for seed pickup. *Transactions of the ASAE, 10*(5), 607–614.

Hunt, D. (2008). *Farm power and machinery management.* Waveland press.

Kepner, R. A., Bainer, R., & Barger, E. L. (1978). *Original from the university of Michigan.* Avi Publishing Company.

Klenin, N. I., Popov, I. F., & Sakun, V. A. (1970). *Agricultural machines . USDA and NSF by amerind.* Publishing Co. Pvt. Ltd, USDA and NSF by Amerind Publishing Co. Pvt. Ltd.

Lovegrove, H. T. (1968). *Crop production equipment.* Hutchinson press.

Moysey, E. B., Lambert, E. W., & Wang, Z. (1988). FLOW RATES OF GRAINS AND OILSEEDS THROUGH SHARP-EDGED ORIFICES. *Transactions of the American Society of Agricultural Engineers, 31*(1), 226–231.

Smith, H. P. (1965). *Farm machinery and equipment.* Tata Mc-Graw Hill Pub. Co. Ltd.

Srivastava, A. K., Goering, C. E., Rohrbach, R. P., & Buckm)ster, D. R. (2006). *Engineering principles of agricultural machines.* ASABE Publications.

Suggs, C. W., Thomas, T. N., Eddington, D. L., Peel, H. B., Seaborn, T. R., & Gore, J. W. (1987). Self-feeding transplanter for tobacco and vegetable crops. *Applied Engineering in Agriculture, 3*(2), 148–151.

Tong, Z., Yu, G., Zhao, X., Liu, P., & Ye, B. (2020). Design of vegetable pot seedling pick-up mechanism with planetary gear train. Springer *China Chinese Journal of Mechanical Engineering, 33*(1), Article 21928258. https://doi.org/10.1186/s10033-020-00484-w. Available from: https://link.springer.com/journal/10033.

Tsuga, K. (2000). Development of fully automatic vegetable transplanter. *Japan Agricultural Research Quarterly, 34*(1), 21–28.

CHAPTER

Machines for weed control

7

Chapter outline

Weed control consumes the lion's share of the total expenditure of producing crops. Weeds not only compete for nutrients and water with a crop but act as hosts to pests that feed on the crop. The removal of weeds that compete with a crop has always been a challenge in agricultural production systems. Because of the difficulties with mechanical modes of weed removal, especially those between the crop on a row (intra-row), chemical methods of weed control using weedicides was used indiscriminately without realising the damage the chemicals cause the crop environment. But recently, there is a rising trend for adopting organic farming practices. Modern innovations in weeding technology and equipment has made weeding more efficient and complete without adopting weedicides. The removal of weeds by soil working tools (Bell, 2005; Culpin, 2013; Hunt, 2008; Kepner et al., 1978; Lovegrove, 1968; Smith, 1965; Srivastava et al., 2006) has again gained significance in farming. It is now recognized that, even in the early stages of crop growth, tools such as a rotary hoe can be operated directly over the crop rows to uproot weeds at the cotyledon stage.

7.1 Objectives of weed control

Primarily, the objectives of weed control are to remove weeds that compete with a crop for soil nutrients and moisture.

Machines of the Farm. https://doi.org/10.1016/B978-0-443-32906-7.00007-4

The removed weeds, when incorporated into the soil, become plant food.
Leaving dry weeds on the soil surface after weeding retains soil moisture.
Removing weeds aerates the soil to allow oxygen to penetrate the soil.
Removing weeds promotes the activity of microorganisms that enhance the soil's health.

7.2 Cultivators

Cultivators are basically inter-cultivators that can traverse between rows of a crop and remove the weeds between the rows. Different types of cultivators are available, ranging from manually operated ones to wide tractor-drawn units. Many of the tractor-mounted cultivators are centrally mounted or front-mounted so that the tractor operator can guide the weeding shovels of the cultivator accurately between cropped rows without destroying the crop. But rear-mounted cultivators are equally popular since they can be used to till loose soil as well to harrow. Centrally or front-mounted cultivators are available in different sizes and are mounted on tractors with a relatively large ground clearance. These tractors are known as row crop tractors and are fitted with thinner rear wheels that can traverse between crop rows without trampling the crop. The wheel treads of both the front and rear wheels are adjusted to be in line and in multiples of the crop row spacing, so that the tractor, along with the cultivator, can traverse inside the row crop without damaging the crop. Tractors have hitch frames that can hydraulically raise or lower the cultivator as desired. Rubber-covered gauge wheels support very wide gangs of cultivators to maintain a uniform depth of cultivation. While using centrally mounted cultivators, it is customary to use a rear-mounted cultivator that loosens the soil behind the tractor wheels that is compacted by the machine weight. A rear-mounted cultivator is mounted on the conventional three-point linkage of the tractor for operation.

The cultivator gang frames consist of beams to which tines (standards or drops) are attached that have a fixed or adjustable foot set at an angle. Cultivator shovels or sweeps, which are the soil working parts, are fastened to these tines at the tip. The working width and number of such cultivator gangs, each carrying multiple tines, vary according to the size of the cultivator. The tines fitted with the shovels are adjustable and removable over the gang width, so that the desired row spacing between the tines can be achieved. These tines are either rigidly fixed to the frames by clamps as in a "rigid tine cultivator" (Figure 7.1) or they are spring-mounted, as in the "spring-loaded cultivators" (Figure 7.2) so that they can ride over obstacles like heavy roots or rocks. When the shovel hits an obstacle, the tine pinned to the frame at the front end swings back and up against the heavy spring pressure, thus overriding the obstacle. The widths and shapes of the shovels and sweeps used on the cultivators are many, depending on the soil, crop, and weeds to be removed (Figure 7.3). A reversible shovel is a general-purpose shovel that is used for cultivation as well for light ploughing and harrowing. These shovels are used for the shallow ripping of open fields with shallow, rooted weeds. These are usually made to be thick and can be reversed and fitted when one

FIG. 7.1 Rear-Mounted Rigid-Tine Cultivator (India Agrovision Implements Pvt. Ltd).

FIG. 7.2 Rear-Mounted Spring-Loaded Cultivator (HD series, India Agrovision Implements Pvt. Ltd).

end is worn out. Sweep shovels are basically for weeding and are operated as shallowly as possible to prevent damaging the sidewise roots of the crop on the rows. Sweeps are, hence, set almost flat with their outer tip of the wing elevated from the ground by a small suction clearance of 6 mm. Half-sweeps are used when weeding is to be done very close to the crop stem.

Figure 7.4 illustrates a typical centre-mounted cultivator setting (Bernachi et al., 1972) in which each set of three sweep shovels weed between each pair of crop rows. As an instance, it may be noted that shovels 'A' and 'B' mounted on the centre frame along with shovel 'C' on the rear frame cover the complete width between that pair

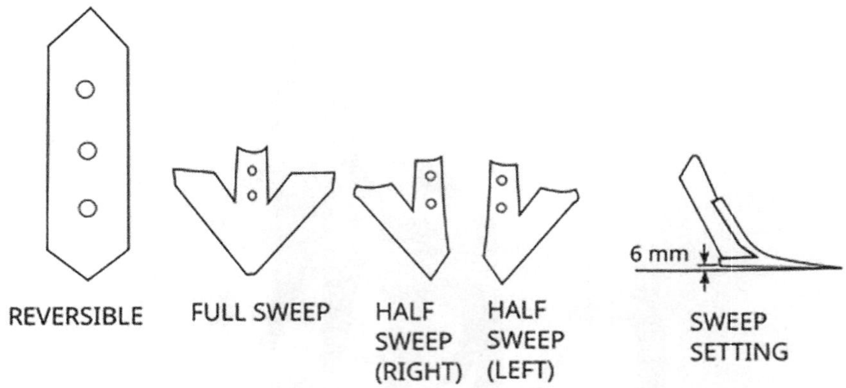

FIG. 7.3 Types of Cultivator Shovels and Sweep Shovel Setting.

of crop rows. It may be seen how the front and rear tractor wheels are set to travel between the crop rows. The shovels at the rear not only weed the row but also loosen the soil compacted by the wheels. Self-propelled toolbars that give the operator a clear view of all the rows being cultivated are available on the market. The example problem in Box 7.1 calculates the power requirement of a sweep cultivator.

7.3 Rotary hoes

Rotary hoes consist of a row of spiked wheels of about 45 cm diameter, placed about 15 cm apart on an axle, and rolled on the soil surface. Two such gangs of wheels are set as staggered in the rows (Figure 7.5) so that they operate with a row spacing of 7.5 cm. The rotary hoe's star wheels break up the soil crust by digging and throwing up soil with their flattened or spoon-shaped star tips and uproot shallow, rooted weeds. The uprooted weeds are thrown on the soil surface and dry out.

Some modern designs of rotary hoes use plastic wheels and are pulled at an inclination for improved weeding. Rotary hoes (Figure 7.6) are excellent tools that can break the upper soil crust as well as uprooting small weeds between established crops. The rotary hoe works over the tops of rows as well as between them, without injuring the crops. The rotary hoe, when operated in reverse rotation, acts as a subsurface packer in mulch tillage (Klenin et al., 1970).

The possibility of high working speeds of rotary hoes, to the extent of 20 $kmh^{-1,}$ provides more field coverage in a short time. Unlike the tine weeder, a rotary hoe works in moist soil, and its operation after a light rain is more effective in breaking the soil crust. A gang of rotary hoes provides sufficient contact pressure on the soil surface to do its job. Gauge wheels provided on the frames are helpful to control the depth of operation in friable soils. On harder soils, hydraulic means are used in some weeders to provide more downward pressure on the star wheels.

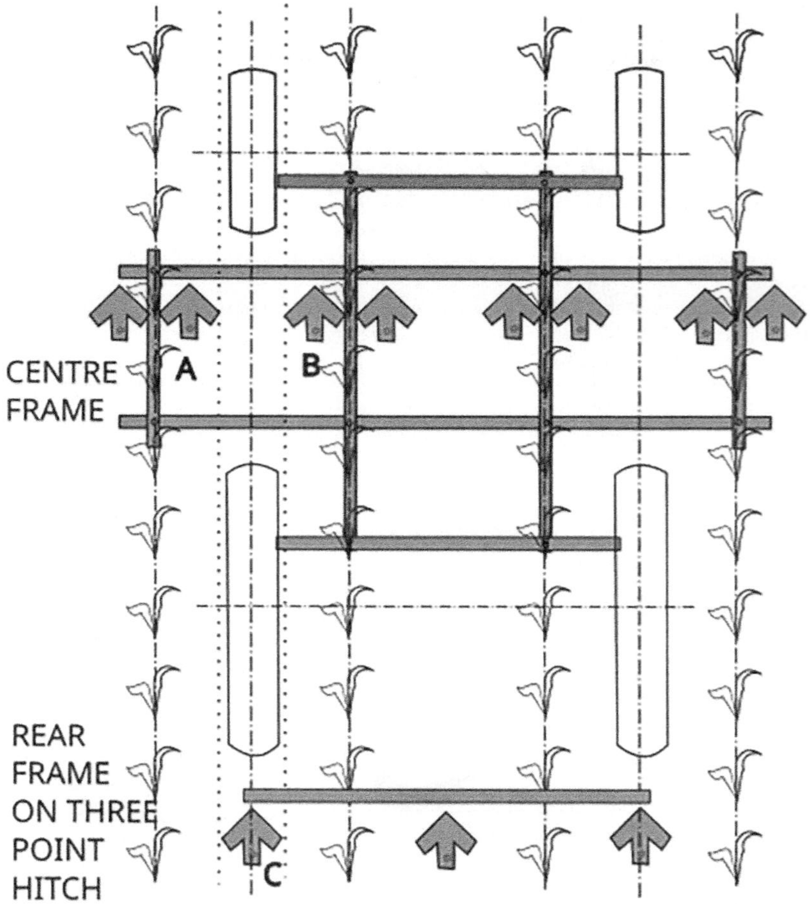

FIG. 7.4 Typical Settings of Sweeps on a Cultivator.

7.4 Finger weeders

The finger weeder has a rotating fingered wheel (Figure 7.7A) made of polymer and is designed to have flexible fingers that follow the ground contour of the crop row. The fingered wheels carry a rigid, metal, spider wheel underneath; this wheel has lugs, which drive the finger wheel to rotate by interacting with the soil below. The spider's lugs rotate the finger wheels at a relatively higher peripheral speed in relation to the forward speed of the cultivator. The finger wheels flex upward, pushing down on the soil before flicking the weeds that are in a tender root hair stage of growth without damaging the planted crop. Obviously, the weeds should be young for these weeders to work. The fingers run very close to the crop rows without damaging them but break up the soil and throw loose soil into the row to cover and kill very young sprouting weeds. They also

Box 7.1 Example 7.1

Example 7.1. Find the power required for a tractor to operate the sweep cultivator arrangement shown in Figure 7.4 at a speed of at 5 kmh $^{-1}$. The width of the sweep blade is 15 cm, and the depth of operation is 5 cm. The sweep blades are set at a rake angle of 15°. The soil parameters are soil density = 15 kN m^3, cohesion = 20 kPa, angle of internal friction of soil = 35°, and soil-metal friction angle = 23°. The soil may be assumed to be adhesionless. Use the McKyes and Ali model (section 5.15) to compute the horizontal soil reaction of each sweep as a simple flat blade. (The computer code of the model is available for the computation accompanying the book.)

Horizontal soil reaction is computed as 220 N for the given parametric values of soil using the software.

Since we can assume this force as a draft requirement for one sweep shovel, the total draft having 11 such shovels = 2420 N

Power required = draft × speed = 2.420 kN × (5000/3600) ms^{-1} = 3.36 kW.

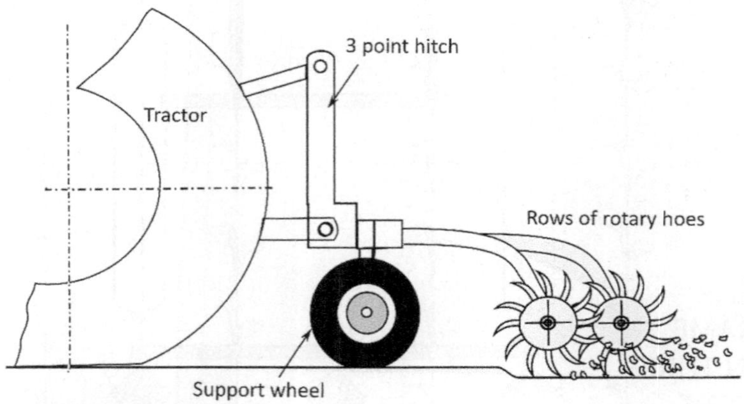

FIG. 7.5 Working of a Rotary Hoe.

heap soil on the crop row, smothering very young weeds. The advantages of using these weeders, especially in vegetable cultivation, is that a thorough intra-row weeding with more loose soil thrown intra- and inter-row is achieved. Finger weeders, attached to a cultivator frame, are used as a set on each crop row (Figure 7.7B). These are intra-row weeding tools that remove weeds present between the plants on the crop row. The weeds present between crop rows (inter-row) are removed by conventional sweeps mounted on a cultivator frame (Figure 7.7 and Figure 7.8).

7.5 Torsion weeders

Steel rods (Figure 7.9) with a torsion spring at the mounting end are used in pairs, mounted on either side of the crop. They are angled downward and back, toward the crop row. The spring allows the tool tips to flex upward with the soil contours as well

FIG. 7.6 Mounted Rotary Hoe Arrays (Helios-6m, Kverneland Group, Norway).

around rooted crops, uprooting intra-row weeds in the white root stage. These devices work as a pair of opposing arms with each arm working alongside and pushing against one side of the row area just beneath the soil surface. Torsion weeder tool tips, which are able to move vertically to conform with the soil contour, are gentler than finger weeders. Much like finger weeders, these are also used on guided toolbars so that the intra-row weeding can be made precisely.

7.6 Tine weeders

Tine weeders are row independent, which means that they can run over a crop without damaging it. They use wirelike spring rods with torsion back ends, very similar to torsion weeders. As this weeder is row independent, rods are used all over the weeder frame, with an effective spacing of 3 cm to 4 cm irrespective of the crop's rows (Figure 7.10). The tines are assembled at a close spacing on the tine shafts, and rows of such tined shafts are used to attain an effective tine spacing of 3 cm. The tines are staggered over the tine shafts on the whole frame in order to achieve this close effective spacing. The tine shafts are inter-connected by parallel links controlled by a common hand lever with a locking arrangement. When the lever is moved, the tine angles are adjusted simultaneously all over the implement frame. The depth of penetration is adjusted by the level of the implement frame and by the support wheel. Row independence is a big advantage because no precise guidance of the tines into the row crop is needed, and, hence, it is faster and more efficient. Here, weed control is performed by removing or burying the weeds in soil at their emergence (cotyledon stage). The working depth and, therefore, the torsion pressure are adjusted to match the age of the crop, its root depth, and the seedling size. The right depth and tine pressure on the weeding tines are crucial for the

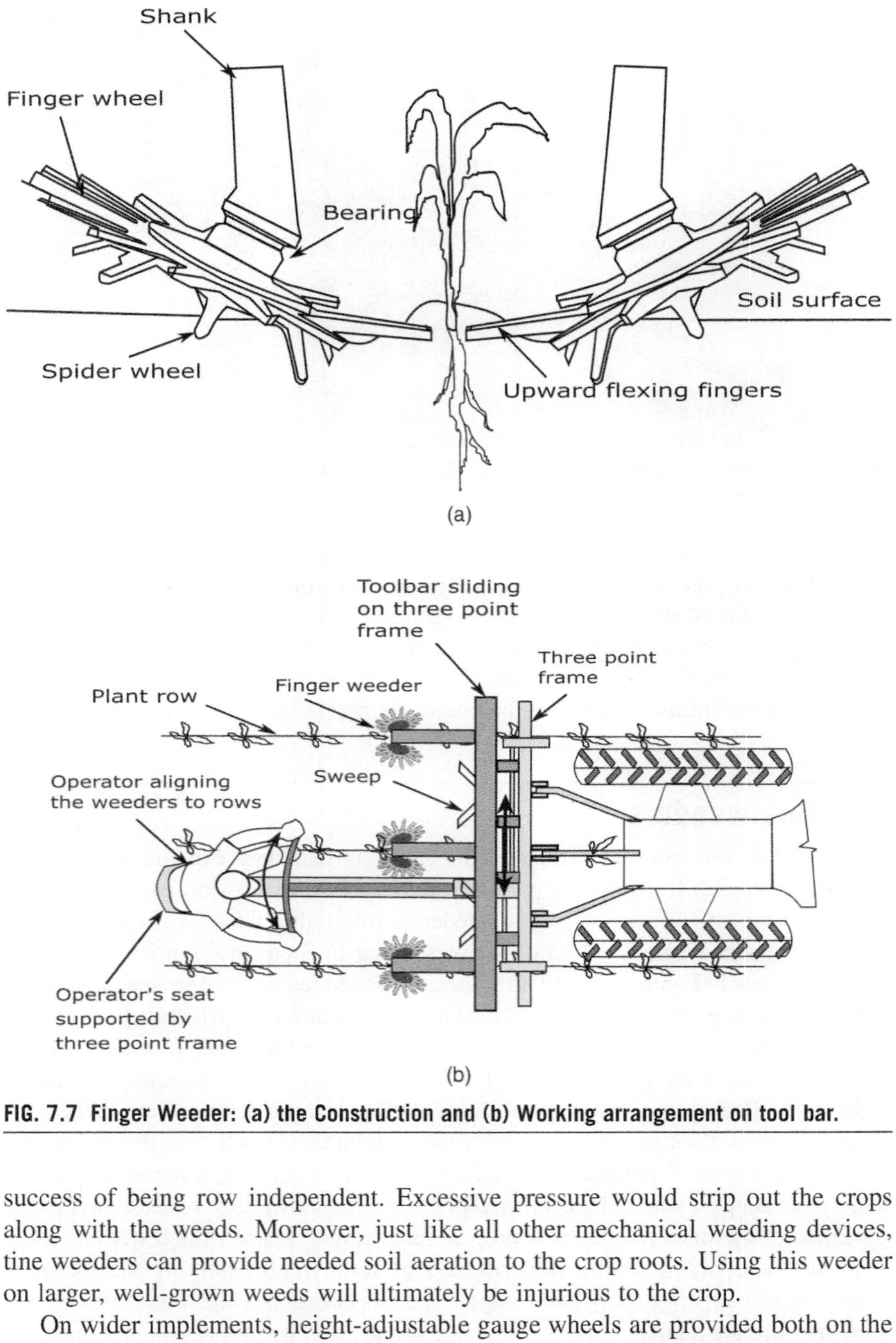

FIG. 7.7 Finger Weeder: (a) the Construction and (b) Working arrangement on tool bar.

success of being row independent. Excessive pressure would strip out the crops along with the weeds. Moreover, just like all other mechanical weeding devices, tine weeders can provide needed soil aeration to the crop roots. Using this weeder on larger, well-grown weeds will ultimately be injurious to the crop.

On wider implements, height-adjustable gauge wheels are provided both on the front and rear of the weeder frame, so that a uniform depth of soil disturbance is

FIG. 7.8 Mounted Inter-Row Cultivator with Sweeps and Finger Weeders (Onyx, Kverneland Group, Norway).

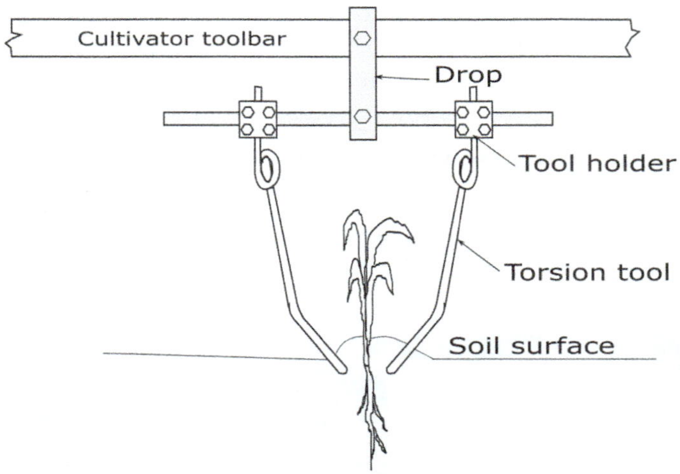

FIG. 7.9 Construction of Torsion Weeder.

achieved. The speed of operation is to be in the order of less than 2 km h^{-1} when the crop is very young, so as to avoid undue damage to the crop. However, it can be increased with the age of crop. An optimally dry soil condition is most suitable for this weeder's operation. This weeder is definitely not ideal for clayey, wet soils. It is suitable for the very early stages of crop growth in which the crop height is just 2 cm to 3 cm. Unlike row crop weeders, in which the tractor tires need to ride in the row spacings, this weeder does not require wider row spacings, but tractor wheel lanes are required depending on the width of the available row crop machines

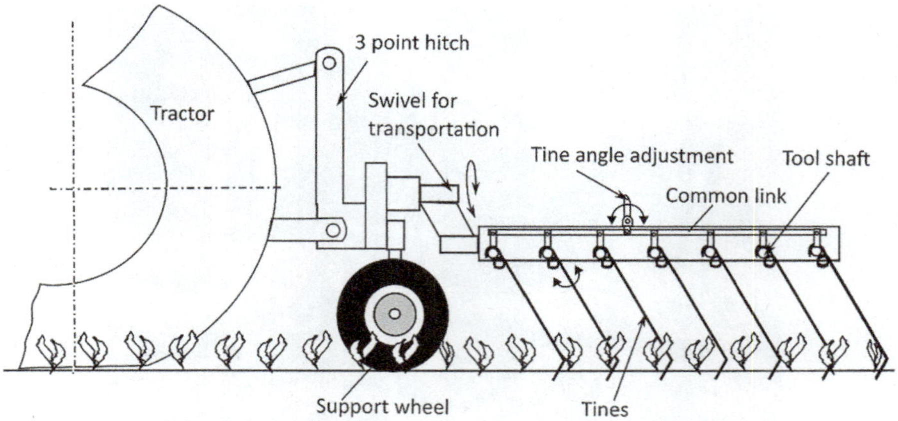

FIG. 7.10 Working of a Tine Weeder.

including the weeder. Figure 7.11 shows a mounted-tine weeder working in a cropped field. The example problem in Box 7.2 calculates the maximum angular deflection of a tine weeder's tine based on the power requirement to tow the implement.

7.7 Steering systems for row crop weeding equipment

The use of sweeps and hoes set to operate between crop rows and the use of finger weeders for weeding the intra-spacing of a crop is the gold standard in modern row crop weeding systems (Figure 7.8). But these systems should be provided with steering

FIG. 7.11 Trailed-Tine Weeder (Arcadia, Kverneland Group, Norway).

Box 7.2 Example 7.2

Example 7.2. The maximum power required for a 3 m wide tine weeder is specified as 22 kW at the maximum operating speed of 6 kmh $^{-1}$. The effective spacing between tines is 2.5 cm. It uses spring steel tines of 7 mm diameter with an effective length of 50 cm from tine tip to its holder. The tine has a built-in torsion spring of 1 coil of 5 cm diameter at a distance of 8 cm from the tine holder along the tine. Find the maximum angular deflection of the tine about the spring centre.

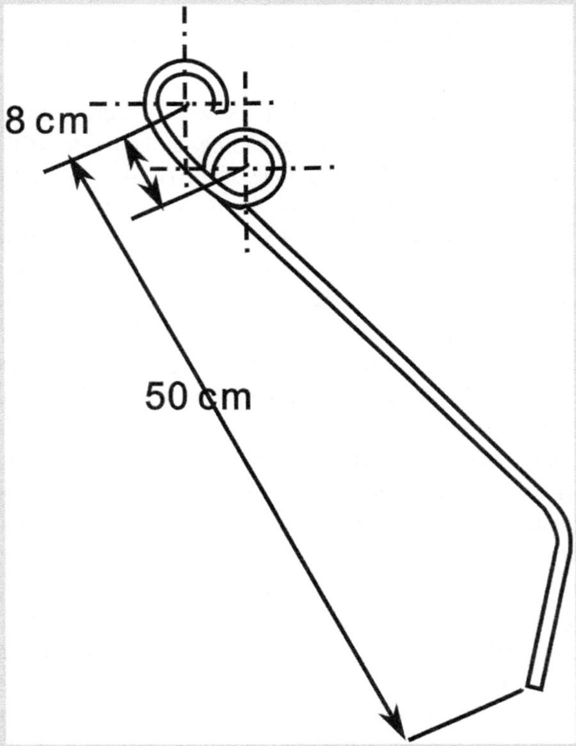

The maximum power exerted on the machine = 22 kW
Effective width of the machine = 3 m
∴Number of tines at 2.5 cm effective spacing = 300/2.5 = 120
Maximum power exerted on a single tine = 22 / 120 = 183 W
Maximum forward speed of the machine = 6 kmh $^{-1}$ = 1.67 ms $^{-1}$
∴Maximum draft on one tine = 183 / 1.67 = 110 N
For a torsion spring,

$$k = \frac{M}{\theta} = \frac{Ed^4}{64Dn}$$

where k = the angular stiffness of the torsion spring

Continued

Box 7.2 Example 7.2—cont'd

M = the moment applied to the tine tip, Nm = 110 N × (0.5-0.08) m = 46 Nm
θ = The angular deflection of the spring, °
E = Young's modulus of the rod material, Pa = 210 GPa = 210 × 10^9 Pa
d = the diameter of the rod, m = 0.007 m
D = the diameter of the torsion spring coil, m = 0.05 m
n = number of torsion spring coils = 1

$$\frac{46}{\theta} = \frac{210e09(0.007)^4}{64(0.05)2}$$

Maximum angular deflection, θ = 0.29 rad = 16.7°

systems independent of the tractor to negotiate between the crop rows as well to align the rows on the finger weeder operation. Depending on the model of the weeding machine, different steering systems are used. One of the possibilities is that the weeding frame is mounted on the front or middle of the tractor on row crop wheels and the tractor driver has a full view of the hoes. This aids the tractor operator to guide the tractor and the weeding system precisely between the crop rows. This is ideal for relatively narrower toolbars. Self-propelled weeder machines with such weeding frames are available on the market, especially for vegetable cultivation.

Rear-mounted weeding equipment always has the weeding frame mounted on a subframe that can move laterally on the main three-point hitched frame. A manually operated steering arrangement was provided in earlier models (Figure 7.8B) in which an operator rides on the seat at the back and guides the weeding tools precisely into the crop rows without damaging them. In modern machines, this steering control is rendered by hydraulic means actuated by camera systems that detect the crop rows in real time and steer the frame precisely in between the rows (Figure 7.12). The automatic camera steering is set by the tractor driver through an operating terminal. The operator sets a suitable grid over a photo image of the crop row, which in turn sets all the required parameters such as the row spacing, number of rows in the camera's view, and plant width and height. The weeder frame that is mounted on a hydraulically actuated slide (inset of Figure 7.12) is moved laterally based on the acquired parameters, cantering the hoes and weeders precisely along the crop rows. More sophisticated machines provide RTK-based GPS guidance to these steering systems at a centimetre-level resolution.

7.8 Machines for organic farming

Organic farming, as we may understand, is the concept of eliminating the use of chemical fertilizers, weedicides, and pesticides in farming so that toxic chemical residues in food and produce are reduced. The demand for organic products and the price tag accordingly have attracted many farmers into organic farming. The use of organic manure and conservation technology such as crop residue management is seen as an alternative to chemical fertilizers. The use of advanced pesticide application methods

FIG. 7.12 Camera-Guided Inter-Row Weeder (Lynx-2000, Kverneland Group, Norway).

is also seen as a part of this venture to keep their use under the threshold and protect the environment. Hence, organic farming is not just about mechanical weed control alone but involves many more strategies, such as crop rotation, growing a catch crop, more efficient seed bed formation, etc., to achieve the desired goals. However, the use of efficient weed control machinery in place of weedicides is gaining a lot of significance in this arena. Internationally, especially in European agriculture, there is a growing trend in the use of precise and exclusive weed control machinery for the organic farming of vegetables. The main hurdle in conventional weed control machines is that they do not deal with intra-row weeds that are present on a row between crops. These are to be either removed or covered by soil to kill them. But the availability of row-independent weeding tools such as the tine weeders has alleviated this impediment.

The incorporation of dead organic matter from uprooted weeds adds to the humus of soil, aiding crop growth. The use of natural organic fertilizers such as lime, manure, and farm waste from livestock and farmyard manure is encouraged, eliminating chemical fertilizers. Soil health and structure is improved by the use of organic fertilizers. Growing catch crops such as legumes in between seasons of the main crop promotes soil humus and structure. Reduced tillage, a very old school of thought, is practiced as a part of organic farming such that the soil is not unduly disturbed, more than necessary, by intensive tillage. As explained previously, row-independent tine weeders and the combination weeding toolbars with inter- and intra-weeding tools are seen as alternatives to chemical weed control and are very much in practice in organic farming.

References

Bell, B. (2005). *Farm machinery.* UK: Old pond Publ.

Bernachi, H., Haman, J., & Kanafojski, Cz. (1972). *Agricultural machines, theory and construction* (p. 1). Warsaw, Poland: Published for USDA and NSF by Foreign Cooperation Center of the Central Institute for Scientific, Technical and Economic Information.

Culpin, C. (2013). *Farm machinery.* Hesperides Press, 144370301X, Hesperides Press. ISBN 144370301 :X.

Hunt, D. (2008). *Farm power and machinery management.* Waveland press.

Kepner, R. A., Bainer, R., & Barger, E. L. (1978). *Original from the university of Michigan.* Avi Publishing Company.

Klenin, N. I., Popov, I. F., & Sakun, V. A. (1970). *Agricultural machines . USDA and NSF by amerind.* Publishing Co. Pvt. Ltd, USDA and NSF by Amerind Publishing Co. Pvt. Ltd.

Lovegrove, H. T. (1968). *Crop production equipment.* Hutchinson press.

Smith, H. P. (1965). *Farm machinery and equipment.* Tata Mc-Graw Hill Pub. Co. Ltd.

Srivastava, A. K., Goering, C. E., Rohrbach, R. P., & Buckm)ster, D. R. (2006). *Engineering principles of agricultural machines.* ASABE Publications.

CHAPTER

Fertilizer and manure application machinery

8

Chapter Outline

The application of fertilizer to a crop has become mandatory for acquiring higher yields. Though the use of fertilizers has been argued to be detrimental to the soil health, especially in the long run, the pressure for food production has sustained the use of fertilizer on some farms. Organic farming and products are being encouraged, but getting higher yields in organic farming is still not achievable. The use of fertilizers in crop production has brought about the creation of many machines for fertilizer application.

8.1 Fertilizer drills

The first of this lot of machinery is the fertilizer drill, which, in construction and principle of working, is quite similar to a seed drill, which is addressed in detail in the chapter on sowing machines. Fertilizer drills usually use bulk metering devices since the fertilizer has to flow continuously into the dug furrow rather than with an intra-spacing on the row.

Machines of the Farm. https://doi.org/10.1016/B978-0-443-32906-7.00030-X

8.1.1 Types and characteristics of fertilizer metering devices

Fertilizer drilling systems (Hunt, 2008; Kepner et al., 1978; Lovegrove, 1968; Smith, 1965; Srivastava et al., 2006) that place fertilizer on rows are those that are used for the deep or shallow placement of fertilizer, either as independent machines or as attachments to seeding machines. They are also used in fertilizer-cum-grain drills, which perform sowing and fertilizer applications simultaneously. Fertilizer metering devices should be capable of dispensing a wide variety of fertilizers of various particle sizes. They are intended to deliver the fertilizer without any change in the rate of delivery as influenced by the depth of fertilizer in the hopper. They are meant to control the delivery rate precisely to the desired level. The fertilizer hopper as well as the metering devices are intended to be easily cleaned since most fertilizers are corrosive to metals. Today, the problem of corrosion is completely eliminated after plasticizing the fertilizer metering bins and devices.

The rotating parts of these metering devices are driven from the ground wheel, similar to the function of seeding devices. The star wheel feed is used prevalently on fertilizing devices. Fertilizer carried between the teeth of the star wheel is pushed into the delivery tube by gravity through the bottom of the hopper. The delivery rate is controlled by the speed of rotation of the star wheel as well by raising or lowering a gate above the star wheel. In many instances, a metering device with a simple agitator over a gated aperture is used. The agitator's rotary speed and the aperture size control the delivery rate. But, its delivery rate is predominantly influenced by gravity and the aperture size rather than the speed of the agitator. Positive feed-metering devices such as the star wheel depend more on the speed of the wheel, although gravity comes into play at larger gate adjustments. The drillability or the ease of flow of a fertilizer into the delivery tube depends on many factors such as the particle size and shape as well as its packing characteristics. Generally, it is observed that the drillability of a fertilizer is inversely proportional to the angle of repose of the material, which is the angle subtended at the base of a heap formed by a freely poured fertilizer on a level surface. Any material that has an angle of repose greater than 55° is usually not drillable. Many types of fertilizer are hygroscopic and form lumps at higher relative humidities, which impedes the free flow of the fertilizer. The fingers of a star wheel device, although being a positive flow type, cause pulsation in its delivery. But, the uniformity of its flow rate is more reliable than an agitated gravity-feed device.

Nowadays, fertilizer drills are not available separately and are always used along with seed-metering systems. These pieces of equipment, known as seed-cum-ferti drills, have separate hoppers and metering systems for seeds and fertilizer. The fertilizer drop tubes place the fertilizer in furrows that are dug by separate furrow openers in the vicinity and a bit deeper than the seed furrows. This avoids the direct contact of the seed and fertilizer and thus prevents seed damage. The parts of the fertilizer drill including the delivery tubes are now almost always plasticized to avoid corrosion by chemical fertilizers and are easily cleaned after use. Similar to seed drills, fertilizer systems can be calibrated appropriately for the desired application rate. The same kind of calculation is applied for calibrating the fertilizer-metering system, too. Simple aperture indexing along with the proper setting of a speed ratio between the ground wheel and the fertilizer-metering system ensure the correct fertilizer application rate.

8.2 Fertilizer distributors

Usually, these fertilizer drills are row crop machines and not used for bulk application in fields yet to be cropped or in bulk sown crops. In those cases, fertilizer distributors or spreaders are used. The centrifugal-type fertilizer distributor (Hunt, 2008; Kepner et al., 1978; Lovegrove, 1968; Bernachi et al., 1972; Klenin et al., 1970) (Figure 8.1) is one popular choice. The machine is usually mounted on the rear of a tractor. It consists of a hopper in the shape of an inverted frustum of a cone with a shuttered aperture from which the fertilizer can flow. At the bottom of the hopper, a rotating disc with a number of deflecting vanes is provided. The disc is rotated, taking a drive from the PTO of the tractor. The drive is also coupled to a simple rod-type agitator on a central axis revolving vertically inside the hopper. The agitator prevents any bridging of fertilizer flowing from the aperture. The aperture is shuttered and can control the opening size, in turn influencing the application rate. The fertilizer flowing through the apertures hits the centre of the revolving disc at the bottom and is flung outside by a centrifugal force imparted to the fertilizer particles through the vanes. This causes the distribution of the metered fertilizer in a circular area around the centre of the revolving disc. As the tractor moves forward, it distributes the fertilizer on a rectangular swath of a width equal to the diameter of the circular distribution pattern. The density of the fertilizer distributed near the disc center is higher than that on the periphery, causing non-uniformity in the distribution of fertilizer over the swath.

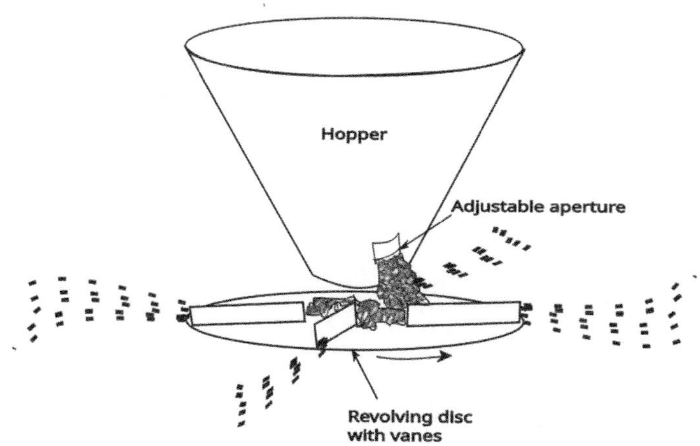

FIG. 8.1 Centrifugal-Type Fertilizer Distributor

No Permission Required

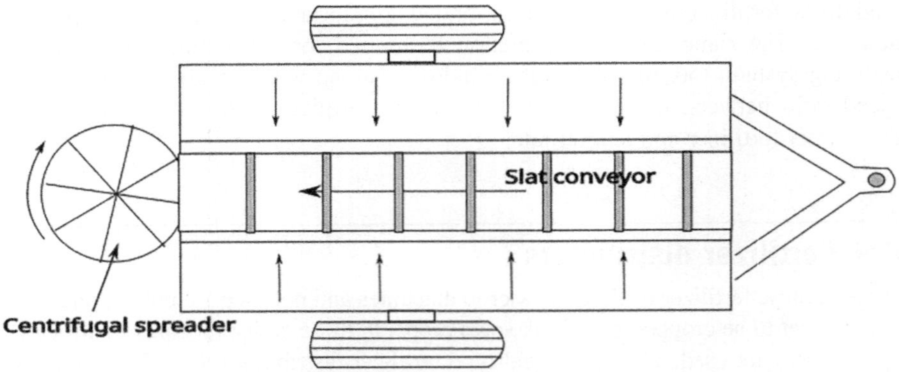

FIG. 8.2 Moving Floor Fertilizer Distributor

No Permission Required

Laying out collection cups in a grid to catch the distributed fertilizer in the circular pattern of distribution can give a measure of the samples collected. Calculating the uniformity coefficient of these samples collected over the grid would yield this variability. If needed, the swath could then be overlapped to acquire a more uniform distribution.

The moving floor fertilizer spreader (Figure 8.2) is a trailed fertilizer bin on a set of wheels with a moving floor. The lateral sides of the bin are inclined toward the central portion, where a slat conveyor of a relatively narrower width is placed. The fertilizer loaded into the trailed bin slides down toward the slat conveyor while the conveyor moves the bottom layer of the fertilizer toward the back of the trailer. At the back, where the conveyor ends, a centrifugal broadcaster disc is positioned. The spreader as well as the conveyor receive power from the PTO of the tractor through a propeller shaft and appropriate transmission elements. The fertilizer falling on the rotating disc of the centrifugal broadcaster, is flung out and distributed. The application rate of the fertilizer is regulated by changing the speed ratio between the slat conveyor and ground wheel. A continuously variable transmission makes this adjustment possible to achieve the desired application rate. This machine is appropriate also for relatively dry powdered manure and digested compost since a centrifugal spreader can work efficiently on such media.

8.2.1 Determination of velocity of throw of a centrifugal distributor

The theory of centrifugal particle distribution has been studied by many earlier works (Glover and Baird, 1973; Martínez-Rodríguez et al., 2021; Mennel and Reece, 1963; Ritter et al., 1980; Cunningham, 1963; Davis and Rice, 1973). For simplicity, let us consider a centrifugal distributor that has a flat disc rotating at an angular velocity of ω. Consider a seed or fertilizer particle of mass m deposited on the rotating flat disc at the centre of the disc, the spot where the radial vane starts on the disc (Figure 8.3).

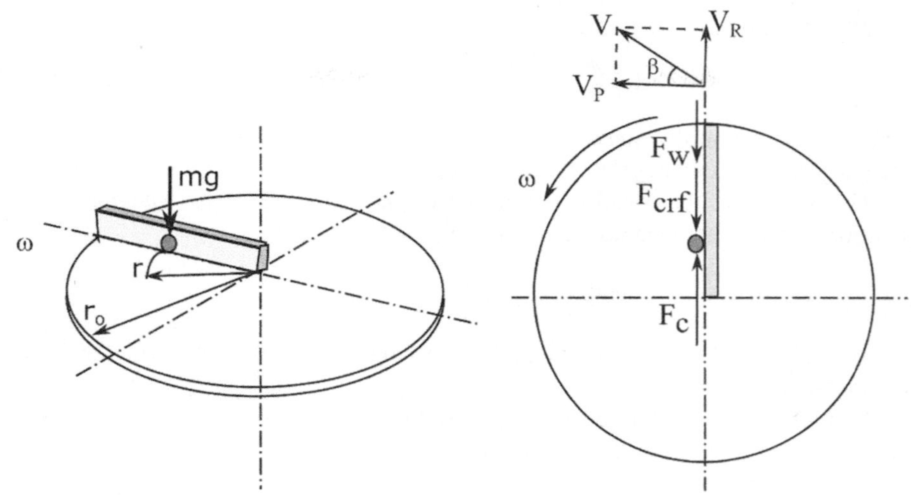

FIG. 8.3 Particle Velocity Determination

No Permission Required

The particle is then subjected to the following forces, namely:

1. The main centrifugal force, F_c, (Figure 8.3) acting radially out pushes the particle outward, along the edge of the radial vane and the disc, $= m\omega^2 r$, where r is the radius at which the particle is on the disc at a time instant, t.
2. A Coriolis force, F_{cr}, (see Figure 8.3) on the particle acting perpendicularly to the centrifugal force, pushing the particle against the radial vane, $= 2m\omega\frac{dr}{dt}$, where r is the actual displacement variable of the particle moving from the centre of the disc to the outer-most radius, r_o, (peripheral edge) of the disc, and $\frac{dr}{dt}$ is the relative velocity of the particle at the position of the seed path along the radius, r. This force is obviously opposite to the direction of rotation and presses the particle against the vane, causing a frictional force radially acting toward the centre, $F_{crf} = 2m\mu\omega\frac{dr}{dt}$, where μ is the frictional coefficient between the particle and the vane.
3. A frictional force is between the particle and the flat disc below it due to the weight of the particle acting downward. This frictional force, $F_w = \mu(mg)$, (see Figure 8.3) acts opposing the particle's outward movement radially toward the centre of disc.

The equation of motion of the particle with the inertial axis rotating with the disc, is then:

Inertial force of the particle $= F_c - F_{crf} - F_w$, which is

$$m\frac{d^2r}{dt^2} = m\omega^2 r - 2m\mu\omega\frac{dr}{dt} - \mu mg.$$

Simplifying,

$$\frac{d^2r}{dt^2}+2\mu\omega\frac{dr}{dt}-\omega^2 r=-\mu g,$$

which is a second-order non-homogeneous differential equation of the variable r of the particle's movement with respect to time, t.

The solution of the equation will have two parts as $r = r_h + r_p$, where r_h is the solution of the homogeneous equation,

$$\frac{d^2r}{dt^2}+2\mu\omega\frac{dr}{dt}-\omega^2 r=0,$$

and r_p is the particular solution part for the constant in the equation on the RHS, which is $r_p = -\mu g$.

Going by the procedure of solving the ordinary differential equation of the second order, the solution of the homogeneous equation will be:

$$r_h=C_1e^{\lambda_1 t}+C_2e^{\lambda_2 t}.$$

The solution of the characteristic equation above of the homogeneous part of the equation will be:

$$\lambda^2+2\mu\omega\lambda-\omega^2=0.$$

The roots of this equation are $\lambda_1=\omega\left(-\mu-\sqrt{1+\mu^2}\right)$ and, $\lambda_2=\omega\left(-\mu+\sqrt{1+\mu^2}\right)$.

As for the particular solution of $r_p = -\mu g$, we can differentiate it, that results in $\frac{dr_p}{dt}=0$, since $-\mu g$ is a constant.

Substituting the derivatives in the motion equation, we get:

$$\frac{d^2r_p}{dt^2}+2\mu\omega\frac{dr_p}{dt}+\omega^2 r_p=-\mu g.$$

$$0+2\mu\omega(0)+\omega^2 r_p=-\mu g.$$

$$r_p=-\frac{\mu g}{\omega^2}.$$

Combining the solutions, we have:

$$r=r_h+r_p=C_1e^{\lambda_1 t}+C_2e^{\lambda_2 t}-\frac{\mu g}{\omega^2},$$

where $\lambda_1=\omega\left(-\mu-\sqrt{1+\mu^2}\right)$ and,$\lambda_2=\omega\left(-\mu+\sqrt{1+\mu^2}\right)$

Applying the initial conditions at the instant t=0, $r=0$, and the relative velocity $\frac{dr}{dt}=0$,

we get $C_1 + C_2 = \frac{\mu g}{\omega^2}$
and $\frac{dr}{dt} = C_1\lambda_1 + C_2\lambda_2 = 0.$
Solving for C_1 *and* C_2,

$$C_1 = \left(\frac{\mu g}{\omega^2}\right)\left(\frac{\lambda_2}{\lambda_2 - \lambda_1}\right) \text{and} C_2 = -\left(\frac{\mu g}{\omega^2}\right)\left(\frac{\lambda_1}{\lambda_2 - \lambda_1}\right).$$

We get the final solution as:

$$r = \left(\frac{\mu g}{\omega^2}\right)\left(\frac{1}{\lambda_2 - \lambda_1}\right)\left(\lambda_2 e^{\lambda_1 t} - \lambda_1 e^{\lambda_2 t} - 1\right)$$

where $\lambda_1 = \omega\left(-\mu - \sqrt{1+\mu^2}\right)$ and,$\lambda_2 = \omega\left(-\mu + \sqrt{1+\mu^2}\right)$.
Then, the relative velocity of the particle is:

$$v_R = \frac{dr}{dt} = \left(\frac{\mu g}{\omega^2}\right)\left(\frac{\lambda_1\lambda_2}{\lambda_2 - \lambda_1}\right)\left(e^{\lambda_1 t} - e^{\lambda_2 t}\right).$$

If we know the disc radius, i.e., the outer radius, r_o, the exit point of the particle, one can evaluate the time at which the particle exits the disc, which is known as the resident time of particle on the disc.

Substituting the time at which it exits the disc in the equation for v_R, the relative velocity can be calculated, which is along the vane.

The velocity, v, of the particle leaving the disc is the resultant of v_R, the relative velocity, and v_P, the peripheral velocity of the disc:

$$v = \sqrt{v_P^2 + v_R^2} \text{ where } v_P = \omega r_o.$$

The example problem given in Box 8.1 may be studied to grasp how this is useful in the design of a centrifugal distributor.

8.3 Manure application equipment

The machine applying the manure is meant to place it in a location where crops have easier access to draw in the nutrients available in the manure. Generally, a surface application causes an undue loss of nutrients from the manure due to their interaction with the atmosphere, largely affected by weather conditions. Nitrogen content in the manure in the form of ammonia is mainly susceptible to such losses. Nutrients such as potassium and phosphorus remain on the soil surface, remain largely unavailable to the crop, and may be lost by leeching due to runoff. Manure placed below the soil surface, that is near the crop rows, is utilized efficiently by the crop. The uniformity of a surface application, though, can be improved adequately by using the right applicators; but surface application is always inferior to sub-surface application. The latter method of application makes the nutrients available uniformly to the crop.

Box 8.1 Example 8.1

Example 8.1. A centrifugal fertilizer broadcaster with a flat disc has its vane's inner radius starting at the centre and the outer radius at 20 cm. Find the magnitude and direction of throw of the disc when it is rotating at 500 rpm. Assume that the frictional coefficient between the broadcast fertilizer and the disc's material is 0.25.

Angular velocity, $\omega, = \frac{2\pi N}{60} = \frac{2\pi(500)}{60} = 52.3$ rad s^{-1};
Frictional coefficient, $\mu, = 0.25$;
r_o = outer radius of the disc = 0.20 m.

We know that $r = \left(\frac{\mu g}{\omega^2}\right)\left(\frac{1}{\lambda_2 - \lambda_1}\right)\left(\lambda_2 e^{\lambda_1 t} - \lambda_1 e^{\lambda_2 t} - 1\right)$

where $\lambda_1 = \omega\left(-\mu - \sqrt{1+\mu^2}\right)$ and $\lambda_2 = \omega\left(-\mu + \sqrt{1+\mu^2}\right)$ and t is the resident time.

Substituting, $\lambda_1 = 52.3\left(-0.25 - \sqrt{1+0.25^2}\right) = -66.94$,

$\lambda_2 = 52.3\left(-0.25 + \sqrt{1+0.25^2}\right) = 40.83$.
Substituting in the equation for r with a known value of $r_o = 0.20m$,

$$0.2 = \left(\frac{0.25 \times 9.81}{52.3^2}\right)\left(\frac{1}{40.83 + 66.94}\right)\left(40.83e^{-66.94t} + 66.94e^{42.83t} - 1\right),$$

$0.2 = 8.04 \times 10^{-6}\left(40.83e^{-66.94t} + 66.94e^{40.83t} - 1\right)$.
Solving by trial and error for t (*easier to do on an Excel sheet*),
Resident time, t, = 0.1385 s;

Relative velocity, $v_R, = \left(\frac{\mu g}{\omega^2}\right)\left(\frac{\lambda_1\lambda_2}{\lambda_2 - \lambda_1}\right)\left(e^{\lambda_1 t} - e^{\lambda_2 t}\right)$

$= (0.000867)\left(\frac{40.83 \times 66.94}{107.71}\right)\left(e^{-66.94\times 0.1385} - e^{40.83\times 0.1385}\right)$
$= 6.27$ m s^{-1};
Peripheral velocity, $v_P, = \omega r_o = 52.3 \times 0.2 = 10.46$ m s^{-1}.

The velocity of the particle leaving the disc, $v, = \sqrt{v_P{}^2 + v_R{}^2} = \sqrt{10.46^2 + 6.27^2} = 12.2$ m s^{-1}.

The angle at which the particle is thrown, $\beta, = tan^{-1}\left(\frac{v_R}{v_P}\right) = 31°$ to the tangent of the disc.

In any type of manure application system, whether solid or liquid, the application rate is relatively high, necessitating that the applicators carry huge volumes of the manure. The huge weight of manure resting on the wheels causes considerable soil compaction that can impede the root development of a crop, resulting in reduced crop yields. It is better not to use manure applicators when the soil is wet. A drag hose liquid applicator is a system in which a long hose supplies the slurry from a central tank to the moving applicator, thereby avoiding soil compaction by the laden vehicles. But this method is not very popular because of the inherent difficulties in the system. Loading the bulk of manure into the applicators and dispensing is a time-consuming affair and takes a long time to complete.

8.3.1 Manure spreaders

The application of farmyard manure to enhance soil health is integral to farming. Farmyard manure happens to be the richest source of organic matter and is recommended at different stages of crop growth. Apart from improving the soil health in terms of nutrients, it restructures the soil to be conducive for crop growth. The first application of manure during field preparation, known as top dressing, is a labour-intensive operation, wherein the volume of manure handled is huge per unit area. The equipment that handles this much volume of manure and distributes it to the field is available as manure spreaders.

A manure spreader (Figure 8.4) is suitable for relatively wet manure, too, since it has beater drums to shred even wet manure. The machine has a slat conveyor at the bottom of the trailed bin. The manure loaded in the bin is moved toward the rear of the spreader; it is first shredded by the beater drums and then passed on to an auger spreader at the extreme back. The spreader has left and right turn flights rotating at relatively high rotary speed. This spreads the shredded manure out on the field quite uniformly over the width of the machine. The shredder drums consist of knifes, which break the lumps of manure and make it amenable for spreading. Unlike a centrifugal fertilizer spreader, the uniformity of distribution is far superior here. Moreover, the quantity of manure output per distance run is higher, though

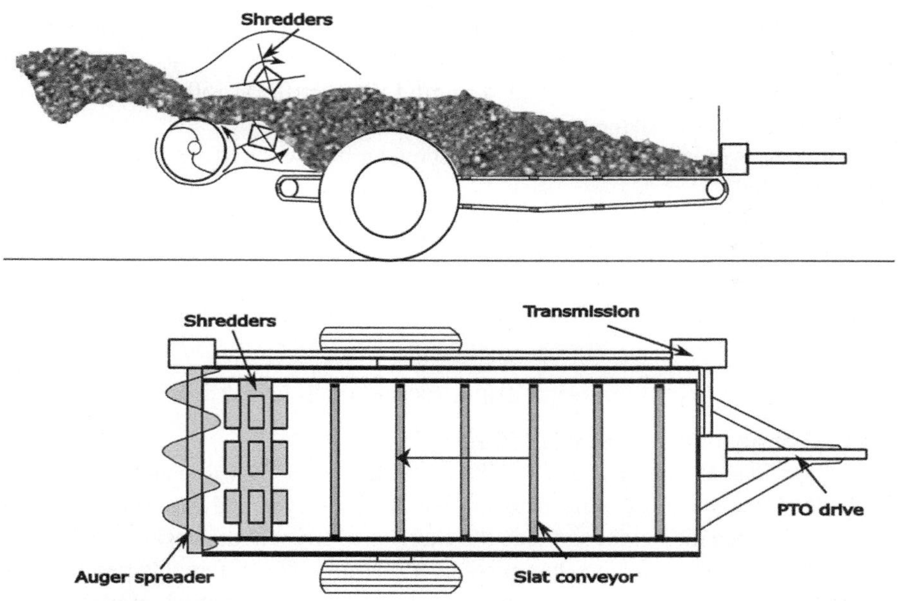

FIG. 8.4 Manure Spreader

No Permission Required

the swath of manure application is smaller than a centrifugal spreader. These machines are available with bin capacities of 1.5 tonnes to 8 tonnes of manure. The wheels are set a bit toward the back end so that the loaded front is supported more by the tractor. The rubber wheels are broader to provide better floatation in the harrowed field. Manure spreaders are usually loaded by front-mounted, tractor-operated shovel loaders. After the application of manure, the field is harrowed again to mixing it with the soil.

8.3.1.1 Calibrating the manure spreader

Similar to a seed drill, a spreader can be calibrated by collecting the manure output from the machine for a predefined distance travelled. A large, plastic sheet spread on the ground can be used to collect this output; the operator can then adjust the output to the desired application rate. But, for manure being applied in bulk, the above method is not usually adopted except under laboratory conditions. A popular method of calibration is to fill the manure spreader with a known depth of the bin and then operating it in the field until the bin is completely emptied. The distance covered for emptying this volume is measured by counting the number of revolutions the spreader's wheel is making. The distance multiplied by the width of application provides the area on which the manure was spread, from which the application rate can be calculated by dividing the volume applied by the calculated area. As an example, if the spreader's manure bin is of a size 4 m × 1.5 m × 0.7 m and it is filled to a full level of 0.7 m without heaping above it, the volume of manure = 4 × 1.5 × 0.7 = 4.2 m^3.

Let us say that the manure spreader's wheel diameter is 0.75 m and the revolutions it made to empty the spreader is 150.

The distance the machine travelled = $\pi d(150) = \pi(0.75)\ 150 = 353.3$ m
where d = wheel diameter.

The area over which the manure was spread = the width of the spreader × distance travelled:

$$= 1.5\ \text{m} \times 353.3 = 530\ \text{m}^2.$$

The volume applied = 4.2 m^3,
∴, the volume applied per m^2 = volume applied/area over which the manure was spread:

$$= 4.2/530 = 0.008\ \text{m}^3\ \text{m}^{-2}.$$

The density of dry manure is 750 kg m^{-3},
the application rate = 750 × 0.008 = 5.94 kg m^{-2} = 59.4 tonnes ha^{-1}.

If need be, the speed ratio between the slat conveyor and the travel speed is altered to change the output and redo the exercise to arrive at the desired application rate. Generally, the machine has predefined settings for different types of solid manure, which will approximately provide the required application rate.

8.3.2 Liquid manure applicators

Liquid manure can be applied on the soil surface or can be directly injected into the soil using various methods and tools. But liquid manure needs to be stored under anaerobic conditions since its decomposition produces unbearable odour. The methods of directly injecting the liquid into the soil has this problem solved to a certain extent. Moreover, direct injection reduces nutrient loss from the manure, enhancing its utilization efficiency in the soil.

Applicators with liquid tanks (Figure 8.5) are the most common machines that apply liquid manure. The injection system uses soil openers that are similar to the furrow openers of grain drills, such as discs, shovels, and knives, but are set to open very narrow cleavages in the soil. They are mounted on a common toolbar with a manifold supplying them the liquid manure from the tank. A pump forces the liquid slurry down to a series of hoses, each connected to one opener on the toolbar through the distributing manifold. These hoses are connected to the injecting spouts mounted on the back of each opener. The injectors apply the manure beneath the crop canopy, closer to and beneath the soil surface. This kind of application is thereby reducing ammonia loss and odour and improving application uniformity. Similar to a sprayer tank, mechanical or hydraulic agitation is provided inside the odour-sealed tank to prevent the liquid manure from settling in layers of different densities. These agitator pumps are often used for loading the liquid slurry into the tank before application.

The surface application of liquid manure is also performed by some machines that, instead of injecting the slurry into the soil, discharge it at a high velocity onto a splash plate that deflects and spreads the liquid onto the soil surface. This application system, although not as nutrient efficient and odourless compared with direct injection systems, offers higher coverage because of the higher travel speed; and the operation can be carried out in less time with these machines. But controlling the uniformity in these machines is difficult. The example problem worked out in Box 8.2 throws more light onto the concept of calculating the application rate of fertilizer and manure applicators using these machines.

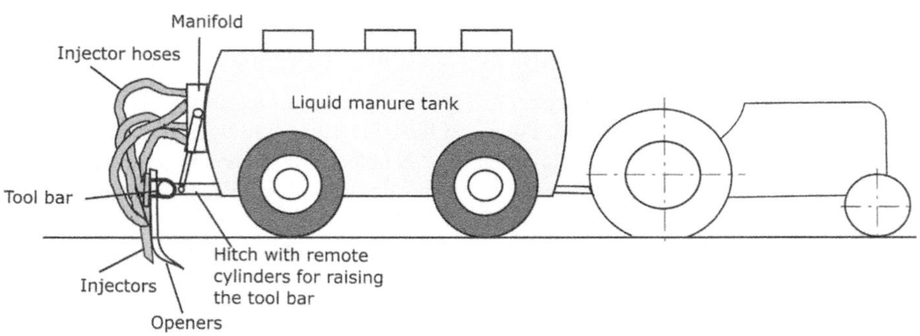

FIG. 8.5 Injector-Type Liquid Manure Applicator

No Permission Required

Box 8.2 Example 8.2

Example 8.2. The recommended application rate of manure for a particular crop is 12 tonnes per hectare. A 3-tonne manure spreader has an output width of 2 m. Calculate the rate of application of manure per m run of the machine. Five kg of manure is fed to the spreader per revolution of the slat conveyor's drive sprocket. The diameter of the ground wheel is 60 cm. Find the ratio between the ground wheel and slat conveyor's drive sprocket for the required application rate.

Recommended application rate = 12 t ha^{-1} = 12000/10000 = 1.2 kg m^{-2};

∴ Application rate per m run = 1.2 × output width = 1.2 × 2 = 2.4 kg m^{-1}.

The output of the slat conveyor to spreader per revolution of the drive sprocket = 5 kg;

Ground wheel diameter = 0.6 m;

Ground distance covered by the ground wheel per revolution = π d = 1.88 m;

Ground distance covered by the ground wheel per revolution of sprocket = 1.88x m;

Output of the machine per m run $= 5\ \text{kg} / (1,88\ \text{x})\ \text{m} = 2.4\ \text{kg}\ \text{m}^{-1}$;

∴, x = 5/ (1.88 × 2.4) = 1.1.

The speed ratio between the ground wheel and slat conveyor sprocket = 1.1:1.

Exercise problems

1. On calibrating a 1.5 m-wide manure spreader with a wheel diameter of 70 cm, it was found that the number of revolutions made by the wheel was 100 to empty a quarter of the manure bin. Find the application rate offered by the machine that is set at a speed ratio of 1:1 between the ground speed and slat conveyor speed. If the recommended application rate is 10 tonnes per ha, what would be the speed ratio of the slat conveyor?
2. An available stock of liquid manure has 6.3 kg of nitrogen per 1000 litres. A standing crop, having a row spacing of 50 cm, requires 200 kg of nitrogen per hectare. What would be the application rate to be set on the injection-type liquid manure applicator with six rows? If the applicator is traveling at a speed of 2 km h^{-1}, find the discharge of the slurry pump that is discharging the manure to the furrow openers.

References

Bernachi, H., Haman, J., & Kanafojski, C. (1972). *Agricultural machines, theory and construction* (p. 1). Warsaw, Poland: Published for USDA and NSF by Foreign Cooperation Center of the Central Institute for Scientific, Technical and Economic Information.

Cunningham, F. M. (1963). Performance characteristics of bulk spreaders for granular fertilizer. *Transactions of the ASAE, 6*(2), 108–114.

Davis, J. B., & Rice, C. E. (1973). Distribution of granular fertilizer and wheat by centrifugal distributors. *Transactions of the ASAE, 16*(5).

Glover, J. W., & Baird, J. V. (1973). Performance of spinner type fertilizer spreaders. *Transactions of the American Society of Agricultural Engineers, 16*(1), 48–51.

Hunt, D. (2008). *Farm power and machinery management*. Waveland press.

Kepner, R. A., Bainer, R., & Barger, E. L. (1978). *Original from the university of Michigan.* Avi Publishing Company.

Klenin, N. I., Popov, I. F., & Sakun, V. A. (1970). *Agricultural machines.* USDA and NSF by Amerind Publishing Co. Pvt. Ltd.

Lovegrove, H. T. (1968). *Crop production equipment.* Hutchinson press.

Martínez-Rodríguez, A., Gómez-ÁguilalI, M. V., & Escobar, M. S. (2021). Model and software for the parameters calculation in centrifugal disk of fertilizer spreaders. *Revista Ciencias Técnicas Gropecuarias, 30*, 82–101.

Mennel, R. M., & Reece, A. R. (1963). The theory of the centrifugal distributor, part II - particle trajectories. *Journal of Agricultural Engineering Research, 43*(1), 78.

Ritter, D. W., Griffis, C. L., & Mathews, E. J. (1980). *Computer simulation of rotary spreader distribution patterns.* ASAE Paper No. SO-1504.

Smith, H. P. (1965). *Farm machinery and equipment.* Tata Mc-Graw Hill Pub. Co. Ltd.

Srivastava, A. K., Goering, C. E., Rohrbach, R. P., & Buckmster, D. R. (2006). *Engineering principles of agricultural machines.* ASABE Publications.

CHAPTER

Plant protection machinery

9

Chapter Outline

Farmers are continuing to apply pesticides and herbicides because chemical control happens to be the most cost-effective way of countering yield loss due to insects and diseases and because of the effects of weed competition on crop production. The global demand for pesticides and weedicides has continued to grow particularly in commercial crops. This is despite the concerns of pesticide residue in food and the environment and the outcry for chemical-free agriculture that can save natural resources. Avoiding the use of agricultural chemicals, including inorganic fertilizers and pesticides, and growing crops organically has been proven time and again to be not sustainable when high yields are expected. The awareness of integrated pest management (IPM) as a method, using different control measures in a balanced

Machines of the Farm. https://doi.org/10.1016/B978-0-443-32906-7.00009-8

manner, to avoid the harmful effect of chemicals on the environment has encouraged the continued use of pesticides. Similarly, the use of herbicides has increased in leaps and bounds because of acute labour shortages and high costs in developing countries. It means that the use of pesticides and herbicides is here to stay and machines to apply these chemicals in the right quantity are very much in demand.

The objectives of spraying and dusting are to apply:

- insecticides to control insects on plants;
- fungicides to control plant diseases;
- herbicides to kill or prevent the emergence of weeds while preparing the field itself or to selectively kill weeds in a cropped field without harming the crop;
- chemicals to defoliate or condition crops such as cotton for mechanical harvesting;
- growth-regulating hormone sprays to increase the fruit set or prevent the early dropping of fruits.

9.1 Spraying equipment

The function of a sprayer is to break the liquid into droplets of a desired size and deposit them uniformly over the target plants to be protected, decimated, or managed. The spraying machine (Bell, 2005; Kepner, Bainer, & Barger, 1978; Lovegrove, 1968; Smith, 1965) should also be capable of regulating the amount of chemical to be applied in the desired quantity to cause the required insect/weed mortality avoiding harm to the environment by excessive application.

9.1.1 Types of sprayers

Depending on the physical principle based on which spraying is achieved, they may be classified as:

- Hydraulic sprayers – they pressurize the spray fluid directly using a positive displacement pump and use the pressure to atomize the fluid into small particles using hydraulic nozzles.
- Hydro-pneumatic sprayers or air-assisted sprayers – they use pneumatic pressure created by pumping air over the surface of the spray liquid inside a sealed container in order to generate the fluid pressure required for atomization.
- Blower sprayers – they use high-speed air streams generated by blowers and fans in which the spray fluid is metered in a desired quantity to be atomized into fine particles.
- Centrifugal sprayers – they use discs rotating at a high speed on which the spray fluid is applied and atomized by the centrifugal force imparted to the fluid.

Based on the spray volume or application rate of the atomized spray generated, sprayers are also classified as in Table 9.1.

Table 9.1 Classification of sprayers based on application rate

Type of sprayer	Field crops, l ha^{-1}	Trees, l ha^{-1}
High volume	>600	> 1000
Medium volume	200 to 600	500 to 1000
Low volume	50 to 200	200 to 500
Very low volume	5 to 50	50 to 200
Ultralow volume	<5	<5

9.1.1.1 Hydraulic field sprayers

Most of the field sprayers work as hydraulic types (Figure 9.1) wherein the fluid pressure is created by a positive displacement pump. The pressurized spray fluid is let out through the hydraulic nozzles, which atomize the liquid into finer-sized droplets. Depending on the type of hydraulic nozzle used, the spray of the atomized

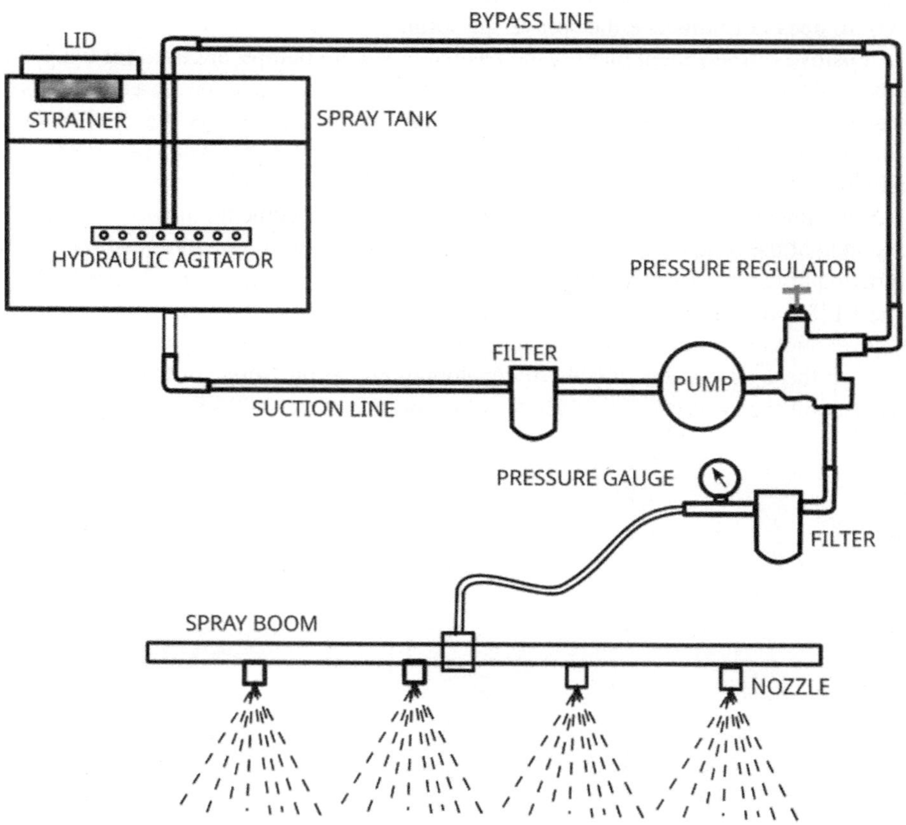

FIG. 9.1 Diagram Showing the Basic Components of a Field Sprayer

particles emanates from the nozzles in a spray pattern or shape of the dispersal desired. The components of a typical hydraulic power sprayer are the pump; air chamber, if it is a reciprocating pump; tank containing an agitator; hydraulic control elements such as the pressure regulator, relief valves, pressure gauge, filters, distribution system with an array of control valves; nozzles; pipes; and fittings.

Pump: Generally, hydraulic sprayers use positive displacement pumps, such as reciprocating pumps, diaphragm, or rotary systems such as vane pumps. Here, the discharge rate is proportional to the speed of the actuator and is not influenced by the opposing fluid pressure.

Figure 9.2 presents a typical performance chart of a reciprocating pump. It shows that the discharge is increasing linearly with the shaft speed since the shaft drives the reciprocating plunger that is displacing a constant volume of fluid with each revolution. It also shows that the discharge is falling slightly with an increase in pressure at the discharge port. This fall is greater at higher pressures due to the leakages in the plunger-cylinder interface. But this effect is very moderate, such that reciprocating pumps can be used even as high-pressure sprayers. Since the sprayer pump is generally driven at a constant speed, the pressure at which the fluid is delivered into the system does not influence the sprayer performance.

Positive displacement pumps, such as reciprocating pumps, are used for the purpose of developing the required pressure in the hydraulic sprayer under a closed system with a controlled opening. The discharge of these pumps is matched to the number of maximum nozzles that is to be used in the sprayers. A pressure-relieving bypass valve is part of the circuit to maintain the fluid circuit at a constant pressure under operation and to protect the positive displacement pumps from an undue load at the time of shutting down the spray. These pumps are mounted onto the spray equipment carried by the three-point linkage of the tractor and driven by the rear PTO shaft. An air chamber is provided on the discharge of a reciprocating pump to level out the cyclic discharge of the pump and provide a constant nozzle pressure. Though these pumps are capable of developing higher pressures, of 5000 kPa, they are used in the sprayers at a lower pressure, of 150 kPa to 400 kPa.

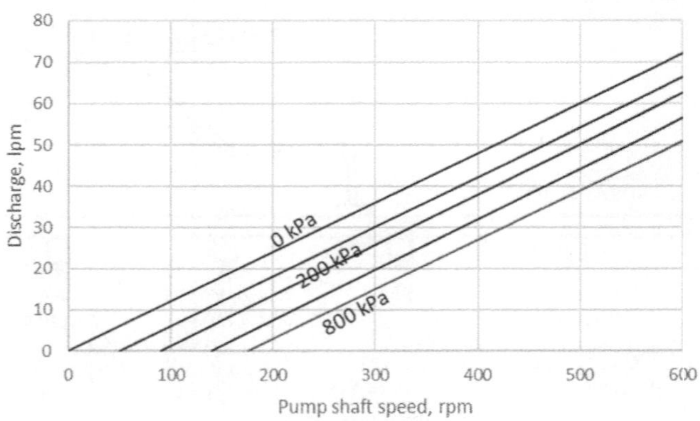

FIG. 9.2 Performance Characteristics of a Reciprocating Pump

Centrifugal pumps push the liquid by centrifugal force and are not positive displacement pumps. They are usually not used in hydraulic sprayers that are required to work in relatively higher pressures. Moreover, they are not self-priming, meaning that they always need a column of fluid in their casing to start pumping. Fluid power (watts) in hydraulic systems is given by the expression Q × P, where Q is the discharge rate in m^3s^{-1} and P is the pressure in Pa. This can be applied to the performance of a pump, and the mechanical power requirement can be obtained by dividing the fluid power by the efficiency of the pump, which is about 50% to 60%. The example problem in Box 9.1 calculates the discharge to be bypassed to the spray tank, with the aim of maintaining the desired discharge pressure.

Tank: Corrosion-resistant plastic or fibre-based materials are used to make the tanks. Tank capacities of 1000 *l* and above are used. They have openings that allow the easy cleaning of the spray residue after spraying. They also have a top opening with a strainer for filling water. The spray chemical poured into the tank of water is agitated continuously to prevent the chemicals from settling at the bottom. The agitation is essential to keep the spray concentration uniform, right from the mixing to the end of the operation. Hydraulic agitation is the process of letting the

Box 9.1 Example 9.1

Example 9.1. The performance characteristics of a reciprocating pump used in a boom sprayer are given below. The pump is set to run at 300 rpm on a machine serving a boom with 20 hollow cone nozzles, each with a discharge of 0.4 l min^{-1}, when operating at a pressure of 300 kPa. Neglecting the pressure losses on the hoses and valves, calculate how much discharge from the pump must be bypassed and returned to the spray tank to set the discharge pressure at the desired level. Also, find the size of the bypass valve opening to the spray tank considering the outlet as a simple orifice.

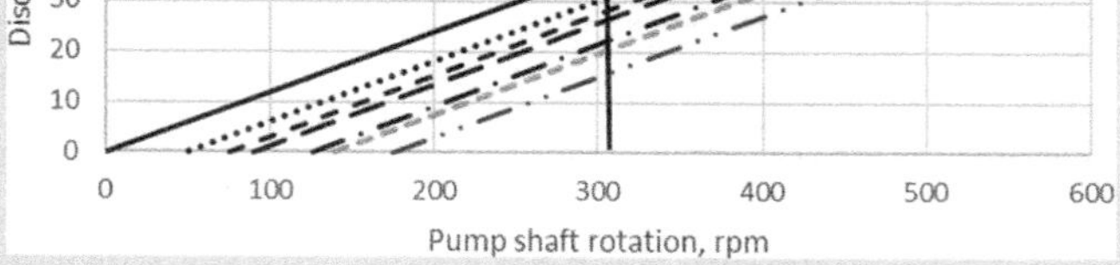

From the performance characteristics, let us first get the pressure vs. discharge relationship at the shaft speed at which the pump is operated on the sprayer. Dropping a vertical line at the pump speed of 300 rpm, the pressure vs. discharge curve drawn is as follows.

Continued

Box 9.1 Example 9.1—cont'd

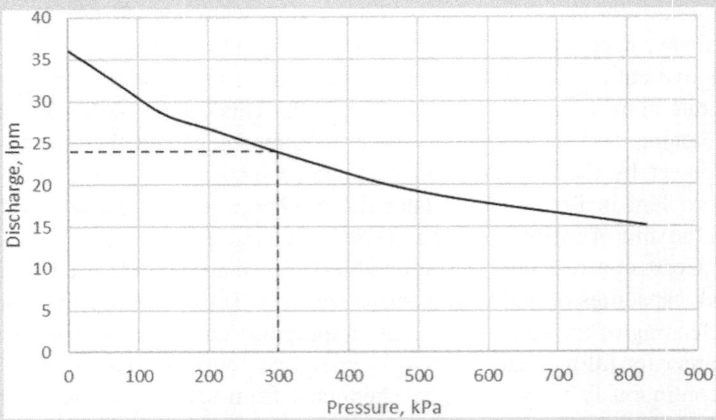

No. of nozzles on the boom = 20

Discharge of each nozzle operating at 300 kPa = 0.4 lpm

Discharge from the boom = 0.4 × 20 = 8 lpm

Discharge of the pump at 300 kPa as determined from the discharge vs. pressure plot = 24 lpm

∴, Discharge to be bypassed to the spray tank = 24 − 8 = 16 lpm = (16/60000) $m^3\ s^{-1}$

Considering the bypass valve opening to the tank as an orifice, the discharge through the bypass outlet

$$Q = C_d A\sqrt{\frac{2P}{\rho}}$$

where Q is the flow rate in $m^3 s^{-1}$,

P is the pressure in Pa,

A is the area of the orifice opening = $\pi d^2/4$ in m^2,

ρ is the density of fluid (water) = 1000 kg m^{-3}, and

C_d is the coefficient of discharge = 0.6 for an orifice (or for valve = 0.5).

Then, $A = \frac{Q}{C_d\sqrt{\frac{2P}{\rho}}} = A = \frac{\frac{16}{60000}}{0.6\sqrt{\frac{2(300000)}{1000}}} = 1.8 \times 10^{-5}\ m^2 = 0.18\ cm^2$.

∴, Diameter of valve opening = $\sqrt{\frac{0.18 \times 4}{\pi}} = 0.48$ cm.

pressurized fluid returning from the spray circuit into the tank using a perforated tube inside the tank. This returning liquid creates a turbulence in the tank, thus giving the agitating action. The pumps used are selected to have higher discharges to contend with the requirement of hydraulic agitation.

The hydraulic control elements in the sprayer circuit are:

- the pressure gauge to adjust the fluid pressure in the hydraulic circuit to the required level;

- the pressure regulator, which has a built-in pressure-relief valve (Figure 9.3).

When the fluid pressure in the spray circuit exceeds the set pressure, it discharges the excess back to the tank. The pressure can be set by adjusting the hand wheel and thus the pre-tension of the spring, which is pressing down the valve against its seat. The inlet pressure connected to the discharge of the pump, when exceeding the set pressure, lifts the valve and fluid escapes out to the outlet and into the tank. It also acts as a safety device when the nozzles are closed completely. The entire discharge from the pump is returned back to the tank, thus unloading the pump. In addition to these, there are:

- the control valves fitted onto a manifold that controls the flow to the different sections of the boom, carrying multiple numbers of spray nozzles;
- the online filters on the suction line between the tank and the pump to remove dust and debris that may clog the nozzles. These must be cleaned periodically.

The **spray boom** is a horizontal member mounted on the sprayer in both lateral directions of the sprayer operation. The hydraulic nozzles are mounted onto this boom, with spacings matching the desired row spacing of the crop. The boom, along with the nozzles, can be raised and lowered to suit the crop height. The boom is made structurally sound so as not to swing or oscillate while in operation. It is divided into three or more sections that are foldable and suitable for on-road travel of a tractor with a sprayer.

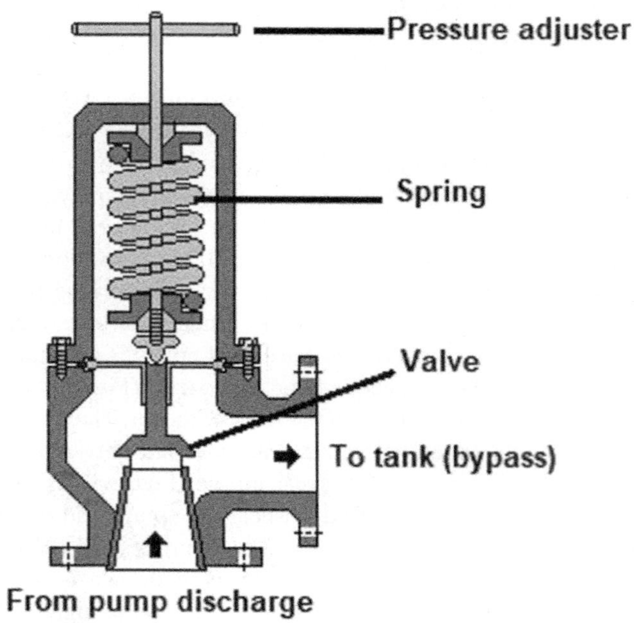

FIG. 9.3 Pressure-Relief Valve

The nozzles are mounted directly onto the boom or to the lower ends of the vertical drop pipes, which position the nozzles between the crop row for spraying the sides of the crop canopy of tall row crops. Field boom sprayers (Figure 9.4) are available either as tractor rear-mounted units or trailed units with a separate set of wheels. The latter possesses large-capacity spray tanks and are used in large farms.

FIG. 9.4 Tractor Trailed and Mounted Boom Sprayers (iXter B and IXTrackT3, Kverneland Group, Norway

9.1.1.2 Hydraulic spray nozzles

Hydraulic spray nozzles may be classified as hollow cone, solid cone, fan, impact, and solid stream. In hollow cone nozzles (Figure 9.5B), the spray liquid under pressure is sent into a swirl plate or a spiral passage to give it a rotating motion. The rotating fluid, when exiting the orifice placed very near the swirl plate, emerges as a hollow, conical sheet of fluid and then atomizes into droplets. These are general-purpose nozzles used in field sprayers for row crop spraying.

The solid cone nozzle is basically the same as the hollow cone nozzle except that an additional axial exit is provided for the fluid to fill the hollow portion of the spray. In some models (Figure 9.5A), a plain, outwardly tapering orifice creates the solid cone. A fan-type nozzle (Figure 9.5C) has a thin, rectangular hole or channel across the outside face of the orifice plate, and the liquid emerges as a fan-shaped fluid sheet before atomization. The impact or flood jet nozzles (Figure 9.6A) are simple jet nozzles and have a deflection plate on the front of the jet. The fluid jet, on hitting the plate, forms a flat stream and atomizes. Fan-type and impact nozzles are used mainly for weedicide spraying. The air shear nozzle (Figure 9.6B) is one that has a high-velocity stream of air and is used to shear the spray fluid dripping into the air stream and atomize it into fine particles. These nozzles are used in motorized knapsack sprayers and orchard sprayers. Each type of nozzle has a specified discharge at the recommended discharge pressure. The use of higher or lower pressure will cause an undesirable change in the spray droplet sizes generated. The recommended discharge pressure for a typical cone nozzle is from 300 kPa to 500 kPa,

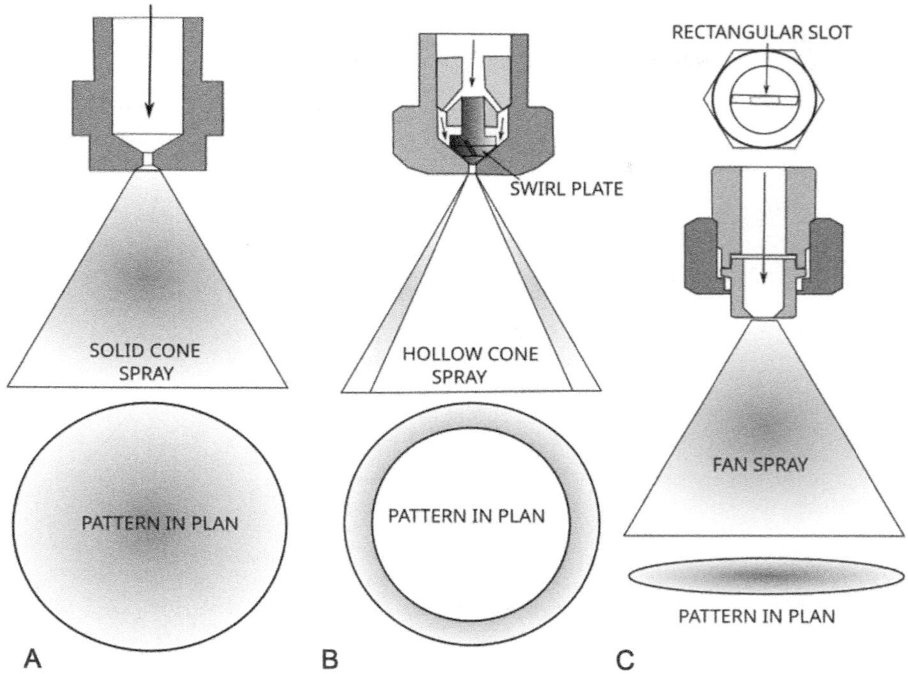

FIG. 9.5 Types of Hydraulic Spray Nozzles

giving a discharge of 0.3 min to 0.7 l min^{-1}. Fan-type nozzles operate at a pressure of 100 kPa to 500 kPa. The nozzles used on field sprayers have spray angles of 60° to 90°, which is the apex angle of the spray cone generated.

Brass was the most common material for nozzles with hardened stainless steel orifice disks to reduce orifice wear. Nowadays, plastic and resin-based nozzles with an appreciable life are available for use.

9.1.1.2.1 Hydraulic spray nozzle as a fluid jet

Theoretically, the fluid flow of a hydraulic nozzle can be considered as that of a fluid jet emanating from an orifice due to fluid pressure and is expressed, as usual, by:

$$Q = C_d A v = C_d A \sqrt{2gH} = C_d A \sqrt{\frac{2P}{\rho}}$$

where Q is the flow rate in m^3s^{-1};
v is the discharge velocity in ms^{-1};
P is the pressure in Pa;
H is the pressure head in m;
g is the acceleration due to gravity, ms^{-2};
A is the area of the orifice opening, m^2;

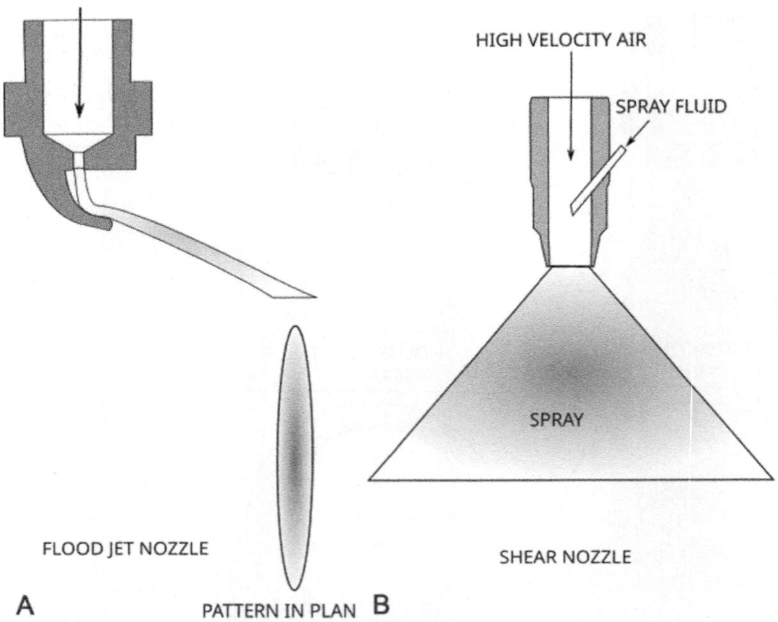

FIG. 9.6 Flood Jet and Shear Nozzles

ρ is the density of fluid (water) = 1000 kg m^{-3}; and
C_d is the coefficient of discharge of the flow process.

However, the coefficient of discharge that is applicable to the spray nozzle cannot be that of an open jet nozzle and has to be ascertained from the relationship between the fluid pressure and the flow rate that yields at that pressure. We can see that $Q = fn(\sqrt{P})$ in the above flow equation with the other parameters being constant. If the relationship between P, the fluid pressure, and Q, the flow rate, is plotted using the manufacturer's data of the spray nozzle, the slope after applying the other constants yields the coefficient of discharge.

It was established by experimentation that the atomization of a jet issuing from an orifice happens only when the flow is as per the empirical expression below (Dombrowski & Johns, 1963; Dorman, 1952; Ford & Furmidge, 1967; Marshall, 1954):

$Re > 280\left(\frac{\mu}{(\sigma\rho d)^{0.5}}\right)^{-0.82}$ where $Re = \frac{\rho v d}{\mu}$ is the Reynold's number of fluid flow

and $\left(\frac{\mu}{(\sigma\rho d)^{0.5}}\right)$, a dimensionless number (Ohnesorge number) relating viscous forces to inertial and surface tension forces where,

ρ is mass density of the fluid, kg m^{-3};
μ is the fluid's viscosity, Pa s;

σ is the fluid's surface tension, N m^{-1};
v is the fluid velocity emanating from the jet, m s^{-1}; and
d is the orifice diameter, m.
From which, the minimum fluid velocity of the jet causing atomization is:

$$v > 280\left(\frac{\sigma^{0.42}\mu^{0.18}}{\rho^{0.59}d^{0.59}}\right)$$

The following examples in Box 9.2 and Box 9.3 use this expression to calculate the minimum discharge velocity for atomization in a hydraulic nozzle as well the discharge pressure required to develop this discharge pressure.

Box 9.2 Example 9.2

Example 9.2. Find the minimum velocity at which atomization of the fluid occurs in a hydraulic nozzle with an orifice diameter of 3 mm spraying a water-based fluid.

The fluid velocity causing atomization, m s^{-1}

$$v = 280\left(\frac{\sigma^{0.42}\mu^{0.18}}{\rho^{0.59}d^{0.59}}\right)$$

where ρ is mass density of water = 1000 kg m^{3},
μ is the water's viscosity =1.0 mPa s = 0.001 Pa s,
σ is the surface tension of water = 0.0728 N m^{-1},
d is the orifice diameter = 3 mm = 0.003 m,

$$v = 280\left(\frac{0.0728^{0.42}0.001^{0.18}}{1000^{0.59}0.003^{0.59}}\right) = 14.05 \text{ m s}^{-1}.$$

Box 9.3 Example 9.3

Example 9.3. The relationship between the discharge pressure and the flow rate of a hollow cone nozzle is furnished below. Its orifice size is 2mm. Find the velocity of fluid flowing from the orifice. If the velocity for complete atomization is 25 ms^{-1}, find the minimum discharge pressure for complete atomization.

Discharge, lpm	**0.3969**	**0.4536**	**0.6237**	**0.6804**	**0.8505**	**1.1907**
Pressure. kPa	**206.7**	**310.05**	**516.75**	**689**	**1033.5**	**2067**

We know that $Q = C_dAv$, for a flow through an orifice such as a spray nozzle, and $v = \sqrt{\frac{2P}{\rho}}$ making the expression as $Q = C_dA\sqrt{\frac{2P}{\rho}}$
where Q is the flow rate in m^3s^{-1}
v is the discharge velocity in ms^{-1}
P is the pressure in Pa
A is the area of the orifice opening = $\pi d^2/4$ in m^2 = $\pi(0.002)^2/4 = 3.14 \times 10^{-6}$ m^2
ρ is the density of fluid (water) = 1000 kg m^{-3}, and
C_d is the coefficient of discharge of the flow process, which is unknown.

Continued

Box 9.3 Example 9.3—cont'd

For finding the velocity of flow at any flow rate, C_d is to be found from the given data.

The above equation can be written as $Q = \left(C_d A\sqrt{\frac{2}{\rho}}\right)\sqrt{P}$ for our convenience, where $\left(C_d A\sqrt{\frac{2}{\rho}}\right)$, the constant component, is the slope of the plot between Q and $\sqrt{P}$.

Let us first convert the pressure and discharge data to our units of requirement.

Q, m^3s^{-1}	0.000006615	0.00000756	0.000010395	1.13E-05	1.42E-05	0.000019845
P, Pa	206700	310050	516750	689000	1033500	2067000
$\sqrt{P}$, $\sqrt{Pa}$	454.6427169	556.8213358	718.8532535	830.0602	1016.612	1437.706507

Then, plot the relationship between Q and $\sqrt{P}$.
From the plot, the slope is calculated $= 1.25 \times 10^{-8}\ m^3s^{-1}Pa^{0.5} = \left(C_d A\sqrt{\frac{2}{\rho}}\right)$

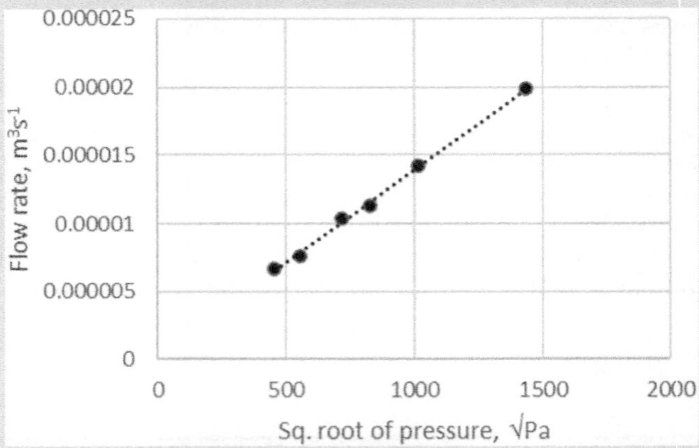

$= C_d \times 3.14 \times 10^{-6} \times \sqrt{\frac{2}{1000}}$
$\therefore C_d = 0.089$
We know that $Q = C_d A v$.
Since the minimum velocity required for atomization is 25 ms^{-1}
the minimum flow rate required $= 0.089 \times 3.14 \times 10^{-6} \times 25$
$= 0.00000698\ m^3s^{-1}$

From the table of data, 0.00000756 m^3s^{-1} corresponding to 310 kPa discharge pressure is greater than the calculated minimum flow rate. This implies that a discharge pressure of over 300 kPa is needed for good atomization.

9.1.1.3 Spray pattern of the nozzle and its measurement

Different types of nozzles, as explained, generate different shapes of spray. When the spray thus generated hits a flat surface, the projected shape of the spray will be circular, annular, a narrow band, etc, according to the type of nozzle. This spray, when traversing in the forward direction, will deposit a rectangular strip equal to the swath of the spray at the surface but with varying quantities of fluid applied relative to the centre of the spray. It can be visualized that more fluid will be collected at the centre for a solid hollow cone nozzle, which tapers toward the edges. This distribution of fluid over the width of the spray is known as the spray pattern of the concerned nozzle. For a hollow cone nozzle, this pattern is a bit flattened at the centre over the swath.

This pattern is measured on an instrument known as the spray patternator (Figure 9.7). It consists of an array of collecting channels leading to fluid-collecting containers. The whole array of channels is set at a slight forward inclination, so that the fluid falling into each channel is led into the collecting tubes against each. The height of the tubes is calibrated for the fluid volume collected, so that the direct measurement of volume is obtained. The nozzle under measurement is mounted on a height-adjustable frame, so that the pattern can be measured for the nozzle at the desired height of the nozzle from the sprayed surface. This setup can hence be used to measure the volume of the sprayed fluid collected on equi-

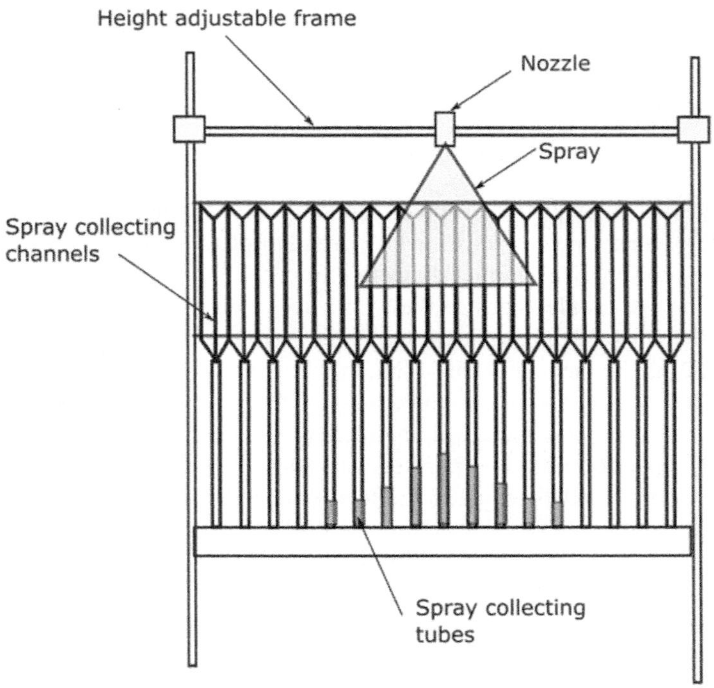

FIG. 9.7 Measurement of Spray Pattern of Nozzles

spaced strips over the spray swath. When a histogram of these volumes is plotted against the swath, the spray pattern of the concerned nozzle is acquired. Generally, the pattern is symmetrical over the centre of the swath.

9.1.1.4 Droplet spectrum of spray

The sizes of droplets deposited on the target plant are crucial, since the volume of the spray applied on the surface influences the mortality of the pest that it intends to kill. Moreover, the sizes determine whether they are deposited on the target or carried away as drift by the wind. Drift is a function of the rate of fall of particles in relation to the horizontal velocity, which in turn is related to particle size and local wind velocity. If the droplet size is too small and wind velocity is high, they drift away from the target and are deposited in unwanted places.

Thedropletsizesofaspraydependuponthetypeandoperatingparametersoftheatomizingdevice(Bode & Butler, 1981; Deysson & Karian, 1978). Low-pressure hydraulic nozzles, such as the hollow or solid cone or the fan types, generate 200 micron- to 100 micron-sized droplets (Matsumoto and Takashima, 1978), whereas the air shear nozzles and centrifugal nozzles generate smaller droplets, in sizes of <100 microns and <50 microns, respectively. The droplet size produced by hydraulic nozzles has an inverse square root relationship with the discharge pressure, and as the pressure drops, the droplet sizes increase. The droplet size of a fan-type nozzle increases from 75 microns to 150 microns when the pressure is reduced from 700 kPa to 200 kPa. However, hydraulic nozzles do not produce uniform droplet sizes but can create a whole spectrum of sizes. The uniformity of spray droplet sizes in this spectrum was experimentally found in a laboratory by capturing suitably dyed fluid droplets on photo paper and measuring their sizes under a microscope. Spread factors, which are multiplication indices, are applied to determine the spherical droplet diameters from the flat deposits measured. After determining the droplet sizes of the whole spectrum, the numerical and volume median diameters of the droplet sizes can be computed. Numerical median diameter (NMD) is the median value of the droplet sizes or that value that divides the droplets into two equal portions on the basis of the number or the frequency of distribution of the droplet sizes. Volume median diameter (VMD) similarly divides the droplet spectrum into two portions such that the total volume of all droplets smaller than the VMD is equal to the total volume of all droplets larger than the VMD. The uniformity ratio, which is the VMD/NMD ratio, when nearing unity, indicates that the spectrum is uniform (the droplet size range is very small). Ideally, the uniformity ratio is unity, when there is only one droplet size in the whole spectrum. Controlled droplet application (CDA) devices, which are basically centrifugal disc sprayers, generate a droplet spectrum that has a uniformity coefficient nearing unity. The following problem, in Box 9.4, illustrates how the NMD and VMD can be calculated from the droplet spectrum of a spray.

9.1.1.5 Nozzle spacing

For the complete coverage of a full swath of spraying and the uniform deposition of the spray on the effective boom width, the nozzle spacing has to be set properly. The height of the boom above the deposition surface is a function of:

(a) the nozzle spacing on the boom;
(b) the nozzle spray angle;
(c) the amount of overlap between the spray geometry required for uniform coverage.

Figure 9.8 shows the typical relationship between these three factors for an example nozzle spacing for hollow cone nozzles with a 70° spray angle fitted on a boom raised to a certain height (Kepner, Bainer and Barger, 1978). It shows the amount of overlap required for uniform coverage and is based on the volume of spray deposited at each unit area on the target level along the width of the boom. It may be seen that the relationship between the volume of liquid deposited by each nozzle against the width is in the shape of a truncated triangle with the volume declining on both sides and constant at the centre (as shown by the thick, dashed line).

Box 9.4 Example 9.4

Example 9.4. The droplet size distribution of a hydraulic sprayer nozzle is given below as measured from a droplet spectrum captured on photo paper. The sizes are classified into bins of size ranges, and the corresponding counts of droplets found in that range are given. Find the NMD, VMD, and uniformity ratio.

Range of droplet sizes, μm	Number of droplets found	Range of droplet sizes, μm	Number of droplets found
1-50	12	251-300	81
51-100	24	301-350	62
101-150	43	351-400	41
151-200	54	401-450	22
201-250	123	451-500	5

The average droplet size of each bin (range) of droplet sizes is calculated.

The total number of droplets measured is found. Based on this, the percent numbers in each bin of droplet sizes is calculated. Then, the cumulative percent of the sequence of droplet sizes is calculated. A plot is made between the average droplet sizes and the corresponding cumulative percent of numbers. Drop a vertical line at the intersection of the curve against the 50% cumulative frequency. This is the number median diameter (NMD), which divides the number-based distribution of droplet sizes in half.

Then, calculate the volume of droplet of each average size. Multiply the volume of each droplet size with that of the number of droplets in that size range to calculate the total applied volume of droplets in that size range. Calculate the sum of applied droplet volumes of all the sizes and then the percent applied volume of each droplet size based on the sum of volumes. The cumulative percentage of these values is calculated next. Plot these cumulative percent values against the droplet sizes. Drop a vertical line at the intersection of the curve against the 50% cumulative frequency. This gives the volume median diameter (VMD), which is the value that divides the volume distribution in half. The table and plot below present the calculations.

Uniformity coefficient = VMD/NMD

Continued

Box 9.4 Example 9.4—cont'd

Range of droplet sizes, μm	Average drop dia, μm	Number of droplets found	Per cent distribution of numbers	Cumulative per cent distribution of numbers	Volume of droplet, μm^3	Applied volume = Volume of droplet x Number of droplets	Per cent distribution of droplet volumes	Cumulative per cent distribution of droplet volumes
1-50	25	12	2.5696	2.5696	8177	98125	0.0019	0.0019
51-100	75	24	5.1392	7.7088	220781	5298750	0.1014	0.1033
101-150	125	43	9.2077	16.9165	1022135	43951823	0.8412	0.9444
151-200	175	54	11.5632	28.4797	2804740	151455938	2.8986	3.8430
201-250	225	123	26.3383	54.8180	5961094	733214531	14.0323	17.8753
251-300	275	81	17.3448	72.1627	10883698	881579531	16.8717	34.7470
301-350	325	62	13.2762	85.4390	17965052	1113833229	21.3166	56.0636
351-400	375	41	8.7794	94.2184	27597656	1131503906	21.6548	77.7183
401-450	425	22	4.7109	98.9293	40174010	883828229	16.9147	94.6331
451-500	475	5	1.0707	100.0000	56086615	280433073	5.3669	100.0000
		467	100.0000		162723958	5225197135	100.0000	

Calculated value of VMD = 310 μ m; NMD = 211 μ m; Uniformity coefficient = 1.47

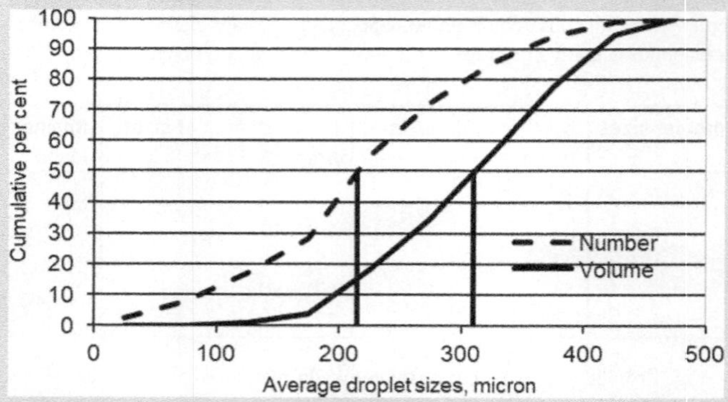

When the nozzles are set at a certain spacing and the volumes deposited by adjacent nozzles are overlapped, the total volume of deposition is constant over the whole width (as shown by the thick, solid line). The top part of Figure 9.8 shows that, for the given configuration of spray geometry (spray angle), nozzle spacing, and overlap, a particular height of the boom from a target surface yields a uniform volume of deposition over the width. If we set this height to be 80% of the earlier setting (bottom half of Figure 9.8) after keeping the nozzle spacing and nozzle angle the same, then the spray overlap is reduced and, thereby, the total volume deposited is not uniform anymore. This illustrates the fact that the height of the boom is to be set according to the three influencing factors to obtain a uniform deposition over the swath width of the sprayer. The example in Box 9.5 shows how the desired nozzle spacing can be calculated from the spray pattern of a given nozzle.

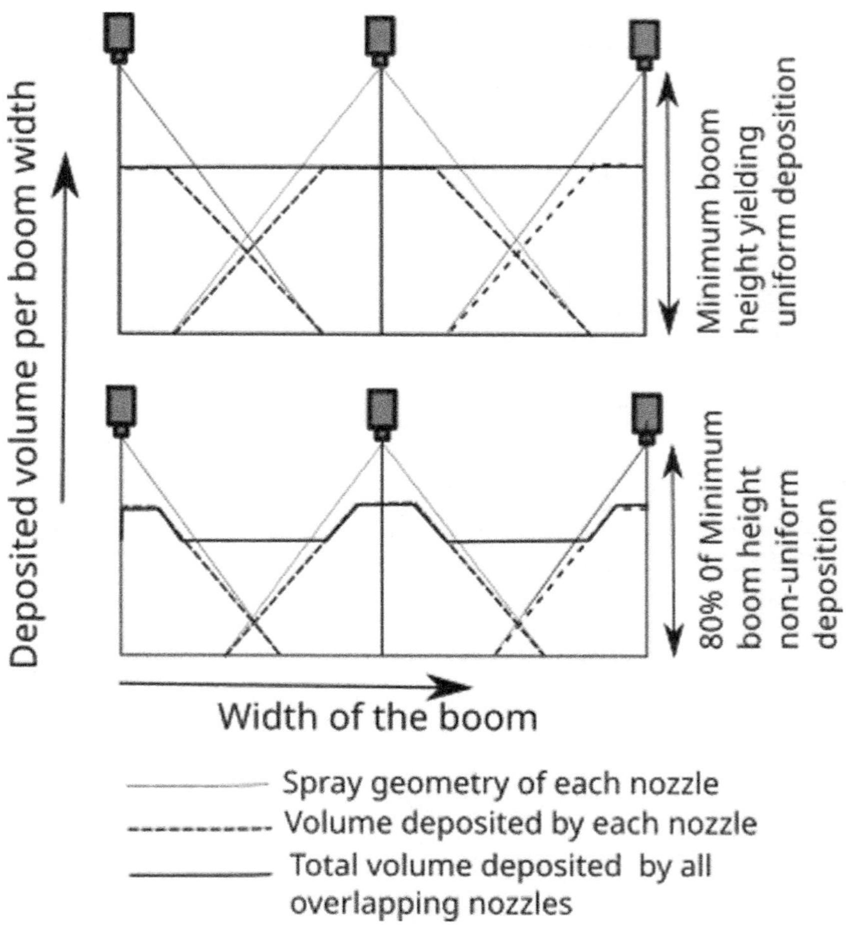

FIG. 9.8 Uniformity of Spray Distribution over the Swath

Box 9.5 Example 9.5

Example 9.5. The data concerning the spray pattern of a single fan-type nozzle are given as the volume of deposit made at each 2.5 cm width of the spray. The nozzle has a 60° spray angle. It discharges 0.4 l min^{-1} at a pressure of 350 kPa. Find the nozzle spacing to be set to acquire a reasonably good uniformity of deposit over the full boom width.

In the table below, the deposited volume of the single nozzle is given against its width of spray swath.

The height at which the nozzle was set to measure this data must be

=Half of the spray pattern width / tan(spray angle/2) = 20/tan(30) = 35 cm.

Eyeballing the data, it can be seen that the left tail of the pattern to a width of 12.5 cm is added to the right of the pattern and vice versa to emulate the overlapping of the spray pattern of one similar nozzle on each side. Then, the addition of the deposits at each width location gives the total deposit at that location. When plotted against the widths, the total deposit is more or less uniform over the width of 40 cm compared with that of a single nozzle. (The student may try to increase or decrease the overlap and find how it affects the uniformity.) An overlap of 12.5 cm between the spray patterns of adjacent nozzles at a boom (nozzle) height of 35 cm will require a nozzle spacing of 27.5 cm, as may be seen in the illustration below. This setting will provide a reasonably good uniform deposition over the width of the boom.

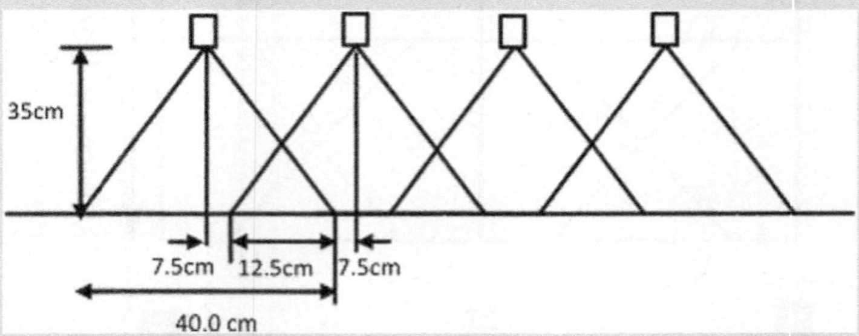

Given		Calculated	
Width of spray pattern, cm	**Deposited volume of single nozzle, ml**	**Overlapping12.5 cm on both sides**	**Total deposit on overlap**
2.5	3	14	17
5	9	11	20
7.5	8	8	16
10	10	8	18
12.5	15	3	18
15	19		19
17.5	21		21
20	24		24
22.5	23		23
25	22		22
27.5	19		19
30	14	3	17
32.5	11	9	20
35	8	8	16
37.5	8	10	18
40	3	15	18

Box 9.5 Example 9.5—cont'd

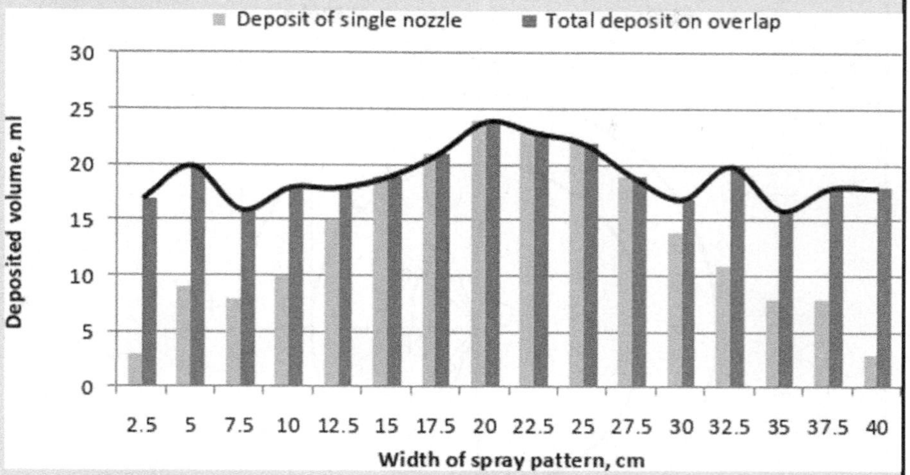

The following example in Box 9.6 illustrates how the deposition pattern of a hydraulic nozzle can be calculated analytically based on its discharge and the shape of spray pattern. And, it also shows how the spray pattern of the spaced nozzles can be adjusted to acquire a better uniformity over the swath.

Box 9.6 Example 9.6

Example 9.6. A hollow cone nozzle, which discharges 0.5 l min^{-1} at 300 kPa pressure, has a 60° spray angle. The nozzle, when set at a height of 43.3 cm from the horizontal target surface, provides an annular deposition pattern, the inner diameter of which is 70% of the outer diameter. Analytically find the pattern of discharge deposited over the width of spray at target level and plot it against the distance from the centre. If two such nozzles are used with an overlap of 50%, plot the total deposition pattern with respect to the distance from the centre.

Half spray angle is 30° and the height of the nozzle above the target = 43.3 cm.

∴, *the* outer radius of the spray distribution = 43.3 tan(30) = 25 cm.

Then, the inner radius = 25 cm × 0.7 cm = 17.5 cm.

Then, draw the spray deposition profile, as in the figure below, and divide it into three zones, Z1, Z2, and Z3. The division was done such that the segments of deposition can be calculated easily.

Continued

Box 9.6 Example 9.6—cont'd

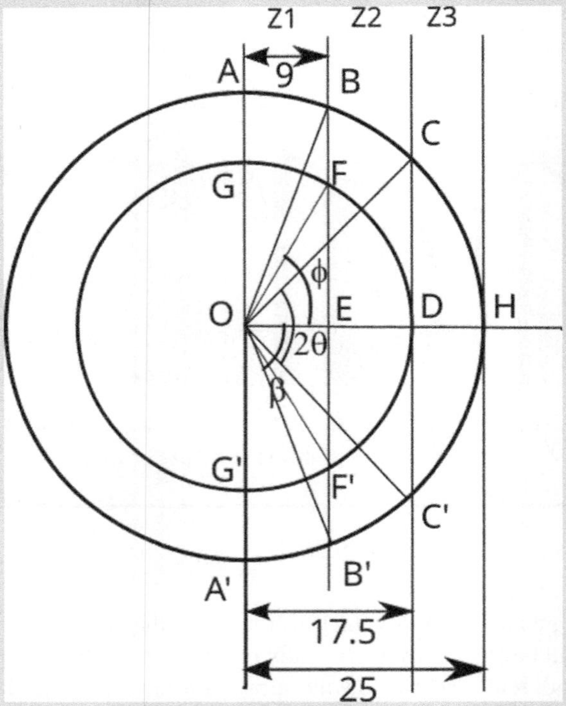

In triangle OCD, OC = 25 cm; OD = 17.5 cm

$$\theta = \cos^{-1}(17.5/25) = 45.57°;\ 2\theta = 91.145°$$

In triangle OBE, OB = 25cm; OE = 9 cm

$$\beta = \cos^{-1}(9/25) = 68.89°;\ 2\beta = 137.8°$$

In triangle OFE, OF = 17.5 cm; OE = 9 cm

$$\phi = \cos^{-1}(9/17.5) = 59.05°;\ 2\phi = 118.1°$$

$$\text{Area of segment CDC'H} = \text{AreaZ3} = \left[\frac{2\theta \times \pi}{360} - \frac{\sin 2\theta}{2}\right] r^2 = (0.794 - 0.5)625 = 183.75\,\text{cm}^2$$

$$\text{Area of segment BEB'H} = \left[\frac{2\beta \times \pi}{360} - \frac{\sin 2\beta}{2}\right] r^2 = (1.2 - 0.34)625 = 538.7\,\text{cm}^2$$

$$\text{Area of segment FEF'D} = \left[\frac{2\phi \times \pi}{360} - \frac{\sin 2\phi}{2}\right] r_i^2 = (1.03 - 0.44)306 = 180.54\,\text{cm}^2$$

Area GFF'G' = Area of sector OGF + Area of sector OG'F' + Area of triangle OFF'

$$\text{Area of sector OGF} = \text{Area of sector OG'F'} = \left[\frac{\eta}{360}\right] \pi r_i^2 = 82.74\,\text{cm}^2$$

where η is angle subtended between OF and OG = between OF' and OG'

$$= (90\phi) = (90–59) = 31°$$

$$\text{Area of triangle OFF'} = 2\left(\frac{1}{2} \times 9 \times 17.5 \sin\phi\right) = 135\,\text{cm}^2$$

Box 9.6 Example 9.6—cont'd

$$\therefore \text{Area GFF}'\text{G}' = 82.74 + 82.74 + 135 = 300.48\,\text{cm}^2$$

$$\text{Area of segment (semi circle) AOHA}' = \pi r^2/2 = \pi(625)/2 = 981.25\,\text{cm}^2$$

$$\text{Area Z1} = \text{Area ABFG} + \text{A}'\text{B}'\text{F}'\text{G}' = \text{Area ABB}'\text{A}' - \text{Area GFF}'\text{G}'$$
$$= \left(\text{Area AOHA}' - \text{Area BEB}'\text{H}\right) - \text{Area GFF}'\text{G}'$$
$$= 981.25 - 538.7\text{–}300.48 = 142.07\,\text{cm}^2$$

$$\text{Area Z2} = \text{Area BCDF} + \text{B}'\text{C}'\text{DF}' = \text{Area BEB}'\text{H} - \text{Area FEF}'\text{D–Area CDC}'\text{H}$$
$$= 538.7 - 180.54 - 183.75 = 174.41\,\text{cm}^2$$

$$\text{Area Z3} = \text{CDC}'\text{H} = 183.75\,\text{cm}^2(\text{calculatedalready})$$

Areas Z1, Z2, and Z3 are the spray-deposited areas. And these zones are symmetrically repeated as Z3, Z2, and Z1 on the LHS of the vertical centre line AA' of the spray deposition profile.
Total spray-deposited area = 2(Area Z1 + Area Z2 + Area Z3)
= 2(142.07 + 174.41 + 183.75) = 1000.5 cm^2
Total spray-deposited area is also = Area of outer circle – Area of inner circle
$= \pi r^2 - \pi r_i^2 = \pi(25^2 - 17.5^2) = 1000.9\ cm^2$ (Calculation verified)
Discharge of the nozzle = 0.5 l min^{-1}
Proportioning the total discharge on the calculated areas of deposition in Z1, Z2, and Z3,
Spray deposited in Z1 = (0.5 / 1000.5) × 142.07 = 0.071 l min^{-1}
Spray deposited in Z2 = (0.5 / 1000.5) × 174.41 = 0.087 l min^{-1}
Spray deposited in Z3 = (0.5 / 1000.5) × 183.75 = 0.092 l min^{-1}
Then, the spray deposited over the width of deposition of the nozzle can be plotted as below.

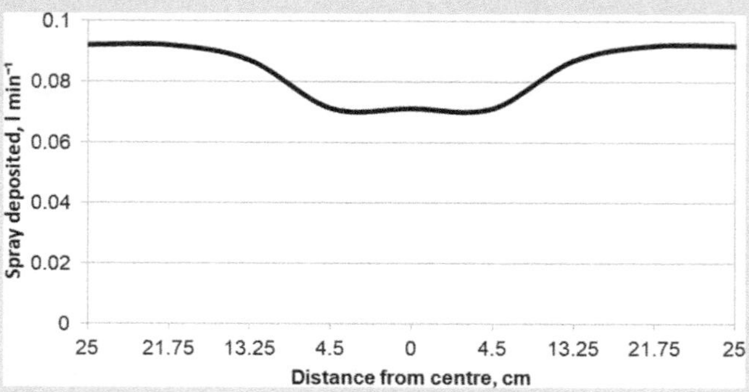

When these nozzles are used with 50% overlap, the discharges are added from the overlapping nozzles at the selected zones, as below, implying that spray rendered by one nozzle to zone Z1 is added to the spray rendered to Z3 of the adjacent nozzle (0.071+0.087 = 0.168 l min^{-1}). And, the spray of both nozzles on Z2 is added together (0.092+0.092 = 0.184 l min^{-1}). It may be noted that the uniformity of the deposited liquid is much better by overlapping the nozzles.

Continued

Box 9.6 Example 9.6—cont'd

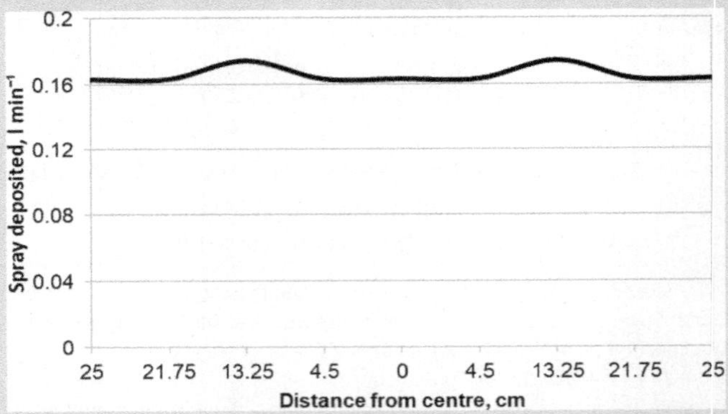

Then, try solving this problem by plotting the spray deposition area on graph paper, and try calculating the area of each segment and verify the results.

Try to think of a better geometrical method for these calculations. A software that can calculate the quantity of deposit per unit time on each of the segments of the deposited area was made on such an algorithm and is available which may be tried by the student to get an idea. It also permits simulating the pattern when two such nozzle patterns are overlpped.

9.1.1.6 Application rate of sprayer

The spray application rate is the quantity of the spray fluid applied to a unit cropped area (l ha^{-1}). Usually, pesticide/weedicide manufacturers prescribe the application rate for each of their chemicals as applied to each type of crop, pest, disease, or weed. As an example, they advise that a particular chemical, namely, "2-4 D," a weedicide, may be applied at 500 l ha^{-1}. This means that 500 litres of spray fluid with the chemical diluted in water at the recommended dilution should be applied to an area of one hectare. The two factors that affect the application rate are the sprayer's total discharge rate (l h^{-1}) and its field coverage or capacity (ha h^{-1}).

Generally, a hollow cone nozzle discharges about 0.3 min to 0.7 l min^{-1}. It depends on the discharge pressure setting of the sprayer. The discharge per nozzle is 0.6 min to l min^{-1} for a discharge pressure of 400 kPa. If the boom has 10 such nozzles, the discharge from the whole boom = 0.6 min × 10 = 6 l min^{-1}, = 360 l h^{-1}. This is the total discharge rate of the boom sprayer.

Field capacity of the sprayer = swath width of coverage × speed of operation.

Width of coverage of the sprayer = swath width of the boom with 10 nozzles = number of nozzles × spacing between nozzles.

If we assume that we have set the nozzles at 40 cm spacing without any spray overlap, then the swath width = 0.4 × 10 = 4 m and then the

field capacity of the sprayer = swath width of sprayer × Speed of operation, (say) 3 km h^{-1}

$$= 4 \text{ m} \times 3000 \text{ m h}^{-1} = 1.2 \text{ ha h}^{-1}$$

Having the total discharge rate of the sprayer as 360 l h^{-1} and field capacity as 1.2 ha h^{-1}, the application rate can be calculated as

total discharge rate / field capacity = 360 l h^{-1} / 1.2 ha h^{-1} = 300 l ha^{-1}.

Fundamental hydraulics teach us that the flow rate for a hydraulic system (here, it is the nozzle discharge) is proportional to the square root of the pressure. This implies that, if we desire to change the discharge of the nozzle by two times to increase the application rate, then the discharge pressure is to be increased 2^2 times. Such an increase in pressure will cause an undesirable change in the spray droplet sizes generated. So, if we require an application rate of 500 l ha^{-1}, the only option is to decrease the speed of the operation to decrease the field capacity and not the total discharge rate of the sprayer.

9.1.1.7 Sprayer calibration

Knowing the above, sprayer calibration is just the calculation of the operating speed of the sprayer to acquire the required application rate.

Since Application Rate = total discharge rate / field capacity and thereby field coverage = total discharge rate / application rate,

the desired field coverage for getting an application rate of 500 l ha^{-1} in the last example = 360 (lit / hr) / 500 (lit ha^{-1}) = 0.72 ha h^{-1} = 7200 m^2h^{-1}

Total application width (swath) was calculated as 4 m for the desired nozzle setting, and so,

speed of operation = (7200 m^2h^{-1}) / (4 m) = 1800 m / hr = 1.8 km h^{-1}.

This means that, only if the boom sprayer moves at a slower speed, of 1.8 km h^{-1}, can the desired application rate of 500 l ha^{-1} be achieved.

Before operating the sprayer in the field, one should calibrate it so that the desired application rate is achieved to make the pest/weed control effective. The following example, in Box 9.7, details the calculations to achieve the desired application rate of a boom sprayer, given its speed and discharge.

Box 9.7 Example 9.7

Example 9.7. Cotton crop planted in 30 ha is to be sprayed with Imidacloprid 17.8% SL using a tractor mounted boom sprayer. The boom has 16 hollow cone nozzles of a rated delivery of 0.4 l min^{-1} of fluid at 400 kPa. The nozzle spacing on the boom is 50 cm and no overlap of spray given. Imidacloprid 17.8% SL is to be applied at a rate of 150 ml ha^{-1}. The recommended dilution of the pesticide is 3.0 ml in 10 l water. At what speed is the sprayer to be operated to achieve the desired application rate when the discharge pressure is set at 300 kPa?

Since Q, the discharge at pressure from an orifice (nozzle) $= A \times \sqrt{(2gh)}$ where h is the discharge pressure and A is the area of orifice. A being constant, $Q \propto \sqrt{h}$ or $Q2 \propto h$

or $\frac{Q_1^2}{Q_2^2} = \frac{h_1}{h_2}$

$h_1 = 300$ kPa; $h_2 = 400$ kPa; $Q_2 = 0.4$ l min^{-1}

$\therefore Q_1 = 0.12$ l min^{-1}

Discharge from the boom = 16 nozzles × 0.12 l min^{-1} $nozzle^{-1}$ = 1.92 l min^{-1} = 115.2 l h^{-1}

Spray swath = 16 nozzles × nozzle spacing = 16 × 0.5 = 8 m

Area covered by boom in unit time (Field capacity) = 8 m × S, speed in km h^{-1}= 8000S m^2 h^{-1}
= 0.8S ha h^{-1}

Required spray discharge per unit area = (Pesticide to be applied / unit area) × dilution rate
= 150 ml ha^{-1} × (10 l water / 3 ml pesticide)
= 500 l ha^{-1}

Then, we can discern that:

Discharge from the boom (l h^{-1}) / Field capacity (ha h^{-1})
= Required discharge per unit area (l ha^{-1})

115.2 l h^{-1} / 0.8S ha h^{-1} = 500 l ha^{-1}

$\therefore$ S = 0.288 km h^{-1}.

9.1.2 Small sprayers

Small sprayers are hydraulic, air compression, or air-assisted sprayers, generally carried manually as a knapsack, and are ideal for spraying small crop areas and for other spraying works such as mosquito eradication work.

9.1.2.1 Hydraulic knapsack sprayers

Hydraulic knapsack sprayers are compact sprayers (Figure 9.9) that work as a hydraulic unit with all the components required for operation. They have plastic spray tanks of about eight litres, so that they can be easily carried when full. A built-in, lever-operated reciprocating or diaphragm pump is provided with an air vessel integrated in it. While the sprayer along with the tank into which the components are built is carried as a backpack by the operator, the pump lever protrudes from the side on the operator's front. The operator can oscillate the pump lever up and down, and this reciprocating motion is transmitted to the pump's plunger or diaphragm by way of a bell crank. The pump's plunger operates inside a cylinder or the diaphragm moves up and down to draw and exhaust the fluid through a set of suction and delivery valves. Similar to any other reciprocating pump, the fluid inside the tank is drawn at the suction stroke through the suction valve and delivered into the air vessel and delivery passage at the delivery stroke.

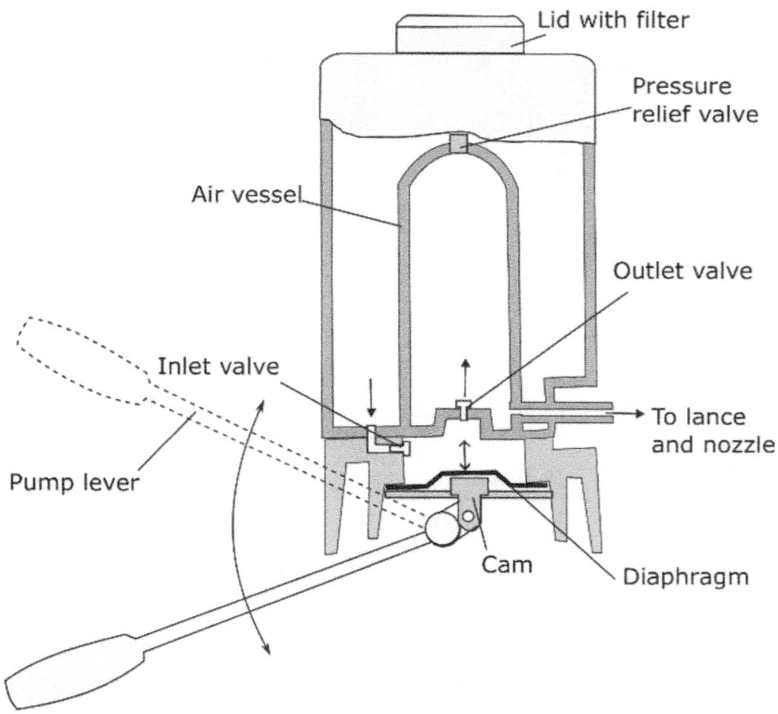

FIG. 9.9 Hydraulic Backpack Sprayer

The one-way valves alternate accordingly to make this possible. The air vessel (pressure chamber) delivers a continuous flow to the delivery passage by discharging stored fluid under pressure at the suction stroke. The air vessel has a pressure-relief valve to safeguard the pump from pressure overload. The delivery passage is connected through a hose on to a lance with a hydraulic nozzle at the end. The pressurized liquid is thus sprayed out by opening the trigger valve on the lance handle. There is a lidded-top opening for pouring in the spray fluid, and a strainer in its mouth to filter the poured fluid.

9.1.2.2 Backpack air compression sprayers

Backpack air compression sprayers are small backpack sprayers (Figure 9.10) that work by using air pressure to pressurize the spray liquid. The spray tank is generally made of brass or durable plastic; a reciprocating pump, similar to the tyre pump, is mounted inside the tank. The air pump's handle comes from the tank's top. The tank has a lidded opening, which is sealed by a rubber gasket so that the tank is air sealed. The tank is filled with spray fluid, with about one-quarter tank of air space on the top. When the air pump is operated after placing it on the ground, the air inside the tank is pressurized. There is a non-return air valve at the pump's delivery end, and pumped

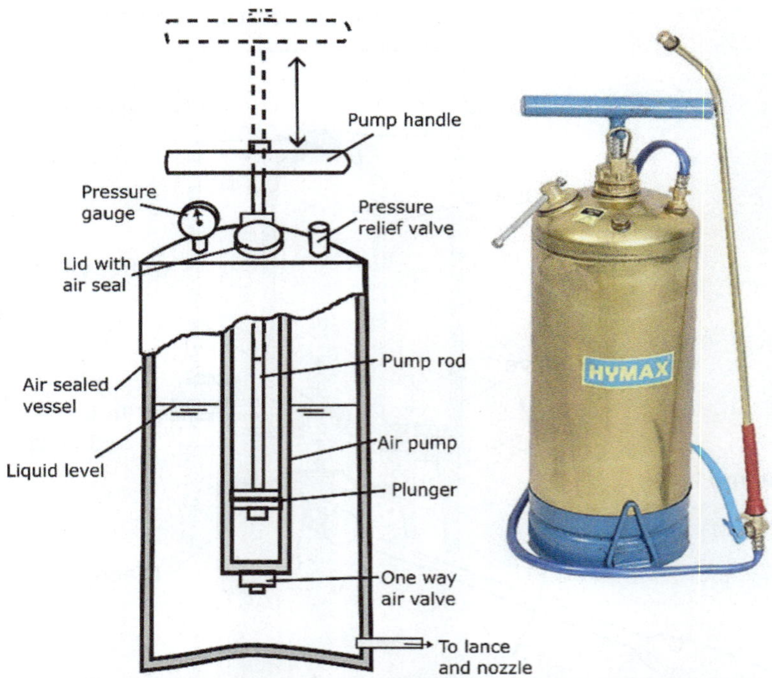

FIG. 9.10 Backpack Air Compression Sprayer (HX-203, Hymax Agro India)

air cannot return while pumping is occurring. The air pressure inside the tank is thus raised, and this pneumatic pressure acts on the surface of the spray liquid inside, ready to push it from the delivery passage. The delivery passage is connected through a hose on to a lance with a hydraulic nozzle at the end. It also has an anti-dribbling, lever-type, cut-off valve, which closes the delivery positively.

The sprayer is lifted and slung as a backpack by the operator, ready for operation. Upon opening the cut-off valve, the pressurized liquid is atomized by the nozzle. When the stored air pressure declines, it is built up again by pumping more air. This type of sprayer is popular not only for small nurseries but also in municipalities for insect and disease control. A pressure gauge is provided to monitor the operating pressure. A safety valve is also provided to prevent pressure overload. The example problem given in Box 9.8 explains the theory of building the air pressure inside the air compression sprayer.

9.1.2.3 Knapsack mist blowers

Knapsackmistblowersareair-assistedsprayers(Figure 9.11) that are very popular for field spraying rice and other commercial crops in Asian countries. A knapsack mist blower has a plastic tank with two compartments. The larger is of eight-litre capacity for spraying fluid; a smaller compartment is for petrol. The heart of the sprayer is a centrifugal blower driven by a 0.8-1 hp, single-stroke, petrol engine with all the

Box 9.8 Example 9.8

Example 9.8. A backpack air compression sprayer has a cylindrical backpack canister (air-sealed vessel) of 16 cm diameter and 45 cm height. An air pump of 5 cm diameter and 40 cm height is fitted concentrically inside the backpack canister. A spray fluid is put into the canister to a three-fourths level to allow free air space over the fluid in the canister. When air is pumped into the canister, the air above the fluid is pressurized and delivers a spray through the nozzle. Calculate how many strokes must be given to the air pump to build a 350 kPa over the fluid. Neglect the wall thickness in the calculations and assume the stroke of the air pump as 10 cm less than the air barrel height.

Volume of the canister $= \pi D^2H/4 = \pi(16)^2 \times 45/4 = 9043.2$ cc
where D = 16 cm, H = 45 cm.
Outer volume of the air barrel $= \pi d^2h/4 = \pi(5)^2 \times 40/4 = 785$ cc
where D = 5 cm, H = 40 cm.
Annular volume for the spray fluid = 9043.2 − 785 = 8258 cc = 8.258 *l*
Spray fluid volume filled in the sprayer = 0.75 × 8.258 = 6.2 *l*
Air space volume over the fluid = 8.258 – 6.21 = 2.048 *l*
When air is pumped into this closed air space, the pressure is built according to the expression, $\frac{P}{V}$ = constant.
P_1 = Existing air space pressure over fluid = atmospheric pressure = 100 kPa,
P_2 = Desired air pressure in space above the fluid = 350 kPa,
V_1 = Free air volume over fluid = 2.05 l.
$\therefore, V_2$ = Volume of air over fluid after pumping in air to desired pressure
$= \frac{P_2V_1}{P_1} = \frac{350 \times 2.05}{100} = 7.21$ l.
Then, air volume pumped in = 7.21 – 2.05 = 5.16 l,
Air pump stroke = 30 cm,
Air pump stroke volume $= \pi d^2h/4 = \pi(5)^2 \times 30 / 4 = 0.59$ l,
Number of strokes to build the desired pressure over fluid = 8.7 = 9 (appr.).

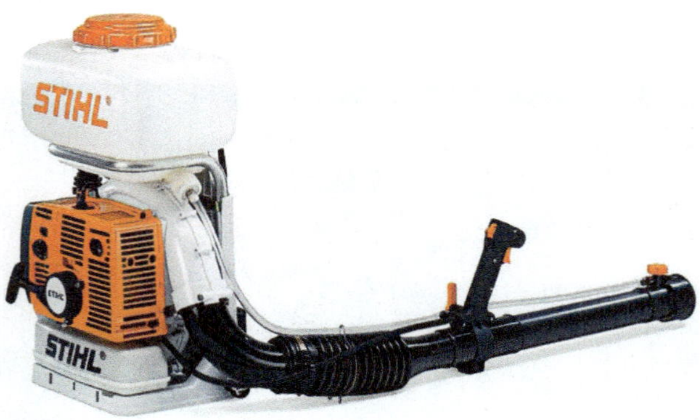

FIG. 9.11 Knapsack Mist Blower (SR420, STIHL India)

necessary parts such as a carburettor, air filter, magneto, spark plug, etc. The engine operates at 3000 rpm and the blower delivers 4000 m^3h^{-1} of air. The blower draws air through its central suction hole and delivers high-velocity air through a single radial outlet. This stream is led out by a flexible spray boom tube, which protrudes from the side of the sprayer to the front. The 1-metre-long air boom carries an air shear nozzle at the front, through which the air stream is forced at a high velocity. The shear nozzle has a jet at the centre to which the spray fluid from the tank is supplied through a thin transparent plastic tube. The tube is brought out from the bottom of the spray tank with an online strainer. The nozzle also has a control valve, which can set the fluid discharge at 0.5, 1.0, 1.5, or 2.0 $l\ \text{min}^{-1}$. The liquid coming from the jet at the centre of the nozzle is sheared off by the high-velocity stream, producing an atomized spray. The particle size of this spray is 100 microns, and the swath is about 0.75 m at a distance of 1 m. The operator, while carrying the sprayer as a backpack, can hold the spray boom with the nozzle in hand and direct the spray. Some of the air from the blower is sent into the spray tank for pneumatic agitation.

9.1.3 Self-propelled high-clearance sprayers

Self-propelled high-clearance sprayers (Figure 9.12) are machines that are used in row crops that are too high for conventional tractors, whose ground clearance is too low. They have special chassis with large-diameter row crop wheels to yield the desired clearance. These machines are usually low-pressure, low-volume sprayers, like other conventional sprayers. The spray boom is usually mounted on the rear of the machine. The height of the boom can be varied to permit spraying either high- or low-growing crops.

9.1.4 Air-assisted sprayers

Air-assisted sprayers, also known as air-blast sprayers, are generally used in orchards for spraying large areas of fruit trees, large shade trees, vineyards, and certain

FIG. 9.12 Self-Propelled High-Clearance Sprayer (Patriot 50 series, CaseIH, USA)

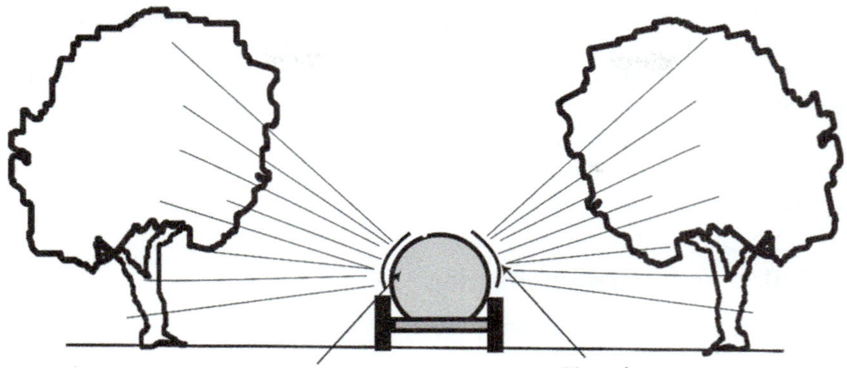

FIG. 9.13 Working of an Orchard Sprayer

FIG. 9.14 Tractor Trailed Air-Assisted Orchard Sprayer (KX600V600, KrishiSpray, AV Agritech Inputs Pvt Ltd, India)

other crops. These machines use a blast of air generated by axial or radial high-velocity fans. These fans deliver an air volume in the range of 10,000 m^3h^{-1} to 40,000 m^3h^{-1} and velocities in the range of 160 kmh to 250 kmh^{-1}. The spray fluid is metered into these air streams using special nozzles (Figure 9.13). The high-velocity air stream atomizes the fluid and carries it deep into the tree canopies. These are generally large, trailed machines and they deliver a high volume of spray to both sides and to larger heights. They generate relatively smaller droplet sizes compared with low-pressure sprayers since the principle of air shear is used for atomization. Figure 9.14 shows a tractor trailed orchard sprayer, on which a high-capacity blower is used to provide a high-velocity air stream. This air stream is radially blown from the sides of the sprayer through air ducts placed with jet spray nozzles. The spray jet is thereby sheared by the air stream, rendering an atomized stream of spray that penetrates the canopy of the orchard trees. Penetration into the tree canopy should be sufficiently adequate such that there is a uniform spray coverage on the tree leaves. Large trees should be pruned to permit the proper penetration of the spray stream deep into their canopies for the effective treatment of pest and diseases.

9.1.5 Drones for plant protection

Unoccupied aerial vehicles, commonly known as drones, are flying machines without a pilot but are operated by remote controls. Their use has become very prevalent in all spheres of activities. Starting as secret military weapons, advancement in technology has made them available for applications such as aerial photography; precision agriculture; the monitoring of crops, forest fires, and environmental surveillance; and many more uses. Their application in agriculture has become extensive (Ozkan, 2024), especially in crop surveillance and spraying. Recently, powerful drones are tending to replace aerial spraying in many countries. Global positioning systems are now able to guide a flying drone precisely over precalibrated paths. The drone market is now set to go forward in leaps and bounds because of their extensive applications including agriculture. China is the greatest user of multi-rotor drones and constantly promoting their availability and use in that country and elsewhere. Newer control algorithms, more powerful rotor drives, and battery technology will enhance their use further. Drones, at present, are more useful for small field plots because of their inherent limitation with the spray liquid payload per fill. Drone spraying minimizes the exposure of chemical toxicity to its operators.

Spraying drones (Figure 9.15) have all the advantages (Ozkan, 2024) of a conventional boom sprayer, despite the fact that they have some limitations regarding

FIG. 9.15 Drone for Spraying (AGRAS T30, DJI inc., China)

the liquid payload they carry. Additionally, they have navigational systems based on GPS and sensors for collision avoidance. Many spray drones marketed now possess dual rotors on each rotor arm. This give these drones more lifting capacity; they can carry a 37- to 56-litre spray tank, providing more liquid sprayed per fill. The spray boom and the nozzles, in particular, are positioned deliberately below the rotors so that the turbulence they cause in the plant canopy improves the spray penetration into the lower parts of the crop. Drones' spray penetration is very similar to that caused by air-assisted ground sprayers. Recent-model drones do not have wider booms since the nozzles on their ends cannot be under the rotors that can provide spray penetration. Some drones use spinning disc atomizers or controlled droplet applicators. The spinning speed of the discs causes the liquid atomization. Moreover, these disc drives can be set up for fertilizer and seed broadcasting, too.

Generally, drones use hydraulic nozzles that can be chosen according to the discharge rate they offer. Drones with hydraulic nozzles usually can fly more closely to the crop canopy, thereby preventing the very small spray particles from being deposited away from the crop boundary. The drawback in using hydraulic nozzles is that their discharge is relatively fast, which empties the spray tank quickly and thus reduces the area covered per fill. On the contrary, spinning disc atomizers are low-volume applicators with uniform-sized particles and do not require a large liquid volume. They have the advantage of controlling their spinning speed, varying the discharge rate and droplet size depending on the flying speed. As previously stated, the major disadvantage of drone sprayers is their limited tank capacity and the need for frequent fills to cover larger areas. The next important hurdle is their battery charge to endure flying for a longer time.

Basically, drone spraying falls under the category of low-volume spraying since the application rate is 14 litres to 20 litres per hectare. As in the case of the ground sprayers, the flying speed, spray swath, number of atomizers on the drone, and the discharge rate decide the application rate. A drone with a 20-litre tank empties the tank in about 3 minutes. Modern drone control systems based on GPS guidance can direct the drone to leave for refilling and then return to the spot it left to continue operation. The typical flying speed of drones varies between 15 km h to 50 km h^{-1}; they usually fly just over the crop canopy or at a maximum height of 2.5 m from the ground. All drone models have a height sensor that can follow the terrain of the land. Modern drones use real-time kinematics (RTK) as GPS assistance and are guided at a centimetre resolution.

The theoretical width of a drone's swath is difficult to assess, unlike that for a ground sprayer. This is because of the height at which it is flown while spraying. So, the best method is to perform a practical observation of the spray droplets' presence on a trial run over the crop and determine the width. Based on this, the planning of spray runs can be made. There has to be a compromise between the choice of the used hydraulic nozzle's discharge and the flight speed. A high flight speed at larger discharge rates may give the required coverage with more time available for tank refills per unit area. But, this may deteriorate the droplet sizes rendered because of the higher discharge. The choice of low flight speeds may cause more tank refills per unit area covered. Too small a droplet size will cause undue drifting.

The problem of battery charging is a major hurdle in adopting drones for large fields because a full charge lasts for only a maximum of 10 minutes in a medium-sized drone. Keeping multiple replacement batteries with a full charge is a viable solution. The fast charging of batteries is now possible so that multiple batteries can be charged quickly. The usage of drones is restricted by government regulations and licenses; that is one reason for their slow adoption on farms.

As stated previously, a drone is an ideal spraying solution for small field sizes but may not replace conventional ground-driven sprayers in the near future in large field areas. Advancements in the sphere of unoccupied and guided farm machines will make huge row crop sprayers popular in large farms. But, large-capacity drones will be indispensable for plantation crops and forests where machinery traffic is difficult.

The specifications of a recent model of a leading Chinese drone manufacturer are given in Table 9.2 for understanding the machine better.

An example problem, as shown in Box 9.9, illustrates the relationship of different factors influencing the spray coverage of a drone.

Table 9.2 Specifications of a commercial spraying drone

Maximum field coverage 16 ha	Weight with battery 36.5 kg
Maximum power consumption 11,000 W	Hovering endurance 7.8 min (@29,000 mAh & take-off weight of 66.5 kg) 20.5min (@29,000 mAh & take-off weight of 36.5 kg)
Maximum operating flight speed 7 m s^{-1}	Maximum flight altitude 4500 m without payload
Maximum tank volume 30 l	Nozzle numbers 16
Maximum discharge of drone on choice of nozzles 7.2 l min^{-1}, 8 l min^{-1}, 3.6 l min^{-1}	Droplet sizes 130 μm to 250 μm and 110 μm to 135 μm
Maximum effective spray width 4 m to 9 m (with 12 nozzles and a distance of 1.5 m to 3 m to crops)	
Battery 29,000 mAh 51.8 V	Battery charging input 100 VAC to 260 Vac Output voltage 40 V to 60 V; Full charging in 9 to 12 minutes
Precision radar, RTK/GNSS (GPS systems) for flight guidance and planning, Precision flowmeter for flow detection, Tank level detector, Omnidirectional obstacle avoidance radar Top-view radar, Drone remote control with Wi-Fi, Bluetooth, and GPS guidance	

Box 9.9 Example 9.9

Example 9.9. The width of a spray swath of a drone, for which specifications were given previously, is 6 m when it is spraying from a 2 m height over a row crop. Find the flight speed of the drone to get an application rate of 20 l ha^{-1}. Use the maximum discharge of the machine. What is the area covered per tank load? Also, calculate how much area can be sprayed with one tank load and one fully charged battery. How many batteries need to be kept to perform a continuous spraying? Use the specifications of the drone given in Table 9.2.

Desired application rate = 20 l ha^{-1}

Spray discharge rate = 8 l min^{-1}

∴, Area covered / unit time =

$$\frac{(Spray\ discharge\ rate)}{(Desired\ application\ rate)} = \frac{8\,l\ min^{-1}}{20\ l\ ha^{-1}} = 0.4\ ha\ min^{-1} = \left(0.4 \times 10000\ m^2\ ha^{-1} \times\ 60\ min\ h^{-1}\right)$$

$$= 240000\ m^2\ h^{-1}$$

· Flight speed $= \dfrac{Area\ covered\ per\ unit\ time}{width\ of\ spray} = \dfrac{240000\ m^2\ h^{-1}}{6\ m} = 40000\ m\ h^{-1} = 40\ km\ h^{-1}$.

Tank capacity = 30 l.

Time required for the drone to empty the tank $= \dfrac{Tank\ capacity}{Discgarge\ rate} = \dfrac{30\ l}{8\ l\ min^{-1}} = 3.75\ min.$

Area covered per tank = Area covered per unit time × time to empty one tank load

$= 0.4\ ha\ min^{-1} \times 3.75\ min = 1.5ha.$

The capacity of the drone battery = 29000 mAh = 29 Ah @51.8 V output,

implying that its maximum power output = 29 Ah × 51.8 V = 1502 Wh = 90132 Wmin.

Maximum power demand of the drone = 11000 W.

Time for which it can sustain at maximum power demand = = 8.19 min.

Kindly note that the hovering endurance is specified as 7.8 min at a payload of 66.5 kg (36.5 kg of machine + 30 kg liquid weight) in the specifications.

Since the time of spraying for one tank load is 3.75 min,

The number of refills per a full charge of battery 7.8 min / 3.75 min = 2.08 = 2 times appr.

Area covered per full charge of battery = 1.5 × 2 = 3 ha.

Since 12 minutes are required to fast-charge the battery, three batteries need to be used to perform a continuous spraying.

9.2 Dusting equipment

In earlier days, insecticides and fungicides in the form of dust were applied to plants by having them in a thinly woven bag and shaking it over the plants. The dust form of application is still in use for certain formulations. A tractor rear-mounted duster, which is operated by the power take-off, basically has an air blower into which the dust is stirred and fed at a uniform rate. The air stream disperses the dust and distributes it over a wide area. But, the dust is highly susceptible to drift in windy conditions and, hence, dusting is done in the early mornings, when wind is not as much of a factor and the morning dew will aid the dust in sticking to the plants. Mechanical or pneumatic agitation is provided in the dust hopper to prevent the bridging of fertilizer.

Exercise Problems

1. A field sprayer has a 10 m boom with nozzles set at 25 cm spacing. If the nozzles have a 70° spray cone angle and the set spacing produces a 50% overlap to provide uniform coverage, find the height at which the boom is to be set above the plant surface. At this boom height setting, the desired application rate is 300 l ha^{-1}. The sprayer discharge pressure is set at 500 kPa. The operating speed of the sprayer is 2 kmh^{-1}. Find the sprayer's pump capacity, assuming 25% of the flow is recycled back into the tank while in operation. Find out the discharge rate per nozzle of the sprayer.
2. A tractor mounted blower sprayer is operated at a speed of 3 kmh^{-1}, and the desired application rate is 50 l per tree. The tree spacing is 10 m × 10 m, and each nozzle delivers 4 l min^{-1} at the operating pressure of 400 kPa. If one half row is sprayed from each side of the machine, find the number of nozzles needed. Also, find the area the sprayer can cover with one full load of spray in its 2000-litre tank.
3. In Example 9.8, solve for the amount of time the sprayer can be operated without losing the desired atomization, which happens when the discharge pressure of the nozzle falls below 200 kPa. Assume the nozzle discharge as 0.4 lpm.
4. The performance characteristics of a reciprocating pump used in a boom sprayer is given in Example 9.1 of this chapter. The pump is run at 500 rpm on a machine supplying a boom with 30 hollow cone nozzles, each with a discharge of 0.5 l min^{-1}, when operating at a pressure of 350 kPa. Calculate the power required to drive the pump and how much liquid is bypassed to the tank to maintain the desired pressure.
5. The coefficient of discharge of a solid cone nozzle, as obtained from experimenting with the discharges obtained at various pressures, is 0.12. Its orifice size is 2 mm. If the velocity for complete atomization is 25 ms^{-1}, find the minimum discharge pressure for complete atomization and the discharge from the nozzle at that pressure.
6. A hollow cone nozzle, which discharges 0.55 l min^{-1} at 350 kPa pressure, has a 70° spray angle. The nozzle, when set at a height of 45 cm from the horizontal target surface, provides an annular deposition pattern, the inner diameter of which is 70% of the outer diameter. If two such nozzles are used with an overlap of 50%, plot the total deposition with respect to the distance from the centre. (Hint: Use the software given in accompanying the book to find the deposition area.)
7. A backpack air compression sprayer has a cylindrical air-sealed spray tank of 15 cm diameter and 50 cm height. Spray fluid is generally put into the canister to a three-fourths level to allow free space over the fluid in the vessel. An air pump of 6 cm diameter and 35 cm height is provided concentrically inside the backpack tank, the stroke of which is 30 cm. The operator pumps air into the

tank by giving seven strokes to the pump handle. Neglect the wall thickness and find the pressure built over the fluid.

8. A spraying drone, the swath of which is 7 m as operated from a 2 m height over the crop, is being used to apply pesticide on cotton. It is operated at a speed 25 km h^{-1}. How many minutes are needed using 16 nozzles with a 0.55 lpm discharge rate each? Calculate the application rate. If its tank capacity is 40 litres, find the area covered per tank load. The drone demands about 10 kW for doing this job with a battery of 40,000 mAh capacity at 60 V. Find the area of field coverage per full charge of the battery.
9. By using mathematical integration, find the segmental areas Z1, Z2 and Z3 of the sprayed area of a single hollow cone nozzle that we calculated in problem 9.6 of the text book.
10. Solve for a generic method of geometrically calculating the strip areas (problem 9.6) that can be easily implemented on computer code.
11. A fan type nozzle generates a spray pattern in the shape of an ellipse as viewed in plan with a major axis of 48 cm and a minor axis of 20 cm. Theoretically arrive at the distribution of areas of 8 equal segments on the right hand side of y axis, based on which the spray pattern of this nozzle can be estimated.

References

Bell, B. (2005). *Farm machinery*. UK: Old pond Publ.

Bernachi, H., Haman, J., & Kanafojski, C. (1972). *Agricultural machines, theory and construction, 1*. Published for USDA and NSF by Foreign Cooperation Center of the Central Institute for Scientific, Technical and Economic Information, Hesperides Press.

Bode, L. E., & Butler, B. J. (1981). The three d's of droplet size; Diameter, drift, and deposit. AA-81-004 *ASAE Paper*.

Deysson, J. Y., & Karian, J. (1978). *1978 1st international conference on liquid atomization and spray systems approximate sizing of single fluid and pneumatic atomizers*.

Dombrowski, N., & Johns, W. R. (1963). The aerodynamic instability and disintegration of viscous liquid sheets. *Chemical Engineering Science, 18*(3), 203–214. https://doi.org/10.1016/0009-2509(63)85005-8, 00092509.

Dorman, R. G. (1952). The atomization of liquid in a flat spray. *British Journal of Applied Physics, 3*(6), 189–192. https://doi.org/10.1088/0508-3443/3/6/305, 05083443.

Ford, R. E., & Furmidge, C. G. L. (1967). The formation of drops from viscous Newtonian liquids sprayed through fan-jet nozzles. *British Journal of Applied Physics, 18*(3), 0508–3443. https://doi.org/10.1088/0508-3443/18/3/312

Kepner, R. A., Bainer, R., & Barger, E. L. (1978). *Principles of farm machinery*. Avi Publishing Company.

Klenin, N. I., Popov, I. F., & Sakun, V. A. (1970). *Agricultural machines. USDA and NSF by amerind*. Publishing Co. Pvt. Ltd, USDA and NSF by Amerind Publishing Co. Pvt. Ltd.

Lovegrove, H. T. (1968). *Crop production equipment*. Hutchinson press.

Marshall, W. R. (1954). Atomisation and spray drying. *Chem Eng. Prog. Monogr. Ser.*

Matsumoto, S., & Takashima, Y. (1978). *1978 1st international conference on liquid atomization and spray systems design criteria of hollow cone nozzle and prediction of drop size distribution*. Tokyo, Japan.

Ozkan, E. (2024). *Agricultural and biological engineering drones for spraying pesticides - opportunities and challenges - pesticide application technology. Department of food*, 2024 2024 https://ohioline.osu.edu/factsheet/fabe-540.

Smith, H. P. (1965). *Farm machinery and equipment*. Tata Mc-Graw Hill Pub. Co. Ltd.

Srivastava, A. K., Goering, C. E., Rohrbach, R. P., & Buckmaster, D. R. (2006). *Engineering principles of agricultural machines*. ASABE Publications.

CHAPTER

Machines for crop harvesting

10

Chapter outline

Harvesting of the matured crop is the final operation in a farm, whether it be a forage crop for livestock, a cereal crop for its grain, or a commercial crop for its fibre. The machines intended for this important operation (Bell, 2005; Culpin, 2013; Hunt, 2008; Kepner, Bainer, & Barger, 1978; Lovegrove, 1968; Smith, 1965; Srivastava, Goering, Rohrbach, & Buckmaster, 2006) should be designed in such a way that they do not damage the final product being harvested and that they collect all the matured products from the harvested crop without any wastage in the machine or field. For instance, a grain combine should handle the jobs of harvesting and separating the grain from the straw effectively without causing much grain loss.

Machines of the Farm. https://doi.org/10.1016/B978-0-443-32906-7.00010-4

10.1 Mowing

Mowing is the process of cutting grass and plant stock using an appropriate cutting mechanism. Lawn mowing is one such operation in which a lawn is cut to the required level using either a rotary or a reciprocating cutting knife. The mowing machine may or may not include a system for collecting the cut grass. A similar cutting mechanism is also used in grain harvesting machines for cutting and collecting matured plants in order to thresh the grain out and separate it.

10.1.1 Reciprocating mower

In a typical reciprocating mower, the cutting of a plant stem is achieved by two shear knives moving in opposite directions, thus causing a shearing action on the plant. Usually, the mower knife sections move linearly over a fixed set of knives. There are double-action mowers available, wherein both the knife units move opposite to each other.

A typical cutter bar (Figure 10.1) consists of a finger bar, which is an assembly of fingers fastened in a linear array. The job of the fingers is to divide the total swath of the standing crop into parts for each mower knife to cut. They guide the crop into the knifes' path. Importantly, these fingers also provide the cutting edge against which the moving knife shears the crop. The mower knife assembly carries triangular steel knife sections, each with cutting edges on adjacent sides of the triangle. The cutting edges of these knife sections may be either smooth or serrated. These knife sections are riveted onto a piece of flat steel in a linear array. One end of the knife assembly has a ball joint to connect with a pitman, which is nothing but a connecting rod driving the reciprocating knife assembly on the cutter bar. The ledger plate is the counterpart of the knife fixed on the guard (Figure 10.1); it is usually serrated on the underside and is replaceable. Wear plates are replaceable and adjustable steel plates at the back of the knife sections to bear the frontal thrust of the crop on the knife sections. Knife clips are holders fixed at regular intervals along the cutter bar; they hold down the knife assembly, at the same time allowing it to slide freely. Apart from the equi-spaced fingers on the cutter bar, inner and outer shoes are provided on the cutter bar's ends. These shoes are bigger in size and carry a skid at the bottom that slides on the ground. These skids are height adjustable and slide on the ground surface so as to alter the cut height of the crop being mowed. The outer shoe carries a swath board at its back; this board is inclined inward toward the cutter bar. The swath board collects and windrows the cut material on a row nearer to the inner shoe.

The whole cutter bar assembly is hinged to the mower sub-frame by a pair of hinge pins provided on the inner shoe of the cutter bar. The sub-frame is carried by the tractor (Figure 10.2) on the side or rear depending on the method of mower mounting. This allows for the swinging of the cutter bar vertically up when it is being transported off the field. The cutter bar is hence overhanging the sub-frame on these hinges. The sub-frame has a crank wheel that receives rotary power from the Power-Take-Off (PTO) of the tractor or a hydraulic motor powered by the tractor's

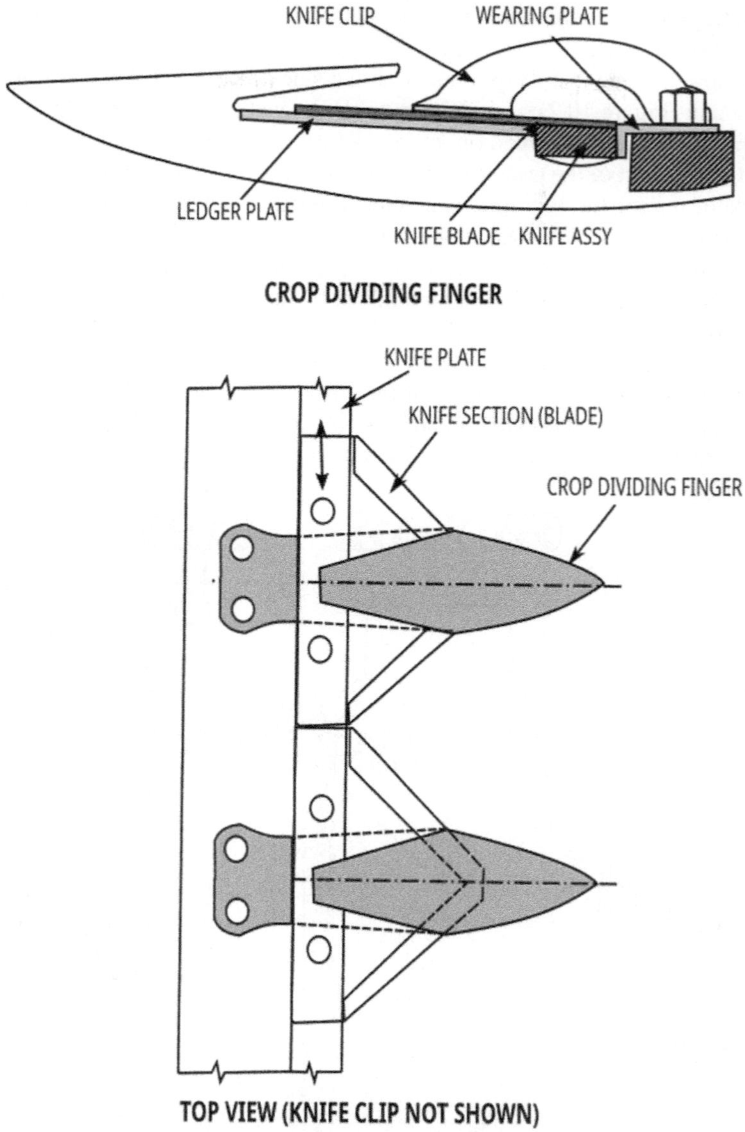

FIGURE 10.1 Parts of a Mower-Cutting Mechanism

hydraulics. The crank wheel drives the reciprocating knife assembly through a pitman connected between itself and the knife bar's ball joint. The pitman is usually made of wood so that it can fail and safeguard the knife sections from damage due to any overloading by stones or hard stalks. Typically, the knife bar is reciprocated at 1000 strokes per minute. The stroke of the knife bar is exactly equal to the pitch between adjacent fingers, which is same as the pitch between the knife sections on the knife assembly.

The mower frame is either mounted on the side of the tractor's chassis as a side-mounted unit (Figure 10.2) so that the cutter bar is well visible to the tractor operator or the mower frame is mounted on a three-point link on the rear and the cutter bar is oriented to the right of the tractor wheels. In both the cases, the cutting swath is on the right side of the tractor, with the inner shoe aligned outside the right wheels.

The first important adjustment on the mower is known as "the registration" of the mower knife. This is the process of exactly aligning the centre of the knife section and the centre of the finger on the same line at the dead ends of the knife stroke. The pitman has some adjustable shims between its ends, which can be added or removed so that its length is adjusted accurately. The crank radius is equal to the stroke of the knife and hence when one dead end of the stroke is registered exactly by adjusting the pitman's length, the other dead end of the stroke registers automatically. Improper registration leads to the inefficient cutting of a crop.

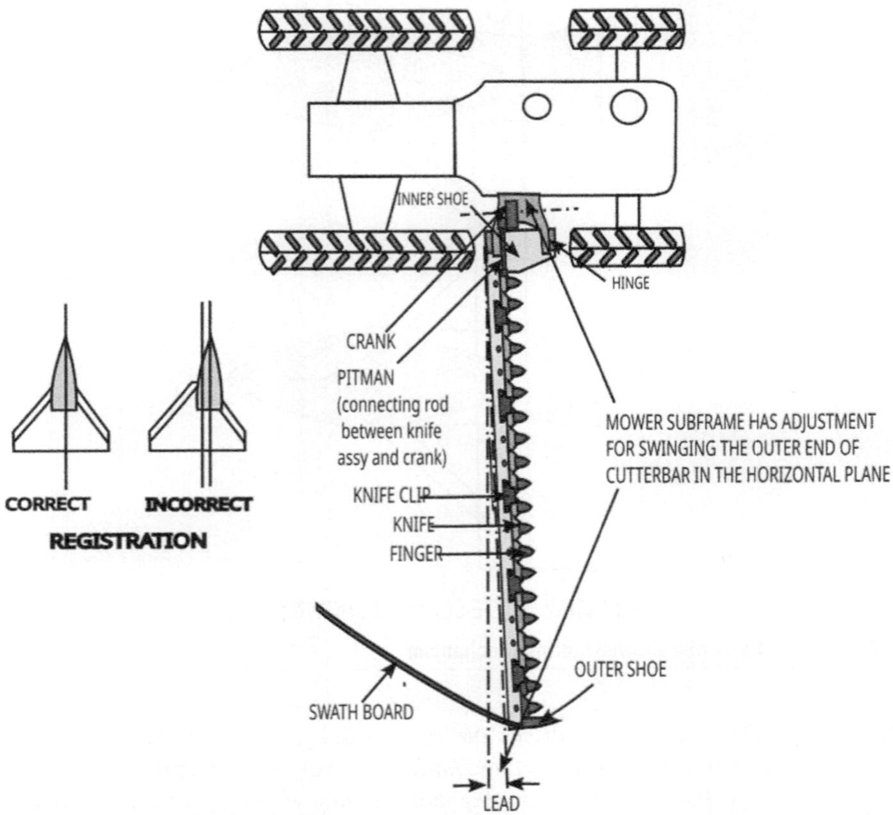

FIGURE 10.2 Tractor-Mounted Mower Cutter Bar Assembly

No Permission Required

The second adjustment concerning the mower is the cutter bar's "alignment" or "lead" with respect to the mower's forward travel (Figure 10.2). When the mower is cutting the crop and moving forward, the ground drag on the outer shoe's skid as well the frontal pressure of the crop on the cutter bar pushes the outer shoe back a bit, since the whole cutter bar is a cantilever beam supported by the tractor. This makes the orientation of the cutter bar not perpendicular to the direction of travel as may be desired. To alleviate this, the cutter bar's outer end is provided a deliberate lead by setting its outer shoe, leading the inner shoe by a certain extent as rotated in the horizontal plane. When the mower is operated, the drag on the cutter bar will straighten the lead and the cutter bar will travel perpendicularly to the travel direction. The mower sub-frame has a provision for rotating the cutter bar, thus setting up this lead.

10.1.1.1 Mower's cutting pattern

When a knife section is moving from left to right (Figure 10.3) between two fingers that are dividing the crop between them, that knife's right edge pushes all the crop stalks between the two fingers enveloped by the shown path toward the finger on the right. The path shown is that of the top-most centre of the knife section (Kepner, 1952), implying that the path of the whole right edge of the moving knife deflects the crop stalks toward the left cutting edge of the finger's ledger plate. It may be noted that the fingers are moving straight ahead along the direction of travel and the knife section is moving forward as well as to the right.

The incremental positions of both the knife section and the left finger are shown in the illustration. But, the actual cutting happens only at the points where the knife's

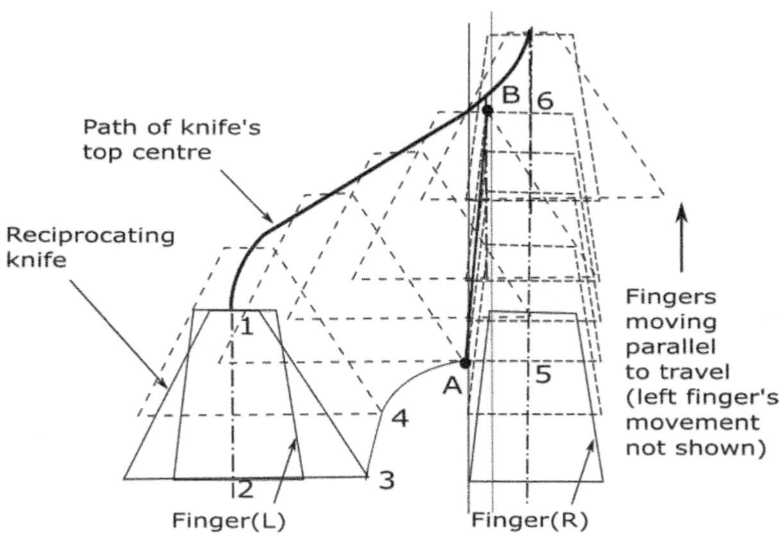

FIGURE 10.3 Cutting Pattern of a Reciprocating Mower

No Permission Required

right edge crosses the points of the left edge of the finger. The line "AB" connecting these points is the line at which all these stalks are sheared. The crop stalks enveloped by points "1234A56B" and the path of the knife's top edge are cut by this stroke. Half of the crop stalks that are in the front of the fingers are also pushed into the knife's path to be cut on this line. The same cutting pattern occurs when the same knife section reciprocates back toward the left finger. Because of the relatively higher speed of the reciprocating knife with respect to the travel speed, all the stalks squeezed between the two fingers are completely sheared.

10.1.1.2 Mechanics of cutting plant stem

Thecuttingofaplantstembyaknifeedgeisacomplexphenomenoninrespecttotheplantmaterial(Persson, 1987). Depending on the sharpness of the knife, which is the cross-sectional edge radius of the knife point, the plant is first compressed, causing the plant fibres along the stem to stretch and fail. Direct shear as well complex stresses are induced in these fibres, making them fail and thus causing the cutting action. Usually, the precut phase consumes more cutting energy involving the compression of the plant stem. Cutting occurs at the knife edge at a pressure exceeding 9 N/mm to 30 N/mm^{-2}. The cutting pressure is the force acting on the frontal cutting area, which is the projected area of the edge on the plant stem diameter.

The forces on the knife are illustrated in Figure 10.4. The force, F_x, in the direction of the knife motion is basically due to the resistance of the plant material to cutting. This involves mainly the frontal forces to the knife edge. It is understood that the cutting force, F_x, will be generated only if it is supported by a reaction. A mower-like mechanism has a counteracting knife on the knife guard, and the shearing of the plant stem happens between the moving knife and the knife edge on the guard. Here,

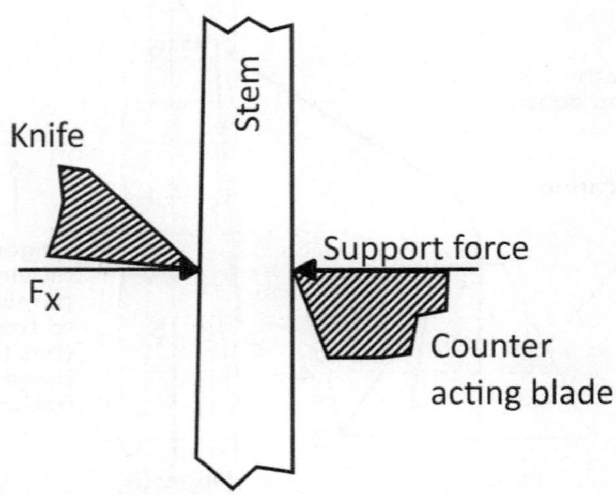

FIGURE 10.4 Forces on Plant Stem Cut by a Knife

No Permission Required

the reaction to the cutting force, F_x, is provided entirely by the knife edge on the guard. But when there is no counteracting knife to give the supporting reaction, as in the case of an impact cutting flail mower, the reaction has to be offered entirely by the plant stem itself through its bending strength of the stump below the height at which the cut is made. The plant's root system embedded in the soil provides the necessary reactive force through a moment about the ground surface (Ige and Finner, 1975). A simple equation for finding the velocity of impact cutting is to use F_b d = ½ mv^2, which is the relationship between the kinetic energy imparted to the cut plant and the work done by the impact force, F_b, traveling through the cutting distance, d, being equal to the stem diameter. Based on this relationship, the velocity of impact cutting is derived, comprising the plant's bending resistance, F_b (N); the mass of the plant stem that is cut, m (kg); F_x, the force required to initiate cutting based on the critical cutting pressure (N); and the diameter of the stem, d (m).

$$v = \left[2d\frac{(F_x - F_b)}{m}\right]^{1/2}$$

When the cut is made nearer to the ground, F_b tends to be nearer to F_x in magnitude, which reduces the minimum required knife velocity. Generally, velocities in the range of 50 to 75 ms^{-1} are required for the reliable cutting of plant stems by impact. An example problem showing the calculation of cutting force required is illustrated in Box 10.1.

Box 10.1 Example 10.1

Example 10.1. A fodder crop having a 5 mm stem diameter is cut by an impact cutting blade at a height of 35 mm above the ground. The mass of a single plant cut apart is 0.015 kg. Cutting occurs at the knife edge at a pressure exceeding 30 N mm^{-2}. The knife's edge radius (sharpness) is 0.25 mm. Find the force needed by the knife to cause the cutting of a single stem and the minimum knife velocity to make the cut by impact. Assume the ultimate tensile strength of the plant fibres as 40 N mm^{-2}.

Let us use the flexural formula $\frac{M}{z} = f$ for the bending of the stem as a cantilever beam
where M is the moment caused on the plant stem due to cutting,
which is the cutting force, F_b × (moment arm from ground to cutting point, L)
z, the section modulus of the circular section of the solid stem = moment of inertia / radius of the stem
and f, the ultimate tensile strength of the plant fibres.
For a circular solid section of the stem, $z = \frac{\pi d^3}{32} = \pi\,(5)^3 / 32 = 12.27\ mm^3$.
Then, F_b = 12.27 × 40 / 35 = 14 N.
The cutting frontal area of the knife edge is calculated as the projected area of the cutting edge on the stem = cutting edge thickness × cutting width = 0.25 × 5 = 1.25 mm^2
where the cutting width is the single stem's diameter.
The force required to initiate cutting will be the critical cutting pressure × cutting frontal area
F_x= 30 × 1.25 = 37.5 N.
The minimum cutting velocity by impact is

$$v = \left[2d\frac{(F_x - F_b)}{m}\right]^{1/2} \quad \text{with d in m}$$

$$v = \left[2 \times \frac{5}{1000} \times \frac{(37.5 - 14.0)}{0.015}\right]^{1/2} = 3.95\ ms^{-1}$$

10.1.2 Flail mowers

Flail mowers cut the crop by impact using flails, which rotate about a horizontal, transverse axis. Flails are steel blades that are hinged to the rotor axis and are stiffened by the centrifugal force on rotation. Flails have a knife edge and cut the crop by impact. The hinging action provides safety for the flails in fields with stones and hard obstructions. Various types of flail knives are used on flail mowers, as shown in Figure 10.5.

Flails are staggered in successive rows on the rotor, and the knife edges being broad provide complete coverage of the cutting swath of the machine as well as a uniform height of the cut. The shroud covering the flail assembly bends the crop forward, so that the height of the cut can be reasonably closer to the ground. This also avoids recutting of the cut crop by the successive flail, which may waste small pieces of cut fodder from the machine. Flail mowers perform badly on a severely lodged crop.

Figure 10.6 shows the loci of the blade tips of a mower axis mounted with two opposing flails 180 deg apart on the rotor axis. As the rotor with flails is rotated and

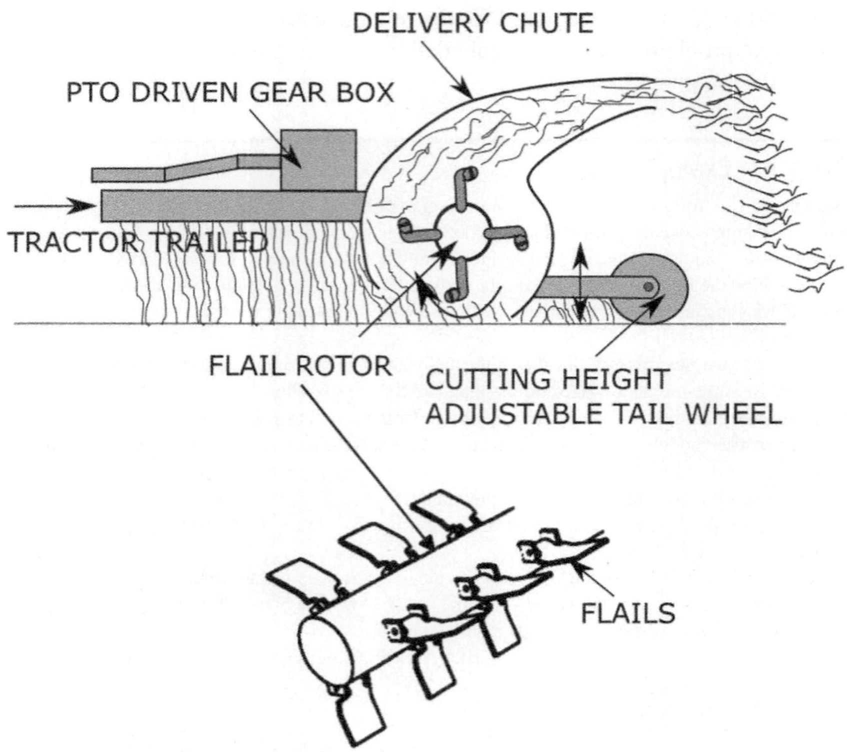

FIGURE 10.5 Flail Mower – Its Operation

No Permission Required

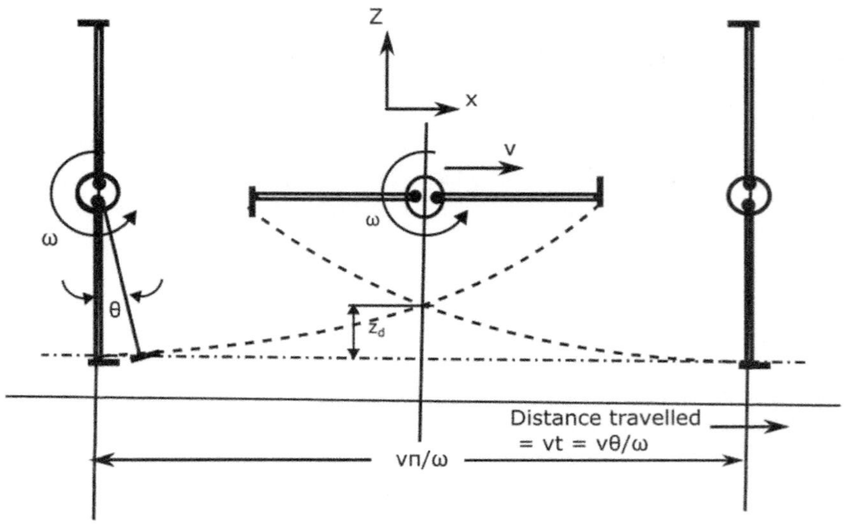

FIGURE 10.6 Loci of the Cutting Tips of a Flail Mower

No Permission Required

the mower moves forward, the tips of the flail blades follow a cycloidal path, as shown. The x and z coordinates of the path can be calculated using the following equations (Bernachi, Haman, & Kanafojski, 1972; Klenin, Popov, & Sakun, 1970):

$$x = v_f t + r sin(\theta) = \frac{v_f \theta}{\omega} + r sin(\theta)$$

$$z = r - r cos(\theta)$$

where x = the displacement of the tip in the x-direction, m;
z = the corresponding displacement of the tip in the z-direction, m;
r = the radius of the blade tip of the flail or rotary mower, m;
v_f = the forward velocity of mower, ms^{-1};
θ = the angular displacement of the rotor, radians = ωt;
t = time, s;
ω = the angular velocity of the flail or rotor blades, radian s^{-1};

Since θ = ωt, it may be noted that $\frac{v_f\theta}{\omega} = v_f t$, which is the distance moved by the mower for the incremental rotation of the rotor.

The locus of the blade tip on flail 1 is shown displacing angularly through 90° from a vertical orientation to a horizontal orientation, covering a 90° rotation of axis as well moving forward with the mower. Similarly, the locus of the second flail is shown for the next 90° rotation of the rotor axis moving the second flail from a horizontal orientation to a vertical orientation. The plot shows the loci of the two blade tips for the mower moving through 180° or π radians of the rotor's rotation given by $\frac{v_f \pi}{\omega}$. The intersecting loci of the two blades show that the cut is not evenly flat and is uneven by a height of z_d where they intersect.

It can be logically understood that the stubble height unevenness, z_d, will increase if v_f increases in relation to the flail tip's peripheral speed, v_p. Since $v_p = \omega r$, the unevenness is influenced by the rotor's rotational speed and the flail radius. The unevenness can, of course, be reduced by adding more flails on the same plane of rotation. That is, increasing the number of flails from two to three or four will reduce the unevenness markedly. Based on this, an approximate relationship has been arrived as:

$$\frac{z_d}{r} = 1 - \cos\left[\frac{\pi}{\lambda_r(1 + C_v)}\right]$$

where λ_r = the number of flail blades on the same plane of rotation

and $C_v = v_p / v_f$ = the velocity ratio of the flail's peripheral velocity to the mower's forward velocity.

Typically, the peripheral velocity is designed to be at about 45 ms^{-1} and the C_v ratio is designed to be more than 10, so that uneven cutting can be avoided. Generally, the flail tips are wide, and, by staggering the rows of flails on the rotor along the axis, the number of tip edges rotating on the same plane is increased, offering the desired evenness of cut. The following example, furnished in Box 10.2, elaborates on the calculation of the height of cut of fodder in relation to the displacement of the flail's tip.

10.1.3 Forage harvesters

The flail mower explained in the last section is a direct cut and throw type of forage harvester that can handle tall grasses. However, the length of cut of the fodder is not precise, as can be visualized from its working mechanism. It is a direct throw-type machine in which the peripheral velocity, v, of the flails is imparted to the cut pieces of fodder, which throw them through a height, h, corresponding to the expression

$$v^2 = 2gh$$

by equating the kinetic energy, $\frac{1}{2}mv^2$, imparted to the cut pieces to the potential energy, *mgh*. The pieces that are thrown out are sent out through the delivery spout and can be deposited only over a short distance.

Another major category of a forage harvester (Figure 10.7) is the type that cuts the fodder into precisely long pieces and then blows them out the delivery spout by using an air blower. These machines (O'Dogherty, 1982) again may be using horizontal flails, vertical cutter blades, or conventional reciprocating cutter bars to cut the long fodder stalks and pass them in bulk to a cylindrical cutter head. The cut stalks are fed by feed rollers and metered precisely to the cylindrical cutter head (Figure 10.8), which has cutting blades arranged on the cylinder's periphery. The stalks are cut into pieces of a precise length, according to the settings made, and passed to another set of feed rollers that convey the cut bulk into a chamber from where an air blower blows them through a delivery spout at greater heights and

Box 10.2 Example 10.2

Example 10.2. Calculate the displacement of a single flail's tip in the travel direction and the corresponding height of cut from the ground surface for a flail mower operating at 5 kmh^{-1}. The rotary speed of the flail with a tip radius of 40 cm is 300 rpm. Use Excel for the calculation, and plot the height of cut in relation to forward travel.

ω = angular velocity of the flail rotor $= \frac{2\pi N}{60} = \frac{2\pi \times 300}{60} = 31.5$ radian s^{-1}

v = forward velocity of mower = 5 kmh^{-1} = 5000/3600 = 1.39 ms^{-1}

Let us calculate x and z for 30° incremental values of θ using the equations explained earlier. The only change is that the cos and sin functions are alternated according to the initial position of the flail from which the displacements are calculated. Here, we assume that the single flail starts to rotate from the horizontal orientation rotating anti-clockwise. And, it should also be noted that the "x" and' "y" displacements of the flail tip are with reference to the rotor axis. Negative values of "x" indicate that the tip is ahead of the axis in the travel direction. The "y" displacement being equal to zero corresponds to the desired cutting height from the ground level. The Excel calculations are as below according to the equations below:

$x = vt + rcos(\theta) = \frac{v_f\theta}{\omega} + rcos(\theta)$

$z = r - rsin(\theta).$

The table and plot below were prepared in Excel.

Angular displacement of rotor	Angular displacement of rotor axis	Distance travelled by axis	Horizontal displacement of rotor tip	Vertical displacement of rotor tip
θ, deg	θ, rad	$\frac{v\theta}{\omega}$, m	x, m	z, m
0	0.000	0.000	-0.400	0.400
30	0.524	0.023	-0.323	0.200
60	1.047	0.047	-0.153	0.054
90	1.571	0.070	0.070	0.000
120	2.094	0.093	0.293	0.054
150	2.618	0.116	0.463	0.200
180	3.142	0.140	0.540	0.400

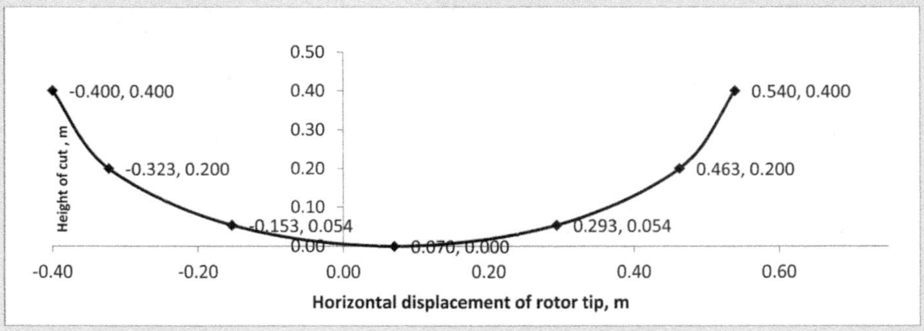

FIGURE 10.7 Forage Harvester with Spinner Discs for Precise Length of Cut (FR920, New Holland, CNHI, USA)

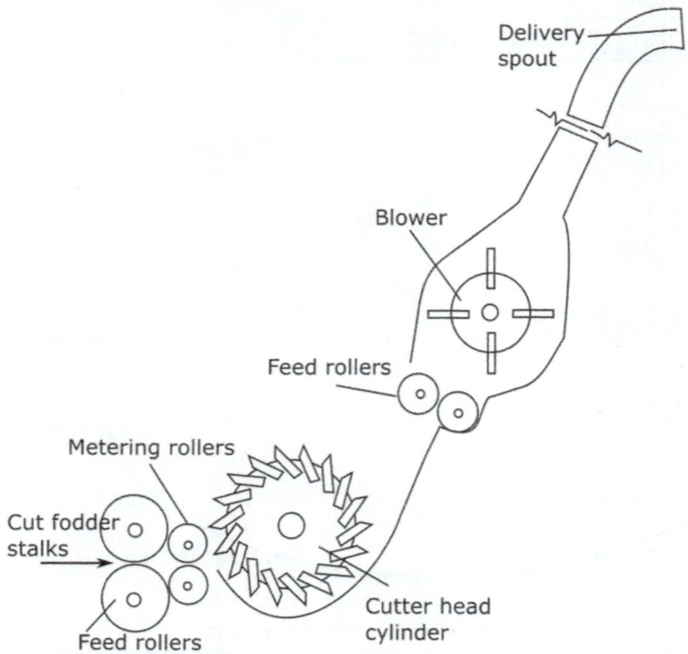

FIGURE 10.8 Flow Path of Fodder in the Forage Harvester

No Permission Required

Box 10.3 Example 10.3

Example 10.3. A 4-m wide forage harvester that uses cutter bars to harvest fodder corn yielding 35 tonnes ha^{-1} is operated at 6 kmh^{-1}. The length of the cutter head cylinder is 60 cm, the number of knives used is 8, and its rotational speed is 900 rpm. The throat height of the feed window to the cutter head cylinder is 10 cm. Fodder cut from the wider swath is conveyed by suitable augers and feed rollers to the cutter head cylinder. Assuming the density of the fodder bulk fed to the cutter head is 400 kgm^{-3}, find the length of cut delivered by the harvester.

Swath of the harvester = 4 m

Area of coverage $= \frac{6000m}{60\ min} \times 4m = 400m^2\ min^{-1}$,

Weight of fodder cut per minute = fodder yield × area covered

$= \frac{35000}{10000} kgm^{-2} \times 400m^2\ min^{-1} = 1400kg\ min^{-1}$,

Volume of fodder cut per min $= \frac{1400kg\ min^{-1}}{400kgm^{-3}} = 3.5m^3\ min^{-1}$,

Width of cutter cylinder = 60 cm = 0.6 m,

Height of throat window = 10 cm = 0.1 m,

Throat area = 0.06 m^2,

Length of fodder bulk fed /min $= \frac{3.5m^3\ min^{-1}}{0.06m^2} = 58.3mmin^{-1}$ = velocity of fed fodder.

∴, the length of cut of fodder $l = \frac{v_{feed}}{nN} = \frac{58.3}{8 \times 900} = 0.008m = 8mm$

onto a transport that travels alongside the forage harvester. Row crop forage harvesters that divide a row crop, as in the case of grain combines, are available. These harvesters cut the fodder stalks using conventional cutter bars. Vertical spinner discs are used in some models to direct even lodged fodder crop into the cutter bars.

The precise length of cut fodder pieces is influenced by the number of cutter head knives, the rotational speed of the cutter cylinder, and the linear speed of the fodder being fed into the cutter head cylinder.

The length of cut $l = \frac{v_{feed}}{nN}$ in m

where v_{feed}, in m min^{-1}, is the fodder feed speed, n is the number of knives on the periphery, and N is the rotational speed of the cutter head in rpm. The example problem worked out in Box 10.3 illustrates the relationship between the length of cut and its influencing factors.

10.2 Grain harvesting

Most of the cereal crops such as wheat, rice, and millet are now harvested with combined harvesters known as grain combines. Throughout the world, direct combining has become the standard for harvesting these crops. The operations of a typical self-propelled combine (Figure 10.9) are, respectively, reeling a standing crop into a cutting mechanism, conveying the cut crop with the ear head to the threshing drum, threshing the grain off the ear heads, separating the grain and chaff from the straw, winnowing the chaff and other debris from the grain, storing the cleaned grain, and, finally, ejecting the straw and chaff from the machine to the field.

The combine (Figure 10.10) has a reel at the front, which rotates at a slow rotational speed and pushes the standing crop toward the cutter bar at the front of the

FIGURE 10.9 Self-Propelled Grain Combine (CR 10.93 Grain Combine, New Holland, CNHI, USA)

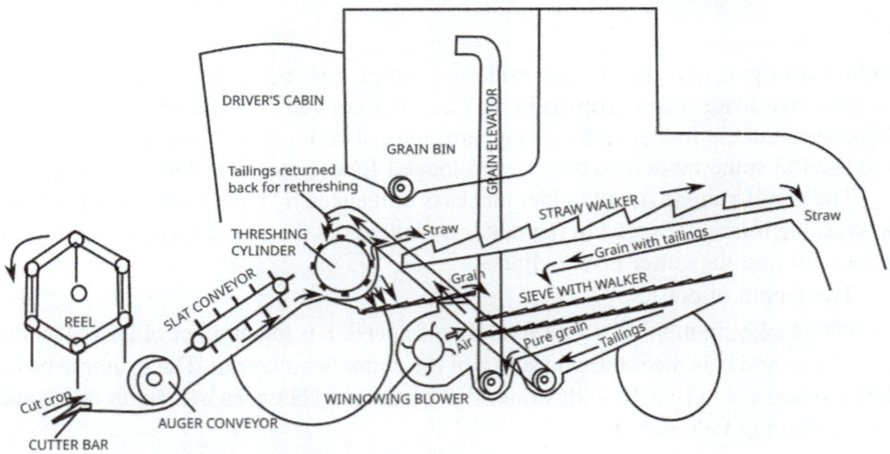

FIGURE 10.10 Flow Path of a Self-Propelled Grain Combine

No Permission Required

machine. The cutter bar is something very similar to that of a mower and shears off the standing crop at an adjustable height of cut from the ground level. The cut crop thus falls into the auger conveyor, which is arranged to cover the whole width of the cutter bar. The conveyor's job is to convey the cut crop from the left and the right to the centre of the machine. Usually, the cutting swath of the combine is much wider than the machine itself. All the mechanisms following the cutter bar are fit into the machine's width. So, the cut material is conveyed to the centre of the machine and then conveyed up to the machine level by a slat conveyor.

The combine header is the unit that has the aforesaid gathering and cutting mechanisms. Pickup reels are the most common type of gathering mechanism; they have radial fingers that can reach even below the lodged crops to help the cutting mechanism cut the crop effectively. The reel is usually power driven and has four to eight slats or arrays of fingers on radial arms. The fingers are made to orient in the proper sweep angles, controlled by cams and parallel bar linkages. The proper orientation of these fingers that push the crop toward the cutter bar is important in reducing the gathering losses of the machine. The approach speed of the fingers or slats also influences these losses due to the shattering of the grain from the crop. The reel index is the ratio between its peripheral speed and the machine's forward speed and is expected to be from 1.25 to 1.5. The cutter bar of a combine operates at 800 to 900 strokes per minute, which is usually lower than that of a conventional mower to reduce the grain shattering from the ear head. The cutter bar, while cutting the crop, also causes considerable grain shedding, but much of it is collected by the header since the grain head is effectively bent into the machine header by the reel. The height of cut as set by the header also affects the performance of gathering and cutting of the crop.

The slat conveyor in turn feeds the crop into the threshing cylinder. The threshing cylinder is the heart of the machine, and it beats the cut crop between itself and a matching concave below it, which separates the grain and chaff from the ear heads. The conveying auger is absent in small, straight-through combines, the working width of which is less than 1.5 m. Here, the cut material is directly taken into the machine since the machine's width is the same as that of the cutter bar.

Threshing is achieved when the cut crop with grain heads passes through the clearance between the rotating threshing cylinder and the concave grating positioned below it. The concave assembly is pivoted at the rear, whereas the front is adjustable vertically to control the amount of clearance between the cylinder and concave. In a conventional threshing cylinder, the crop is fed in a direction transverse to the cylinder's rotation; in an axial flow cylinder, it is fed parallel to the rotating cylinder. Three different configurations are used on the surface of conventional threshing cylinders, namely, the rasp bar, angle bar, and spike tooth. All these configurations work against grated concaves. The most commonly used rasp bars are cast bars with shallow grooves, fastened parallel to the axis of the rotating cylinder; they rub off the grain from the ear heads against the concave grate made of parallel bars running transverse to the cylinder axis. The angle bars are typically rubber-coated angle sections placed helically on the threshing cylinder's surface; they thresh the crop against the rubber-coated concave grates. They perform quite well on small-grained crops such as barley. The spiked configuration is less common; it uses spikelike projections on the cylinder to tear and rub the grain against a grated concave. The threshing performance or efficiency of the threshing cylinder is the percentage of grains threshed from those on the ear heads that are fed to the cylinder. The speed of the cylinder is vastly variable, between 150 rpm to 1500 rpm, and set depending on the type of crop, its maturity, and its moisture (Srivastava, Mahoney, and West, 1990). The peripheral speed of the cylinder is the real influencing factor directly affecting the threshing performance; higher speeds result in more grain damage (Arnold, 1964). Threshing cylinder speeds are adjustable by means of the appropriate drives. The operator's manual of the combine provides details on how and at what level these adjustments should be for each crop. The adjustable concave clearance is crucial and influences the threshing performance and grain damage (Cooper, 1971). The smaller the clearance is, the greater is the threshing efficiency but at the expense of more grain damage. The cut crop's feed rate into the threshing cylinder is another factor influencing the threshing performance and depends mainly on the forward speed of the machine. Excessive feed rates cause clogging and reduce the threshing performance. The number of rasp bars or angle bars does affect the threshing performance, too. Moist crops are hard to thresh and result in excessive grain losses (Srivastava, Mahoney and West, 1990).

The grain and chaff separated by the threshing cylinder fall through the concave's grate directly onto an oscillating grain pan, also acting as a conveyor and dust separator. The threshing cylinder, after striping the ear heads from the stalk, moves the straw over to a straw walker. The oscillating straw walker has ramplike conveyor slats, which move the straw toward the back of the machine for ejection at the back. The oscillations help in shaking off the grain and grain with tailings, which adhere to the straw and are conveyed along with the straw. Although most of the grain is threshed from the ear heads, grain that goes along with the straw after threshing is separated from the straw by the vibratory movement of the straw walker. In axial flow threshers, rotary separators that use centrifugal force to shake the loose grain from the threshed straw are provided. The rate of grain separation is found to be exponentially decreasing with the distance the threshed straw travels on the straw walker. Most of the separation happens within a 2 m length of the straw walker, even at a higher mass rate of material flow (Reed, Zoerb, and Bigsby, 1974). The separation efficiency of the straw walker or rotary separator is the ratio of grains separated to the amount that is entering it. The amount of grain that sticks with the grain even after the straw walker and leaves the combine at the rear is termed the separator loss. As may be visualized, the separating efficiency is influenced by the length of the walker, the walker's crank displacement, the crank speed, and the straw feed rate.

The grain and tailings fall through the straw walker onto another oscillating sieve-cum-conveyor below, through which the winnowing fan blows air. The high-velocity air from the winnower blows out all the chaff and short stalks of straw. The relatively heavier tailings with grain and pure grain fall through to another sieve known as a chaffer sieve at the bottom, where they are again winnowed by the same stream of air. Since the terminal velocity of the chaff is about 1 to 2 m s^{-1}, the chaff is blown away by the air stream, leaving behind the heavier grain and tailings. Here, the sieve openings are designed such that only the pure grain falls through and the grain with tailings moves toward the back and falls over the back edge into a tailing conveyor. The tailing conveyor is a screw conveyor-and-bucket elevator set, which collects all these unthreshed tailings with grain and cycles it back into the threshing cylinder for threshing. The tailings are once again threshed and winnowed to separate the pure grain and tailings. Meanwhile, the pure grain falling from the second oscillating sieve mentioned earlier is collected by yet another screw conveyor and elevated by a bucket conveyor into a grain storage tank. The front portion of the combine, which carries the cutter bar, the screw conveyor, and the slat conveyor, is called the combine's header and is typically hinged to the combine's chassis. The header can be raised and lowered through hydraulic cylinders such that the height of cut can be adjusted and the header can be raised for road travel.

Pure grain is lost in the combine in each step of the combine's operation (Nyborg, McColly, and Hinkle, 1969). Header losses include the grain lost during the cutting and conveying operations. The action of the reel as well as the cutter bar can shatter the grain from the ear head, tossing it to the ground, and depends very much on the maturity and variety of the crop being harvested. Rack or walker losses are the threshed and unthreshed grain carried along with the straw over the straw walkers and ejected from the machine. Shoe or cleaner losses consist of threshed grain and unthreshed grain that escape from the cleaning unit along with the chaff winnowed by an excessive air blast. Cylinder losses are the unthreshed grain that escapes with the straw discharged from the rear of the machine, either along with the straw or the material blown away by the winnower fan. The cleaning efficiency depends on the sieves provided in the cleaning section and the air blast velocity of the winnower fan. The air flow rate is controlled by means of adjustable shutters on the fan's inlet and by changing the fan speed. These are also set according to the recommendations in the manual for a particular crop. The calculation of efficiencies at each stage of operation of a grain combine is illustrated by an example problem worked out in Box 10.4.

10.3 Baling

Rice and wheat straw are left on the field after harvesting the crop using a combine. Except for using a minimal part of it as cattle fodder, the disposal of the rest is a huge problem for farmers in Asia. particularly in India. So, the straw is usually burnt in the field to clear it off for the next sowing season. This has been causing a huge air pollution hazard. Straw-handling machines such as balers, which are in vogue in India now and already prevalent elsewhere, can bale the hay as collected from the field. These straw bales are successfully used for generating energy in gasifiers and pollution-minimizing furnaces and as fodder for cattle.

Baling is essentially a packaging operation for straw and fodder. Baling makes the handling, transporting, and storing of fodder and straw easier. This is because baled straw (0.0065 m^3kg^{-1}) reduces the volume and space requirements by about three times compared with long, loose straw (0.018 m^3kg^{-1}).

10.3.1 Round-bale machines

The schematic view of a round baler (Figure 10.11) is shown in Figure 10.12A. The straw pickup unit picks the straw from the ground and transfers it to the feed rotor and then to the floor roller. Pickup units are simple fixed-spring teeth devices that pick loose straw from the ground and take it up to the machine level. Pickup units

Box 10.4 Example 10.4

Example 10.4. A rice combine having a cutting width of 3.5 m was tested for its performance. Grain collected at each stage of the flow path was weighed over a time span of 20 s in which the combine harvested crop over a forward travel of 12 m. The free threshed grain and unthreshed grain on the straw walker (rack) were 70 g and 55 g, respectively, of the total material, including straw on the walker being 8.5 kg. The total material on the cleaning sieve was 4 kg, of which the free threshed grain weight was 200 g and the unthreshed grain was 75 g. The total cleaned grain collected in the final grain outlet was 16 kg. Calculate the cylinder, walker, and cleaning losses as a percent of the total yield. Also calculate the grain yield and total losses in kg/ha.

Total crop material in the flow path as throughput = 8.5 + 4 + 16 = 28.5 kg.

Threshed grain and unthreshed grain on the straw walker = 70 + 55 = 125 g.

Threshed and unthreshed grain over the cleaning sieve = 200 + 75 = 275 g.

Total unthreshed grain on the cleaning sieve and walker = 55 + 75 = 130 g.

Total threshed grain on the cleaning sieve and walker = 200 + 70 = 270 g.

Total grain available = final grain output + total unthreshed grain on cleaning sieve and walker + total threshed grain on cleaning sieve and walker
= 16000 + 130 + 270 = 16400 g.

Cylinder loss = unthreshed grain in the flow path / total grain available = 130 × 100 / 16400 = 0.79%.

Walker loss = threshed grain and unthreshed grain on the straw walker / total grain available = 125 ×100 /16400 = 0.76%.

Cleaning loss = threshed grain and unthreshed grain on the cleaning sieve / total grain available = 275 × 100 /16400 = 1.67%.

∴, the total losses = 2.43% (cylinder losses are not included since they were already accounted for in the other two losses).

Overall recovery of grain from harvested crop = final grain output / total material throughput = 16 /28.5 = 56%.

Swath width = 3.5 m.

Grain output ha^{-1} = grain collected in test area / test area
= 16 kg / 12 m × 3.5 m = 0.38 kg/m^2 = 3810 kg ha^{-1}.

Total losses ha^{-1} = total loss in grain / test area = 0.4 kg / 12 × 3.5 = 0.0095 kg/m^2 = 95 kg ha^{-1}.

are driven by a ground wheel or from the PTO. A ground drive maintains a constant speed ratio between the picker drum and the forward travel. The peripheral speed of the picker fingers are typically 10% to 20% greater than the forward speed. The floor roller receives the straw from the picker and, in turn, feeds the collected straw to the bar and chain conveyor, which acts as the circular bale-forming chamber. A conveyor is shown in the illustration as a centre line with a few visible roller chain links. The inner surface of the conveyor is in the shape of a cylinder and is driven by a drive sprocket on one end and a driven sprocket on the other. The mechanical drive is obtained from the PTO drive to the machine from the tractor. The floor roller and the bar conveyor, working together, begin to roll and form the bale inside the bale chamber. At this time, the tailgate is closed and latched by the tailgate latches provided on both sides of the bale chamber walls. The bale density springs are retracted and hold the rubber stop firmly against the tailgate frame that can open outward and upward on a top hinge. As the baler moves forward, more and

FIGURE 10.11 Tractor-Operated Round Baler (Pro-Belt Baler, New Holland, CNHI, USA)

No Permission Required

more straw is picked up and the bale chamber rolls and curls the incoming straw layers freely under the combined action of the floor roller and bar conveyor. As more straw is fed in, the floor roller, conveyor, and stripper roll continue to roll and compact the straw layers until the desired bale density is reached. The compacted bale thus exerts pressure on the tailgate. Restraining this pressure to open the tailgate, the tailgate is hooked by the latches attached to the mainframe and the density springs attached to the tailgate. The density sensor springs are adjustable and sense the bale density in the chamber. Once the bale has reached the desired density, the operator gets an indication, stops the forward travel of the baler, and engages the bale wrapper mechanism. The tailgate remains hooked in the tailgate latches provided on both sides of the chamber.

During standstill, the rotating bale is wrapped with jute twine fed from a twine spool. The twine arm mechanism moves the twine arm with the twine down to the middle behind the duckbill. The duckbill is similar to the shuttle in a sewing machine. It moves down to pick up the twine that runs through the twine holder and feeds the twine to the rotating bale through the slot between the conveyor drive shaft and the stripper roll. The rotating bale picks up the twine and the duckbill retracts. The twine mechanism puts wraps around the bale. When the twine arm reaches the end of the bale, transiting from the other end, the duckbill moves up to the position where the bale is formed and pulls the twine cutter up to sever the twine. The twine wrap arm's movement is programmed to follow a definite path along the axial length of the bale, providing the appropriate dwells so that twine wrapping is achieved from end to end. The overlapping rows of twine on the bale lock the twine on the bale and require no knotting. When the wrap cycle is finished, the operator

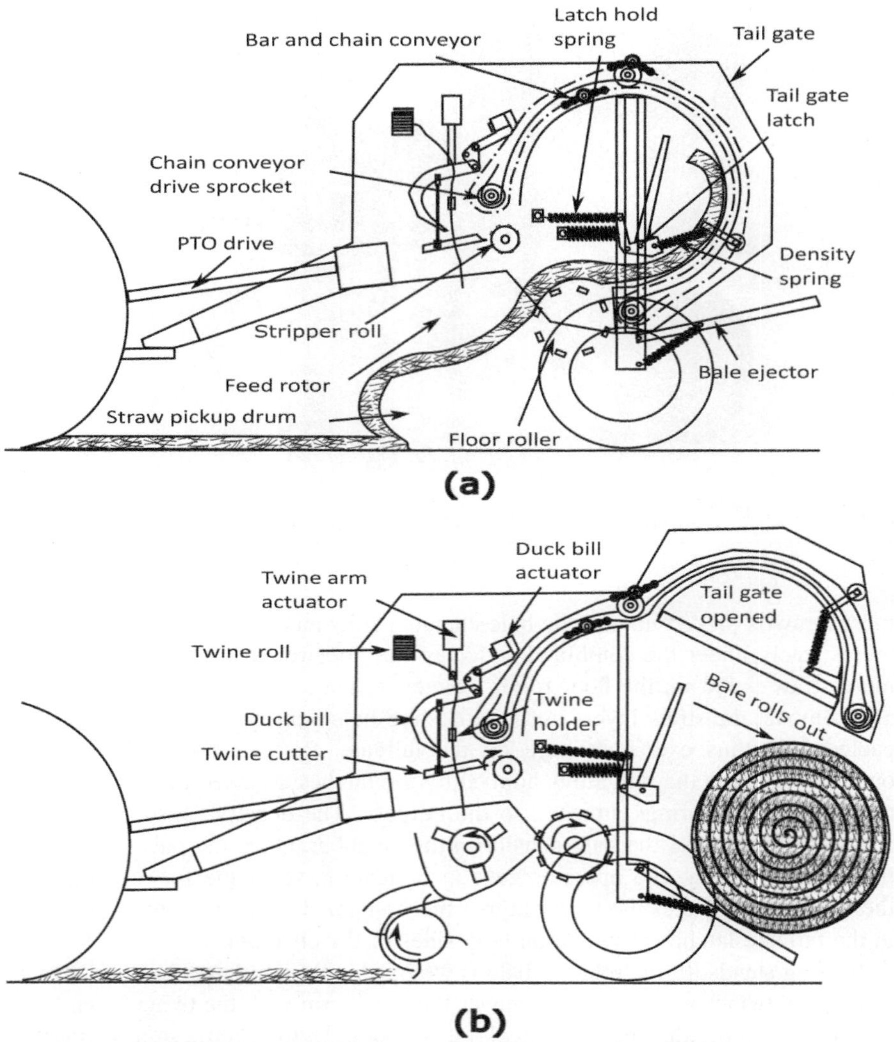

FIGURE 10.12 Components and Working of a Tractor-Operated Round Baler

No Permission Required

knows that the finished bale is wrapped and ready to be ejected. While the PTO is still engaged and the formed bale is rotating in the chamber, the operator opens the tailgate by actuating the tractor hydraulics connected to the baler. During bale ejection, the single-acting tailgate cylinders on either side of the chamber release the tailgate latches and then lift the tailgate up and outward, as shown in Figure 10.12B. The bar and chain conveyor provide for parting the chamber into halves, so that the tailgate section of the conveyor can be moved up with the

Box 10.5 Example 10.5

Example 10.5. A round baler capable of making 120 cm × 120 cm bales operates at a speed of 5 km h^{-1} to bale rice straw after harvesting with a combine. The typical straw yield of rice is 5000 kg ha^{-1}. The expected bale density of the machine is 150 kg m^{-3}. Calculate the distance at which each bale is ejected. Also calculate the theoretical number of bales made by the baler per hour.

Volume of bale $= \frac{\pi d^2 l}{4} = 1.35\ m^3$ where d = bale diameter = 1.2 m and l = bale length = 1.2 m.

Expected bale density = 150 kg m^{-3}.

∴, bale weight = 150 × 1.35 = 200 kg.

Collection swath of the baler = bale width = 1.2 m.

Straw available per m^2 area of collection = 5000 kg ha^{-1} = 0.5 kg m^{-2}.

Straw collected by the baler per m run = straw available per m^2 area × swath width $= 0.5 \times 1.2\ m = 0.6\ kg\ m^{-1}$.

When 200 kg is collected to form one bale, the distance run by the baler $= \frac{200 kg}{0.6 kg m^{-1}} / = 333\ m$.

If the baler is operated at 5 km h^{-1}, then the theoretical number of bales $= \frac{\text{speed of baler}}{\text{distance run by the baler to make one bale}}$

$= \frac{5000 m h^{-1}}{333 m bale^{-1}} = 15\ \text{bales}\ h^{-1}$.

Assuming a field efficiency of 70 percent, the actual number of 200 kg bales made = 11 bales h^{-1}.

tailgate. Upon the opening of the tailgate, the bale rolls from the bale chamber, over the bale ejector, to the ground. The operator then closes the tailgate control, and the tailgate comes down by its own weight, retracting the single acting tailgate cylinders. When the tailgate contacts the front half of the baler frame, the tailgate latches are re-engaged by the latch springs. The baler is now ready for the next baling cycle, and the operator moves the baler forward. The following example worked out in Box 10.5 calculates the throughput capacity of a round baler at a given speed of operation.

10.4 Corn combine

A corn combine (Figure 10.13) is usually a multi-row machine that snaps corn cobs from the standing crop stalks, dehusks the cobs to remove the cob sheath, threshes the cobs, and, finally, cleans the corn kernels for storage. The sheath and threshed cobs are ejected at the back of the combine along with the cleaned debris after winnowing the kernels.

The machine (Figure 10.14) has crop dividers at its front, the job of which is to follow the row of corn stalks, directing the row of plants into the corn header. These dividers are made tall, from sheet metal, and cover almost the height of the corn stalks, thus directing the crop effectively into the machine. All the dividers are hinged independently to the machine frame such that they can follow the furrows on which the corn crop is usually planted. Thus, the dividers pick up the stalks that have lodged close to the ground. As the stalks are guided into the throats between the dividers, gathering chains feed the rooted crop stalks in between the snapping rolls on the corn header. The snapping rolls pull the stalks between themselves

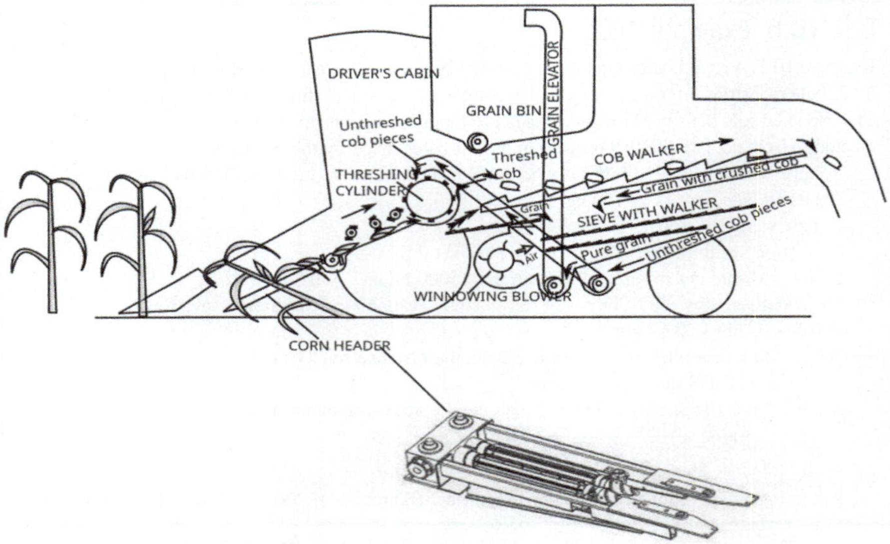

FIGURE 10.13 Flow Path of a Self-Propelled Corn Combine

No Permission Required

FIGURE 10.14 Self-Propelled 12-Row Corn Header and Combine (New Holland, CNHI, USA)

and pinch off the cobs from the stalks. The gap between the snapping rolls could be adjusted so as to match the average stalk thickness for effective picking. The snapping rolls extend to form the dehusking roll. Pairs of dehusking rolls, held together under spring pressure, de-sheath the corn cobs. One of each pair of dehusking rolls is made of rubber. The corrugations and projections on the dehusking rolls grasp the sheaths (husk) and pull them through the rolls so that the sheaths are stripped clean off the cobs without causing any threshing effect. One of the rolls is set a little higher than the other so that the dehusked cobs fall to the side, clearing the roller assembly, and fall into the conveyor, which moves the cobs forward into the machine's thresher.

The thresher consists of a rasp bar or peg-studded cylinder rotating inside a cylindrical concave. The threshed kernels drop through the holes in the concave, while the husks, cobs, and fine trash are ejected and blown out at the end of the thresher. After threshing the dehusked cobs, the kernels fall through the concave onto sieves and are winnowed, similarly to the grain combine. Threshing results in crushed cob pieces with kernels still unthreshed on these pieces. These cob pieces are transported on the sieves and drop at the back of the sieves to be recycled back into the threshing mechanism. The cleaned kernels are elevated and dropped into a storage tank. Figure 10.14 illustrates a self-propelled multi-row corn combine.

The main factors that influence the performance of corn pickers are the condition of the crop stand, timeliness of the harvest, stiffness of the corn stalks, height of the cobs on the stalks, size of the cobs, hardness of the cobs, affecting the threshing, and other machine characteristics such as speed of travel.

10.5 Sugarcane harvesters

Sugarcane is grown in the tropical and sub-tropical climates of countries in Asia, Africa, Latin America, and Australia. Sugarcane cultivation practices vary from country to country and within a country. Row-to-row spacing in India varies from 60 cm to 100 cm, while in many other countries, it is from 150 cm to 180 cm.

Sugarcane harvesters (Figure 10.15) are huge machines that completely combine all the operations in the harvesting of cane. The first operation is the severing of sugarcane leaves, known as topping. Then, the rows of cane are divided and directed into the machine. The cane is cut at the base near or a bit under the soil, and the whole stalk is fed into the machine for the removal of sheaths and trash on the

FIGURE 10.15 Self-Propelled Sugarcane Harvester (Case IH, CNHI, USA)

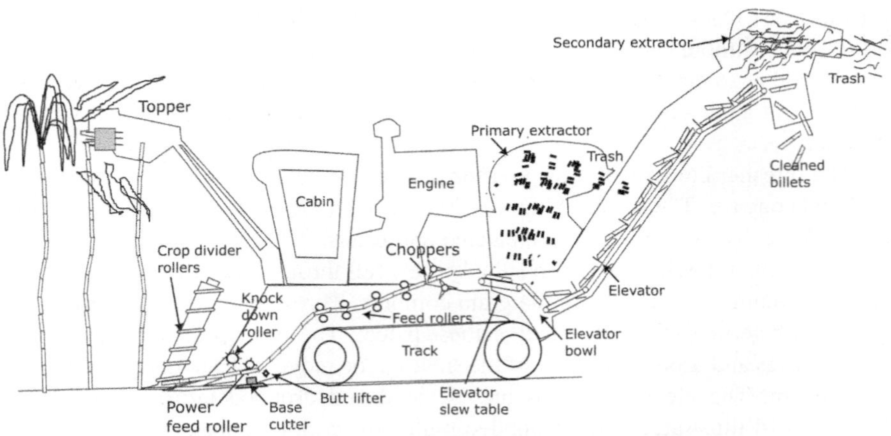

FIGURE 10.16 Construction and Process Path in a Sugarcane Harvester

No Permission Required

surface. Then, the whole cane is distributed laterally and uniformly cut into pieces or billets of about 30 cm to 40 cm.

The billets are elevated and dumped into a tractor trailer that is traveling on the side of the machine while the harvesting is occurring. The trash is removed completely in two stages and dumped well away from the trailer by extractors, the outlets of which can be swivelled away from the billet outlet. Typically, a high-powered engine, of above 250 kW with a turbo charger, is used since the power requirement of harvesting the hardy stem and bunchy crop of sugarcane is very high. The number of operations carried out by the machine is also great, with each demanding a lot of power. All the components are driven by high-powered hydraulic motors drawing hydraulic power from the engine. Hydraulic drives provide flexibility for the components for any orientation and could take a large amount of shock loads in harvesting the cane. The operator has access to all the adjustments from the machine's cabin. Figure 10.16 illustrates the construction and flow path of a typical sugarcane harvester. The primary parts and their purposes are as follows.

The topper cuts the standing sugarcane tops as the first operation, discarding them clear of the harvest area. It has a pair of drums with cutter blades mounted on a height-adjustable mast. Some harvesters have a shredder as an optional attachment to the topper, which shreds the long leaf into smaller pieces before discarding them.

Crop dividers are intricately designed sets of helically ridged rollers, inclined backward and contra-rotating, that raise the bent or lodged canes and separate a row of sugarcane being harvested from the adjacent rows. Side-trim knives are provided in many models and are positioned on the sides of the crop divider rollers, which are hydraulically adjustable. They cut the ends of tangled sugarcane stalks that were not separated by the crop dividers, preventing the canes in the adjacent rows from being pulled out by the crop divider rollers.

The knockdown roller is a slatted roller that tilts the topped canes into the machine for performing the cutting and feeding operation.

The power-feed roller is a finned roller that assists feeding the guided cane to the base cutter. The fins untangle the interwoven cane for easy base cutting.

The base cutter is a set of rugged twin cutter discs positioned at the level of or a bit below the ground. It cuts the sugarcane stalks at a level closer to or below the soil so that the juicy cane bottom is severed at the right spot. It guides the cut canes' lower end to the butt lifter roller. The base cutter is hydraulically adjustable for the height of the cut. The cutter discs consist typically of three heavy cutter blades fastened to the rotating discs.

The butt lifter roller lifts the cut cane stalks and guides them to the feed rollers in the machine.

The feed roller train consists of sets of paired rollers that transport and horizontally distribute the sugarcane stalks to the chopper drums. These rollers, while feeding the canes onward, clean soil from the sugarcane stalks.

Choppers cuts the sugarcane into billets and throw them to the primary extractor chamber. Drums with three chopper knives perform this process of billeting.

The elevator bowl receives the sugarcane billets from the chopper and feeds the elevator.

The primary extractor cleans the billets by removing trash and other impurities on the cane surface. It is a high-powered fan that blows air on the billets while they are being thrown out of the choppers into the elevator bowl.

The flighted elevator carries the billets up to the secondary extractor, where again high-velocity air is blown on the billets through a perforated floor to remove dirt and impurities. Both extractor outlets can swivel independently, which can direct the trash well away from the outlet path of the cleaned billets.

The elevator slew table turns the elevator 180° for unloading the billets on either side of the machine into a tractor-drawn trailer proceeding on the already harvested side of the row being harvested by the cane harvester presently.

10.6 Root crop harvesters

Root crop harvesting of crops such as potato, sweet potato, and yam is generally performed by digging them in bulk and separating the tubers from the soil. Typically, a wide-raked blade (Figure 10.17) is used to dig the soil below the tubers and bring the tubers to the surface. The loose soil, along with the tubers, is then guided up on a slat conveyor with rod-shaped slats, which is also oscillated by the drive itself. The loose soil broken up by the oscillations falls through the conveyor slots to the ground below, and the tubers are transported to the rear of the machine. They are then simply windrowed on a row so as to be picked up manually or by a collecting machine. The conveyor is driven by a gear box that receives power from the tractor's PTO through

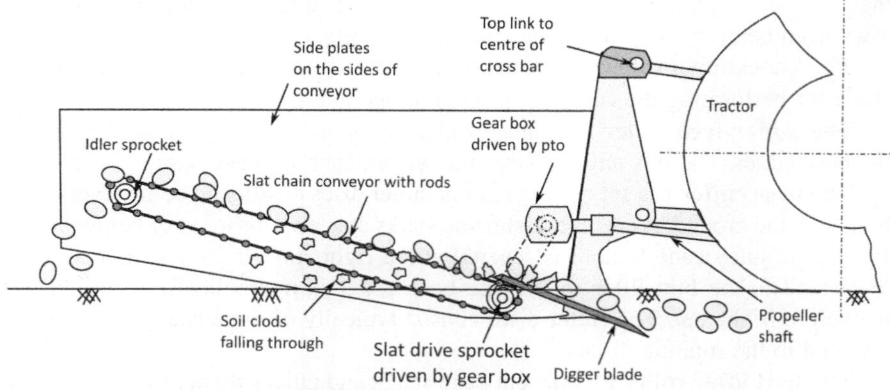

FIGURE 10.17 Working of a Potato Digger

No Permission Required

a propeller shaft. Harvesting practices vary from region to region. In some places, the haulms are allowed to wither and dry, after which the tubers are dug up. In other regions, the haulms are first topped using a topping machine before digging. Some machines have sets of closely spaced corrugated rollers on which a potato dug with the haulms rolls over and the haulms are extracted from the tubers. Modern potato combines have extensive collection systems, which receive the tubers, clean them further, and, finally, elevate them to be loaded on a container traveling on the machine's side. Machines that are intended for tubers such as a cassavas dig more deeply to uproot these crops much more intensively. Moreover, many such crops require topping of the vegetation by separate topping machines or attachments that remove and windrow the stalks before harvesting the tubers. The number of rows these machines dig depend on the power requirement. In many of these machines, apart from the raked blade, which has either a plain edge or toothed configuration, vertical coulters are used ahead of the digging blade, set to operate on the edges of the digging blade's path. The coulters make a precut wall on both sides; this makes the digging easier and can restrain the soil failure front within the intended width of operation. Bigger machines are equipped for sorting the dug tubers into several size ranges.

Root crops such as peanuts and carrots are harvested by an alternate mechanism, wherein a relatively narrow digging blade operating below the roots fractures the root zone adequately, such that pickers can strategically grab the plants ahead of the digging blade and convey them to the rear of the machine (Figure 10.18). The picking mechanism is usually a pair of contra-rotating padded chains, between which the vegetation of the root crop is grasped and pulled from the fractured soil. The conveyor chain carries the crop to the rear, where the vegetation is chopped off by revolving cutter blades to separate the peanuts or carrots from the vegetation.

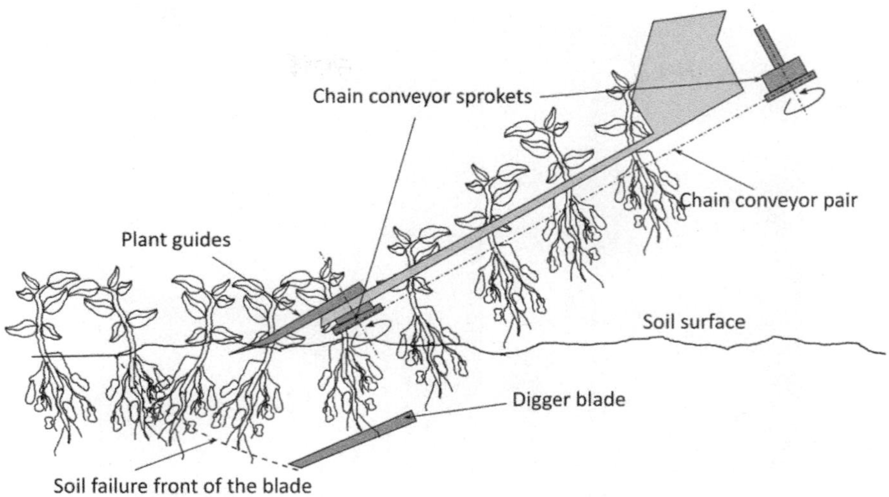

FIGURE 10.18 Digging and Picking Mechanism of a Peanut Harvester

No Permission Required

The produce is then cleaned of the adhering soil and stored in a container on the machine.

An optimal forward speed of root crop harvesters normally ranges from 1.0 km h to 5 km h^{-1} and depends mainly on the soil condition, with higher speeds being adopted for sandy/loamy soils. The conveyor speed ranges from 0.2 m/s to 2.0 m/s, and the digger blade's rake angle ranges from 10° to 24°, again based on the soil type. Unduly higher forward speeds can result in excessive damage to tubers.

Forward speed influences the tuber separation efficiency, too. An optimal digging angle reduces tuber loss as well power requirements.

10.7 Cotton pickers

The presence of trash in machine-picked cotton discolours the cotton fibre while ginning. This is why the laborious hand-picking of cotton is preferred, ensuring the least trash inclusion. But, hand-picking costs are enormous and machine picking is preferred in large cotton belts. Typically, mechanical cotton pickers are multi-row, self-propelled machines (Figure 10.19) and are expected to collect mature seed cotton from the defoliated crop. The collected cotton should have the least leafy trash and damaged cotton fibre possible.

While the mechanical picker moves into the standing crop row, crop dividers and guides lead the defoliated stems with seed cotton into the picking passage of the machine (Figure 10.20). Revolving picker spindles move into this passage and wrap the seed cotton around their axis, thus picking them from the stems. The picked cotton is

FIGURE 10.19 Mechanical Cotton Picker (Case IH Cotton Express, CNHI, USA)

removed from the picking spindles, conveyed, and collected in an on-board storage container.

Two types of spindles are used in modern mechanical pickers. Conical spindles have three or four rows of mild surface serrations to grip the cotton and wrap it around itself while rotating on its axis. The conical shape allows for the easy doffing or removal of cotton from its surface. Straight spindles are longer than the former and are smaller in diameter, with a fluted surface. These spindles are used at higher speeds and depend more on a moistened surface for effective picking. Machines with conical spindles are more popular than the latter.

The picking drum has a row of picking spindles on multiple bars arranged around the drum (Figure 10.20). In the shown machine, 16 spindle bars are on the periphery of the drum and each bar has 20 equi-spaced spindles along the drum axis. Each of the picking spindles is conically shaped with mild serrations on the surface and is fitted on bearings to revolve at a high speed. The spindles on each bar along the drum axis are driven by bevel gear sets from a common shaft behind the bar, and all such shafts inside the drum periphery have a common drive from the drum rotation. This arrangement revolves all the spindles simultaneously at the same speed. The conical shape facilitates the easy removal of the cotton (doffing) after it leaves the picking zone. The spindles rotate at 2000 rpm while the picker travels at a ground speed of 5 km h^{-1}. The spindles' orientation, with respect to the crop row and the doffer cylinders at each angular position of the drum, is achieved by a stationary cam actuating the followers on the spindle bars. The entry and exit of the spindles into the passageway are synchronized with the forward travel of the machine so that the entering or exiting spindles do not move forward or backward with respect to the cotton plant. Each rotating spindle enters and exits the cotton row from the side with a minimum disturbance to the plant, since the relative speed

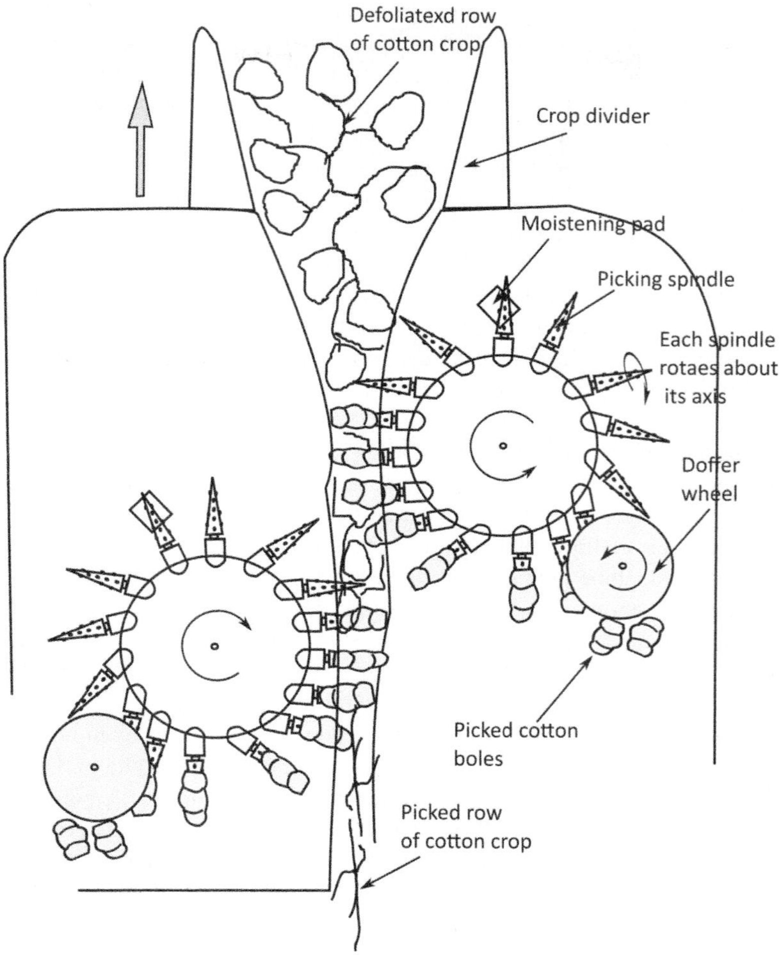

FIGURE 10.20 Construction and Working of the Cotton Picker Drums

No Permission Required

of the machine with respect to plants is zero. The spacing between the spindles is so designed to extract the maximum number of bolls from the plants. The row of spindles on the bars along the drum axis picks the cotton plants over their whole height in order to remove all the bolls from the plants.

The spindles are arranged to enter the picking passage through a grid, thus preventing the plants from being pulled off by the drum itself. The revolving spindles, when incident with opened cotton bolls, wrap them around their axis, extract them from the plants, and exit. Apart from the mild serrations of the spindle, the spindles are moistened by moistening pads before entering the picking passage to ensure a positive picking action. As the drum rotates, the spindles move into the doffer cylinders; they, in turn, strip the cotton from the spindles and make them drop down to be collected by a conveyor. Doffer cylinders have rubber lugs, which, when rotating, remove the wound cotton from the spindles. Typically, two sets of picking drums are used (Figure 10.20) for the complete picking of the bolls. A pneumatic conveyor is usually used to convey the picked bolls into a storage container that can be off-loaded from the picking machine when filled.

Apart from mechanical cotton pickers, cotton strippers are machines that have combing rollers, the fingers or bristles of which comb the open cotton bolls from the plant. But, the main disadvantage of these machines is that trash, green leaves, and stems are also stripped along with the bolls. The stripped cotton is usually separated from foreign matter to acquire the cotton. The machines operate at higher speeds and can cover wider swaths, too; they can cover very large cotton areas in a short time. Stripping also wastes a lot of cotton in the field but becomes economical, owing to its higher field coverage.

The most important prerequisite for using mechanical cotton pickers is that the cotton variety planter should be of a breed that is amenable for single picking. Many of the cotton varieties farmed in Asian countries such as India and African countries such as Egypt, are multiple picking breeds that allow farmers to hand-pick cotton at several stages. Machine picking is not possible in these varieties at all.

Defoliation is another important prerequisite wherein all the leaves are induced to be shed before cotton pickers or strippers are used. Although cotton plants are deciduous and shed their leaves on maturity, particularly in the single-picking varieties, chemical defoliants are sprayed to induce leaf shedding. Defoliation prevents green leaves from obstructing the spindles reaching the open bolls. Moreover, it prevents the inclusion of green leaves and dry trash in the picked cotton, of which the dry trash is very difficult to remove and discolours the ginned cotton.

The example worked out in Box 10.6 shows the relationship between the travel speed and the picker drum's peripheral speed through the calculations.

Exercise Problems

1. A flail mower's axis rotates at 1200 rpm, and the radius of flails' tip is 40 cm. At what height of cut of a 10 mm-thick fodder plant will impact cutting not be possible at all? Static cutting occurs at the knife edge at a pressure exceeding 35 Nmm^{-2}. The ultimate tensile strength of the plant fibres is 38 N/mm2, and the knife's edge radius (sharpness) is 0.20 mm.
2. A 4 m-wide forage harvester is harvesting sorghum fodder at 7 km h^{-1}. The crop population is 100,000 plants per ha, with each fodder plant weighing 300 g. The length of the cutter head cylinder is 60 cm, its rotational speed is 1000 rpm, and the number of knives on it are eight. The throat height of the feed window leading to the cutter head cylinder is adjustable. Assuming the density of the

Box 10.6 Example 10.6

Example 10.6. A conical spindle-type cotton picker has a spindle drum diameter of 40 cm, which has 16 spindles on its periphery with each spindle rotating at 2500 rpm. The picking spindles are 150 mm length from the periphery of the drum. The effective picking zone is about 150 cm long. Find the rotary speed of the spindle drum, when the picker is operated at 4.5 kmh^{-1} and the number of spindles 16 present in the picking zone at any instant. Also calculate the number of revolutions each spindle makes inside the picking zone in the time interval between its entry and exit from the zone. (It may be noted that the spindles enter at an orientation that is perpendicular to the travel direction.)

Circumference of picker drum = π (40) = 126 cm.
Peripheral space between spindles on the drum = 126 /16 = 7.87 cm.
Length of picking zone = 15 cm.
∴, the number of spindles that can be in the zone at any instant = (15 / 7.87) + 1 = 1.9 +1 = 3.
(+ 1 to convert space to number)
Operating speed = 4.5 km h^{-1}.
We know that the speed of the spindles entering the picking zone is synchronized with the travel speed so that the picking is performed at a zero relative speed between the spindle and the cotton plant.
This implies that the drum's peripheral speed = picker's travel speed = 4.5 km h^{-1}
= (4500 / 3600) ms^{-1} = 1.25 ms^{-1}.
∴ the rotary speed of the drum = 1.25 / (πd /60) = (1.25 × 60) / (π × 0.4) = 59.7 rpm.
Peripheral speed of the drum = 1.25 ms^{-1}.
The time taken for the picker to travel through one picker zone length = 0.15 m /1.25 ms^{-1} = 0.12 s.
∴, the revolutions a spindle makes in this time = 2500 rpm × 0.12 s = 41.5 rps × 0.12 s = 5.

fodder bulk fed to the cutter head as 350 kgm^{-3}, find the throat window height to be set, if a length of cut of 20 mm is desired.

3. While testing a wheat combine traveling at 3 km h^{-1} with a cutting swath of 5 m over a period of 30 s, the threshed grain and unthreshed grain found on the straw walker (rack) were 150 and 250 g respectively, out of the total material including straw on the walker being 25 kg. The total material on the cleaning sieve was 7 kg of which the threshed grain weight was 350 g and the unthreshed grain was 200 g. The total cleaned grain collected in the final grain outlet was 42 kg. Calculate the cylinder, walker and cleaning losses as per cent of total yield. Also calculate the grain yield.
4. A round baler capable of making 150 cm × 120 cm bales operates at a speed of 5 km h^{-1} to bale straw after a combine operation that yielded 6000 kg ha^{-1} straw. The number of bales made per hour is 15, with one bale rolling out per meter travelled. Calculate the density of packing of the round bale.
5. A two row spindle type cotton picker has two picker drum on each row and each picker drum has 15 numbers of 150 mm long spindles on its periphery in each row along its axis. The number of rows of picker spindles along the drum's axis is 18 and the effective picking zone per drum is about 20 cm long. Find the diameter of the drum, if 4 numbers of spindles are active at any instant in the

picking zone. Also find the rotary speed of the spindle drum, when the picker is operated at 4.0 km h^{-1}. Calculate the theoretical weight of cotton picked by each spindle if the picker is operated in a crop planted at row spacing of 100 cm. Assume crop yield as 800 kg ha^{-1}.

6. Compare the displacement of a single flail's tip in the travel direction and the corresponding height of cut from the ground surface for a flail mower operating at 300 rpm and 500 rpm. The travel speed of the rotavator with a tip radius of 35 cm is 6 kmh-1. Use Excel for the calculation, and plot the heights of cut in relation to forward travel. Discuss.

References

Arnold, R. E. (1964). Experiments with rasp bar threshing drums. *J. of Agr. Engng. Res., 9*, 99–134.

Bell, B. (2005). *Farm machinery*. UK: Old pond Publ.

Bernachi, H., Haman, J., & Kanafojski, Cz (1972). *Agricultural Machines, Theory and Construction* (p. 1). Published for USDA and NSF by Foreign Cooperation Center of the Central Institute for Scientific, Technical and Economic Information.

Cooper, G. F. (1971). *Cylinder concave performance from laboratory teste. ASAE Paper No. 71-625*. ASAE.

Culpin, C. (2013). *Farm machinery*. Hesperides Press.

Hunt, D. (2008). *Farm power and machinery management*. Waveland press.

Ige, M. T., & Finner, M. F. (1975). Effects and Interactions between factors affecting the shearing characteristics of forage harvesters. *Transactions of the ASAE, 18*(6), 1011–1016. https://doi.org/10.13031/2013.36727, 2151-0059.

Kepner, R. A. (1952). Analysis of the cutting action of a mower. *Agricultural Engineering, 33* (11), 693–697.

Kepner, R. A., Bainer, R., & Barger, E. L. (1978). *Original from the University of Michigan*. Avi Publishing Company.

Klenin, N. I., Popov, I. F., & Sakun, V. A. (1970). *Agricultural Machines. USDA and NSF by Amerind*. Publishing Co. Pvt. Ltd, USDA and NSF by Amerind Publishing Co. Pvt. Ltd.

Lovegrove, H. T. (1968). *Crop production equipment*. Hutchinson press.

Nyborg, E. O., McColly, H. F., & Hinkle, R. T. (1969). Grain Combine Loss Characteristics. *Transactions of the ASAE, 12*(5), 727–732.

O'Dogherty, M. J. (1982). A review of research on forage chopping. *Journal of Agricultural Engineering Research, 27*(4), 267–289. https://doi.org/10.1016/0021-8634(82)90068-3, 00218634.

Persson, S. (1987). *Mechanics of Cutting Plant Material*. ASAE.

Reed, W. B., Zoerb, G. C., & Bigsby, F. W. (1974). Laboratory study of grain-straw separation. *Transactions of the American Society of Agricultural Engineers, 17*(3), 452–460, 00012351.

Smith, H. P. (1965). *Farm Machinery and Equipment*. Tata Mc-Graw Hill Pub. Co. Ltd.

Srivastava, A. K., Mahoney, W. T., & West, N. L. (1990). Effect of crop properties on combine performance. *Transactions of the American Society of Agricultural Engineers, 33*(1), 63–72, 00012351.

Srivastava, A. K., Goering, C. E., Rohrbach, R. P., & Buckmaster, D. R. (2006). *Engineering Principles of Agricultural Machines*. ASABE Publications.

CHAPTER

Performance and cost analysis of field machines 11

Chapter outline

Farm machinery management concerns the optimal use of machines on farms by understanding the economics behind their operations. It has to do with the efficient selection, management, maintenance, and replacement of these machines. The management goal is, of course, to maximize profits, considering the farming system as a factory. The management of machinery in particular deals with the costs involved in the machinery system in respect to the performance desired from the system. The storage of machines on a farm is another area of interest that will protect the machines and provide farmers better trade-in values on their machines. Selection and replacement decisions in respect to the machines are also matters of concern in the proper management of a farm's machinery fleet.

Machines of the Farm. https://doi.org/10.1016/B978-0-443-32906-7.00011-6

11.1 Performance of field machines

Let us define some commonly used terminology in the performance assessment of farm machinery.

11.1.1 Theoretical field capacity

The theoretical field capacity of a machine, expressed in ha h^{-1}, is the rate of field coverage achieved if the machine performs its function 100 percent of the time at a constant forward speed and always covers 100 percent of its rated width. This is an ideal measure of a field machine's performance.

11.1.2 Effective or actual field capacity

The effective or actual field capacity, also expressed in ha h^{-1}, is the actual rate of field coverage yielded by a machine, considering the inefficiencies in maintaining the rated width, speed, and working time.

11.1.3 Field efficiency

Field efficiency is the ratio between the effective and theoretical field capacities in percent.

The effective field capacity (Kepner, Bainer, and Barger, 1978; McKibben, 1930; Srivastava et al., 2006) is always less than the ideal theoretical capacity of the machine because of the time lost due to the inability of covering the rated width of the machine at each pass, inability of keeping the speed of travel constant, time lost in turning, filling seed hoppers/tanks, and idle time of the operator. The rated width of implements, such as row crop planters and seed drills, is the product of the number of functional rows and the row spacing. These machines utilize 100 percent of their rated width because each row must perform its necessary functional role. But, due to operational difficulties in machines such as ploughs, the present ploughing pass overlaps with that of the pass already made; therefore, the machine will not be able to cover the rated width fully. The maximum permissible forward speed is related to such factors as the nature of the operation, the condition of the field, the amount of power available, and the skill of the operator. For instance, the speed of a planter is limited by the maximum seed rate that it can effectively handle. Moreover, the machine speed cannot be maintained as constant all over the operation. So, time is lost in terms of the width being less than the rated width and the lack of uniformity in maintaining operational speed. But, the rest of the lost time is the most difficult variable to evaluate in relation to field capacity. Time is lost in the field while adjusting the machine, performing minor repairs, de-clogging, turning at the ends, loading seed or fertilizer, and unloading harvested crop. However, it may be noted that this lost time does not include the time spent for servicing, preparing, and calibrating a machine before operation. The total time for the operation of the farm machine is, hence, the sum of the actual time the machine is performing its intended function and the lost time discussed earlier.

The theoretical field capacity of the machine, expressed in hah^{-1}, is

W × S, where W and S are the rated width and rated speed of the machine, respectively.

When W is in m and S is in $km\ h^{-1}$,

then the theoretical field capacity in $ha\ h^{-1}$ = {W in m × (1000 S) in mh^{-1}}/10000 = WS/10 and

field efficiency = effective field capacity / theoretical field capacity.

In terms of the time lost in field operation, the field efficiency, 'η_f', can be expressed as

$$= K\frac{T_o}{T_o + T_a + T_h}$$

where K is the percent of the width of the implement actually utilized,

T_o is the theoretical time required for covering a unit area ($h\ ha^{-1}$), which is the reciprocal of the theoretical field capacity in $ha\ h^{-1}$,

T_h is the time lost per unit area ($h\ ha^{-1}$) due to reasons that are not proportional to the field area, and

T_a is the time lost per unit area ($h\ ha^{-1}$) due to reasons that are proportional to the area.

Computing K for soil working machines such as ploughs is cumbersome because the width is reduced only in certain circumstances of the operation and not always. But, for row crop machines like seed planters, K will be 100 percent.

T_a, the time lost due to reasons that are proportional to the field area, depends very much on the field dimensions. While operating a machine, it must be turned quite often at the field corners and edges to make the subsequent passes. The number of such turns depends on the field dimensions and system of operating the machine. If the ploughing is performed along the longer side of the plot, then the number of turns at the field edges will be less. The total number of turns per unit of area with a given implement is inversely proportional to the length of a useful operational pass in the field. The operation may demand the implement to traverse from one pass to the next along the field edge or in the headland without doing useful work, resulting in a loss of time. Similar to the turns, the number of such idle runs also is proportional to the field area. To fully understand how this is relevant, the reader is advised to refer to Section 2.3 of this book, which covers systems of ploughing.

For instance, if W is the width of land being ploughed by a tractor-mounted plough, the number of passes to be made by a plough of width w will be W/w. If the ploughing is performed from one lengthwise edge of the field, then after the first pass, the operator raises the plough, turns the tractor 90°, makes an idle run through a distance of W along the widthwise edge, and turns 90° to enter the field to make the next ploughing pass along the other lengthwise edge. This implies that the tractor does one 180° turn per two passes and one idle run of the distance between the two passes. The width-wise idle run is W at the start and keeps on reducing by W

– 2w for each bout of two passes; until it becomes zero, where two subsequent passes are adjacent to each other. For computational ease, this distance of idle run can be averaged as to be W/2 for the overall operation. The number of turns and idle runs is equal to (W/2w) rounded to a whole number. Half of these runs are performed on one widthwise side and the other half are performed on the other.

∴, the total idle run to complete the operation = (W/2w) × (W/2) = (W^2/4w).

Assuming an average speed of idle traverse, one can calculate the total time lost in an idle run.

Similarly, the time lost in turning can also be found as time per one 180° turn × (W/2w) turns.

Since the time lost is a function of W with the length of the field being constant, this is clearly a function of the area.

To simplify the above computation further, we may neglect the time for 180° turns and consider only the idle runs. Then, the percent of time lost is an idle run per single pass

$$= \frac{W/2}{W/2 + L} = \frac{W}{W + 2L}$$

where W/2 is the average distance of width-wise idle run and L is the length of field; which is same as the length of ploughing pass. This percent loss thus calculated will be the same for the whole field.

Time lost due to reasons that are proportional to the area also includes that is lost in loading and unloading inputs and produce.

T_h, the time lost per unit area due to reasons that are not proportional to the field area, are not easy to compute; rather, they have to be actually observed and recorded in the field. Generally, these losses range from 1 percent to 10 percent for ploughing and cultivation and can rise to 35 percent for planting operations.

11.2 Operational cost analysis

The total cost of any field operation using machines (Fairbanks et al., 1971; Mayfield et al., 1981) will be comprised of two major components, namely, the fixed cost and the variable cost. The fixed cost of a machine includes depreciation, interest on investment, taxes, insurance, and shelter. A fixed cost is incurred on a machine owner regardless of whether the machine is used or not. The variable cost of a machine is that is incurred as a result of the actual machine operation. This includes expenses for fuel, lubrication, and wages for the operator as well as repairs and maintenance. They are the actual operating costs.

The cost per unit time of operation depends not only on the total cost of the machine and the operation itself, but also on the annual use of the machine. Usually, the annual usage of farmer-owned machines is less compared to machines used in industries. A farmer owning a tractor, uses it throughout the cropping period amounting to about 1000 hours per year. But he uses his other field machines for less than 200 h, because those machines are used only for specific operations of crop cultivation. On the other hand, machines that are custom-lent by private entrepreneurs year round have a higher annual usage.

Since the fixed cost per hour of a machine = the total fixed cost / annual usage, the lower annual use tends to raise the overall operational cost. Conversely, an operating cost can be decreased considerably by increasing the annual usage of the machine. The initial investment on the machine as well as the length of service are important factors deciding the fixed cost.

11.2.1 Fixed costs

11.2.1.1 Depreciation

Depreciation is the reduction in value of a machine caused by normal wear and tear. A machine depreciates over its full service life until it is considered for scrapping. The normal service life of these machines depends on the wear and tear and their annual use. But, generally, it is taken to be 10 years for farm machines. Moreover, the model of a machine may become obsolete due to improvements in technology and thereby loose value quite drastically. Depreciation due to obsolescence is not easy to predict. But, various methods are available for calculating depreciation due to wear and tear. They are the straight line, constant percentage, estimated value, and sinking funds methods. The straight line method is the most common and easiest method.

$$\text{Depreciation} = \frac{C - S}{L} \text{Rs year}^{-1}$$

where C is the initial cost of the machine, S is the salvage value of the machine after its service life is over, and L is the service life of the machine. Dividing it by the annual use will yield the depreciation rate per hour.

The constant percentage and estimated value methods depreciate the machine more rapidly than the straight line method in the first few years of the machine and less rapidly later.

11.2.1.2 Interest on investment

The initial investment one makes on a machine, if the person had chosen not to invest, would have accrued an income to the investor by means of interest from a deposit in a bank or elsewhere. This loss of income is, hence, a cost on the machine and is

$$= \left\{\frac{C + S}{2}\right\}\left\{\frac{i}{100}\right\}$$

where i is the percent of interest the investment would have brought the investor,
C is the initial cost of the machine, and
S is the salvage value of the machine after its service life is over.

Here, the investment has been assumed to be the average of the initial investment and the final salvage value it will bring back. This approximates the time value of money that incurs a loss on the real value of the initial investment. The existing bank interest can be taken as the value of i in the expression.

11.2.1.3 Taxes, insurance, and shelter

For the purposes of cost estimation, a constant average yearly charge of 0.5 percent to 2 percent on the investment is assumed as taxes on the machine. This is added if the given machine is taxed as the policy of a government. When the machine is insured by the owner, the yearly cost of insurance is usually assumed as 0.25 percent on the initial investment. The yearly cost of machine housing, as distributed over its lifetime, is related to the size of the machine and usually taken to be from 0.5 percent to 1 percent of the initial cost of the machine. Hence, the total annual charge on taxes, insurance, and housing works out to approximately 1.5 percent of the investment.

All the above components of a yearly fixed cost are summed and distributed over the machine's annual use to compute the fixed cost per hour of machine usage.

11.2.2 Variable cost

Variable cost is the cost that depends on the actual use of the machine in the field.

The major component of a variable cost is the fuel cost incurred per hour. The fuel cost will depend on the machine's actual power utilization. Usually, this cost is for prime movers such as tractors and engine-operated equipment such as back-pack mist blower sprayers.

A practical rule of thumb that can be used for calculating the fuel consumption of a compression ignition engine is to assume a fuel use of 0.12 litre diesel per hp per hour and 0.20 litre petrol per hp per hour for spark ignition engines. The fuel cost per hour is then calculated based on the prevailing fuel cost per litre.

The next important component is the labour cost involved in the operation of the machine. Mostly, this is the hourly wage paid to the operator or tractor driver. Some machines, such as a down-the-row crop cultivator, may need additional labour to direct the machine in its operation.

The cost of repairs, maintenance, and lubrication is another variable cost component and is normally assumed as 5 percent of the fuel cost incurred. This is because this component depends on the amount of time the machine is operated and is related to the fuel being spent.

The sum of all these variable costs would amount to the total variable cost per hour of usage of the machine. The cost per unit area of the field can be easily obtained, if the actual field capacity of the machine is known, as the cost per hour in Rs h^{-1} divided by the actual field capacity in ha h^{-1}.

The whole process of calculating the cost can be best explained as below by taking an example of a 50 hp tractor costing Rs 8 lakhs.

Fixed cost calculations:
Cost of machine (C) = Rs 8,00,000,
Life of machine (L) = 10 years,
Salvage value after life = 10 percent of cost,
Annual usage of machine (AU) = 1000 hours,
Interest on investment (i) = 5 percent.

a. Depreciation:

by straight line method:

$$= \frac{C - S}{L} \text{Rs year}^{-1}$$

$$= (800,000 - 80,000)/10 = 72,000 \text{ Rs yr}^{-1}$$

$$= 72,000/\text{AU} = 72,000/1000 = 72 \text{ Rs h}^{-1} \text{ of usage}$$

b. Interest on investment:

$$= \left\{\frac{C + S}{2}\right\}\left\{\frac{i}{100}\right\} = [(800,000 + 80,000)\ /\ 2] \times 5\ /\ 100$$

$$= 440,000 \times 0.05 = 22,000 \text{ Rs yr}^{-1}$$

$$= 22,000/1000 = 22 \text{ Rs h}^{-1} \text{ of usage}$$

c. Tax and insurance cost:

$$= \text{Assumed 2 percent of cost} = 16,000 \text{ Rs yr}^{-1}$$

$$= 16,000/1000 = 16 \text{ Rs h}^{-1} \text{ of usage}$$

d. Shelter cost:

$$= \text{Usually 1 percent of cost} = 8000 \text{ Rs yr}^{-1}$$

$$= 8000/1000 = 8 \text{ Rs h}^{-1} \text{ of usage}$$

$$\text{Total fixed cost} = 72 + 22 + 16 + 8 = 118 \text{ Rs h}^{-1}$$

Variable cost calculations:

a. Fuel charges:

The fuel consumption of a tractor is 0.12 litre diesel per hp per hour. Say the 50 hp tractor's power utilization for this operation is 40 hp. Then, the fuel consumption = 0.12 x 40 = 5 l h^{-1}

$$= \text{Rs } 100 \text{ x } 5 = \text{Rs } 500 \text{ Rs h}^{-1} \text{ of usage}$$

b. Lubrication and repair charges:

$$= 5 \text{ percent of cost per year}$$

$$= 0.05 \text{ x } 500{,}000 = 25{,}000 \text{ Rs yr}^{-1}$$

$$= 25{,}000/1000 = 25 \text{ Rs h}^{-1}$$

c. Labour charges:

$$= \text{Rs } 800 \text{ per day of 8 hours}$$

$$= \text{Rs } 100 \text{ Rs h}^{-1}$$

Total variable cost $= 500 + 25 + 100 = 625$ Rs h^{-1}

Total cost $= 118 + 625 = 743$ Rs h^{-1}

Say this tractor is used for operating a three-bottom mould board plough with an actual field capacity of 0.40 ha h^{-1}.

Then, the cost per unit area $= 743$ Rs h^{-1} / 0.40 ha h^{-1} $= 1858$ Rs ha^{-1}.

It may be noted that the plough also has to be charged an operating cost, calculated similarly to that of the tractor. But, the variable cost will decrease since there is no fuel or labour costs involved. However, this reduction may be compensated by the enhanced cost per hour of the plough since its annual usage will be much less, in the order of only 100 hours. The student is encouraged to do a calculation and find out.

11.3 Timeliness cost of farm operations

All operations, including soil preparation through harvesting, are to be performed at an optimal time for maximum benefit. As soon as the summer rain begins to fall, a farmer may require the whole area to be ploughed and make it ready for sowing. Otherwise, the soil moisture may be lost. A large area needs to be sown within a certain time frame to utilize the moisture obtained from rain. Pesticide spraying needs to be completed within a time period to prevent excessive crop damage. The timely harvest of a crop is crucial in reducing grain loss during the harvest and so on. When harvesting a crop, yield or produce quality will be affected incrementally by either delaying or harvesting earlier with respect to an optimal time. The non-availability of machine time due to the limited capacity of a machine is the main reason for issues with the timeliness of operations. So, to own machines is advantageous in terms of the timeliness of operations. But, it is evident that expensive machines such as a cane harvester are not economically practical for individual ownership on small farms; these kinds of machines should always be used by means of custom hiring to enhance their annual use so as to bring down the operational cost. But, the custom hiring of machines will also affect the timeliness of an operation since the farmer will have to depend on the machine owner in scheduling operations. The selection of machine capacity, hence, is to account for the cost incurred due to the timeliness of the operation. The timeliness cost (Hunt, 2008) is, hence, important in determining the cost of an operation as well as the amount of annual use that will justify the purchase of a particular machine.

The timeliness cost is based on the fact that there is a parabolic relationship existing between the money loss accrued due to advancing or delaying a field operation away from the optimal time of carrying it out. The y-axis (Figure 11.1) represents the ratio of money lost per ha^{-1} to the crop value (V) per ha^{-1} on a particular day; the x-axis represents the day of operation in respect to the optimal day. The relationship between the two in the form of $y = Kx^2$ has to be standardized for different field operations pertaining to a region and crop similar to those standards created by the ASAE (ASAE D497, 2006). Here, 'K', the timeliness coefficient, is the crop value lost per day and varies with the type of operation. Typical 'K' values for sowing as well as harvesting crops such as corn, rice, and wheat range from 0.006 to 0.010.

Now, D, the daily capacity of the machine, ha day^{-1}

$$= F_c\, G\, (P_{wd}),$$

where F_c *is* the effective field capacity of the used machine (ha h^{-1}),
G is the average working hours per day (h day^{-1}),
and P_{wd} *is* the probability for a working day, as a decimal fraction.

The timeliness cost depends on the total duration of the operation in a particular season and geographical location, as measured in the actual number of contiguous days from the start to the finish of an operation. The probability, $P_{wd,}$ depends on the geographic location of the farm and varies throughout the year. The P_{wd} value will almost be near unity in regions where there is no substantial cold or rain impeding field operations.

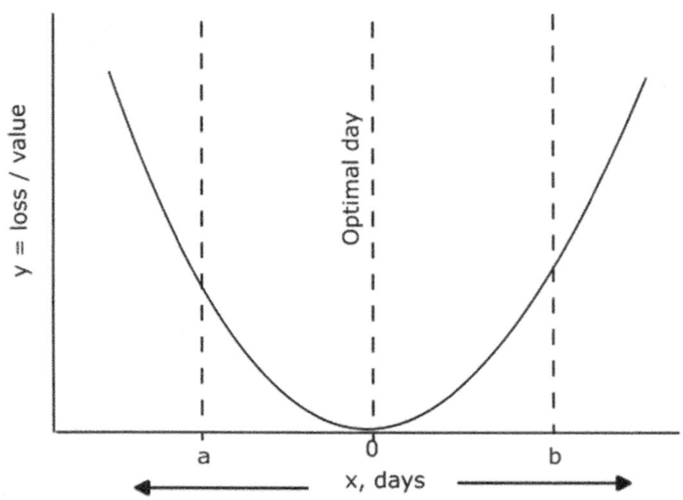

FIG. 11.1 Timeliness Loss Over Day of Operation

No Permission Required

The money loss on a particular day (see Figure 11.1) = $(Kx^2)VD$, as from the parabolic relation.

∴, the total timeliness cost, as integrated over the days of operation from day 'a' to day 'b' in Figure 11.1,

$$= KVD \int_a^b x^2 dx = \frac{KVD}{3}(b^3 - a^3)$$

With 'A' being the total area (ha) to be covered by the operation:

- if the operation is to have the optimal day (origin in Figure 11.1) at the centre of the total duration of the operation (balanced), then substituting -a = b = A/2D,

  the timeliness cost for the covered area (A), $C_t = \frac{KVA^3}{12D^2}$.

- if the operation is advanced before the optimal day or delayed completely after it, with the total duration of the operation, then substituting either a or b = 0, with b or a = A/D,

  then the timeliness cost for the covered area (A), $C_t = \frac{KVA^3}{3D^2}$.

It may be noted that the timeliness cost for the same area, A, is four times less for a balanced operation than the other.

Then, a general expression for the timeliness cost is obtained for the operating period from a to b by substituting for 'D' and dividing by 'A'.

$$C_t = \frac{KVA^2}{3X(F_c G P_{wd})^2}$$

X = 1 if the operation starts or ends at the optimal day or

= 4 if the operation is balanced evenly about the optimal day. The following example in Box 11.1 throws light on how the timeliness cost is calculated for a field operation.

11.4 Selection of farm machine size

Depending on the field coverage required on a farm within a stipulated time available, the machine capacity and the number of such machines to be acquired are to be selected (Siemens, Hamburg, and Tyrrell, 1990). The available models of machines are to be matched with the required capacity. We know that:

$$\frac{A}{HD} = F_c = SWE$$

where

F_c = the required machine capacity, $m^2\ h^{-1}$;
A = the area to be covered by the machine, m^2;
H = the average operating hours per day;

Box 11.1 Example 11.1

Example 11.1. A rice combine in India is in operation in June and is used at an average of 10 hours per day. The effective field capacity of the machine is 1 ha h^{-1}. The cost value of rice is Rs 50 kg^{-1}. Calculate the timeliness cost per ha for harvesting 50 ha in a balanced plan of duration with the optimal day in the middle. Assume the rice yield as 4 tonnes ha^{-1}, P_{wd} as 0.9 for the region in June, and K as 0.004.

A = area harvested = 50 ha;
V = crop value kg^{-1} × crop yield kg ha^{-1} = Rs 50 kg^{-1} × 4000 kg ha^{-1} = 200,000 Rs ha^{-1};
G = average working hours in a day = 10 h day^{-1};
F_c = effective field capacity of machine = 1 ha h^{-1};
P_{wd} = probability of a good working day = 0.9;
X = 4 if the operation is balanced evenly about the optimum time;
K = time loss coefficient for harvesting rice = 0.004;

$$C_t = \frac{KVA^3}{3X(F_c G P_{wd})^2} \text{ Rs;}$$

$$C_t = \frac{0.004(200000)(50)^2}{3(4)[(1)(10)(0.9)]^2} = 2057 \text{ Rs}$$

D = the number of days the machine worked;
W = the required width of the machine, m;
S = the operating speed, m h^{-1};
E = the field efficiency in percent.

Rearranging to solve for the operating width of the machine to be selected,

$$W = \frac{A}{SEHD}$$

Most of the field machinery, including the tillage and sowing machines, are defined by their width and, hence, can be selected accordingly. In the case of harvesters and machinery dealing with other materials such as manure spreaders, the machine capacity in tonne per hour as well as the swath should be considered together.

11.4.1 Variables to be considered for selection of machine

1. The average operating speed is to be selected according to the crop conditions, soil types, weather, etc., and, most importantly, to keep the field efficiency adequately high.
2. Field efficiency, as recommended for each machine, may be considered for the computation of the machine width.
3. The rate of average operating hours per day is, again, seasonal and according to the need for completing the operation in time. Past records of the farm can be used to decide this variable.

Box 11.2 Example 11.2

Example 11.2. A field operation needs to be done on a 320-hectare parcel of land at a speed of 3 km h^{-1}. The effective hours that the machine can work is 10 hours per day, and the work is planned to be completed in 10 days. Assuming a field efficiency of 85%, calculate the width of the machine to do the job.

A = area to be covered by the machine = 320 ha = 320 × 104 m^{2};
H = average operating hours per day = 10 h;
D = number of days the machine worked = 10 days;
S = operating speed = 3 km h^{-1} = 3000 m h^{-1};
E = field efficiency in percent = 85%
W = required width of the machine = $\frac{A}{SEHD}$;
= 320 × 10^4 / (3000 × 0.85 × 10 × 10) = 12.55 m.

4. The available working days may be influenced by:
   a. Farming constraints. For example, only nine days of available working days will be available to use an on-the-row cultivator if the field is irrigated every two weeks, as there may be a need of five days for the field to dry, allowing the cultivator to enter the field.
   b. Environmental conditions such as rainfall or wind, which will affect the trafficability of the machine in the field or the spraying job, respectively.
   c. Contractual limitations. The harvested produce must be delivered within a certain time limit.
   d. Timeliness considerations. This is to keep the timeliness costs at optimum levels by planning the working duration accordingly. A simple example is worked out in Box 11.2 to show the reader how a field machine is selected based on need.

11.5 Break-even analysis for selection of machine

A break-even analysis helps in deciding whether a particular machine is to be purchased or not, in terms of its benefits over time. The break-even point is that point at which the cost incurred by the machine is equal to the revenue generated by it. The concept will be easy to understand, especially with an example of an agricultural machine used for custom hiring. Figure 11.2 shows the total revenue (solid line) generated by a machine on custom hiring against the hours of usage. The revenue is a product of the hourly hire charges the machine generates and the hours of usage. The plot also shows the total cost of the machine (dashed line) as the sum of the fixed costs (FC) and the variable costs as a function of the hours of usage. The fixed cost is the intercept of the cost line and will exist even if the machine is not used at all. The point at which the cost and revenue lines intersect is the break-even point, at which the revenue accrued and the cost incurred are equal. It is obvious that a loss will be incurred if the usage hours are below the break-even point. Based on this simple

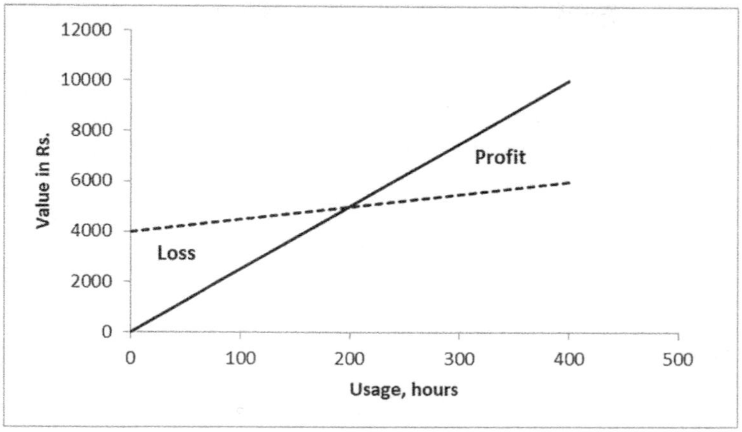

FIG. 11.2 Break-Even Analysis of Cost Against Revenue

No Permission Required

analysis, a custom hirer can decide whether the machine can be hired out sufficiently enough to provide a profitable business. As for a farmer wishing to hire a machine, it will be difficult to find the contribution of the machine to the farm's net profit. But, if the farmer can approximately find it as the revenue made by the machine per hour, a break-even analysis can be used to find the break-even point, based on which a buying decision can be made. As for selecting a machine for purchase, the farmer can use the break-even analysis of costs to compare two machines and decide on the least costly one. The following example problem will throw light on the process.

The following example, shown in Box 11.3, elaborates on and discusses how a break-even analysis can be used effectively to compare the costs of two proposed machines to select the right one.

11.6 Replacement of field machines

The continued use of machines is dependent upon the condition of the machines, and the replacement of obsolete or worn-out machines with newer machines should be decided periodically to prevent undue losses. The replacement decision (Sullivan *et al.*, 2010) uses the terms "defender" and "challenger"; the defender is the existing machine, and the challenger is the best available machine to be procured as a replacement.

Let us consider a replacement problem, wherein a machine costing Rs 2,00,000 was purchased two years ago and is expected to have a three-year life and a salvage value of Rs 50,000. Though the machine is in good condition now, the farmer expects operating costs to run at the rate of Rs 80,000 per year until the end of its life. In addition, the farmer finds that the current market value for the machine is Rs 1,00,000.

Box 11.3 Example 11.3

Example 11.3. A 20 hp, self-propelled, four-row, semi-automatic vegetable transplanter is sold for Rs 400,000. A 30 hp, four-row, automatic vegetable transplanter is available for Rs 850,000. The row spacing is 40 cm for both machines. An entrepreneur plans to buy one of these machines to lease it out on a custom basis. The semi-automatic machine requires four labourers to sit on the machine for its operation. The automatic machine requires only one labourer to do the loading and feeding of nursery plug trays. The labour cost is Rs 100 per hour of work. The operating speed of the semi-automatic machine is 1.5 km $h^{-1,}$ and that of the automatic machine is 3 km h^{-1}. The field efficiency is 70 percent and 80 percent, respectively. Determine the minimum annual use and annual area to be covered that justifies the purchase of the automatic machine. Assume repair and lubrication costs as 15 percent of the fuel cost.

Fixed cost		
Cost of machine (C)	400,000	850,000
Life of machine (L)	10 years	10 years
Salvage value after life	10% of cost	10% of cost
Interest on investment	5%	5%
a. Depreciation: by straight line method =	(400,000-40,000)/10 = 44,000 Rs yr^{-1}	(850,000-85,000)/10 = 76,500 Rs yr^{-1}
b. Interest on investment: =	[(400,000+40,000)/2] × 5 / 100 = 11,000 Rs yr^{-1}	[(850,000+85,000)/2] × 5 / 100 = 23,375 Rs yr^{-1}
c. Tax and insurance cost: assumed 2% of cost	8000 Rs yr^{-1}	17,000 Rs yr^{-1}
d. Shelter cost: usually 1% of cost	4000 Rs yr^{-1}	8500 Rs yr^{-1}
Total fixed cost per year	**67,000 Rs yr^{-1}**	**125,375 Rs h^{-1}**
Total fixed cost per hour	**67,000/AU Rs h^{-1}**	**125,375/AU Rs h^{-1}**
Variable cost calculations		
Fuel consumption of tractor @ 0.12 litre per hp diesel per hour (assuming full power utilized)	0.12 ×20 = 2.4 l h^{-1}	0.12 × 30 = 3.6 l h^{-1}
a. Fuel charges @ Rs 80 l^{-1}	Rs 95 × 2.4 = 228 Rs h^{-1}	Rs 95 × 3.6 = 342 Rs h^{-1}
b. Lubrication and repair charges: 15% of fuel cost	33.75 Rs h^{-1}	51.3 Rs h^{-1}
c. Labour charges: Rs 100 per 8 hour	4 × 100 = 400 Rs h^{-1}	1 × 100 = 100 Rs h^{-1}
Total variable cost	**661.75 Rs h^{-1}**	**493 Rs h^{-1}**
Total cost per hour	**67,000/AU + 661.8 Rs h^{-1}**	**125,375/AU + 493 Rs h^{-1}**
Theoretical field capacity = WS/10	0.4 × 1.5 /10 = 0.6 ha h^{-1}	0.4 × 3.0 /10 = 1.2 ha h^{-1}
Effective field capacity = Theoretical field capacity × Field efficiency	0.6 × 0.7 = 0.42 ha h^{-1}	1.2 × 0.8 = 0.96 ha h^{-1}
Total cost per ha = Total cost per hour / Effective field capacity	**(67,000/AU + 661.8) / 0.42 = 159,523/AU + 1575.7 Rs h^{-1}**	**(125,375/AU + 493) / 0.96 = 130,598/AU + 513.5 Rs h^{-1}**

Box 11.3 Example 11.3—cont'd

Use Excel or an alternate software to calculate and plot:

1. For annual use (AU) in increments of 100 hours each up to 1000 hours, calculate the total cost per hour for each machine using the expression developed in the table above using the spread sheet program. Then plot each row of calculated data against the corresponding annual use.

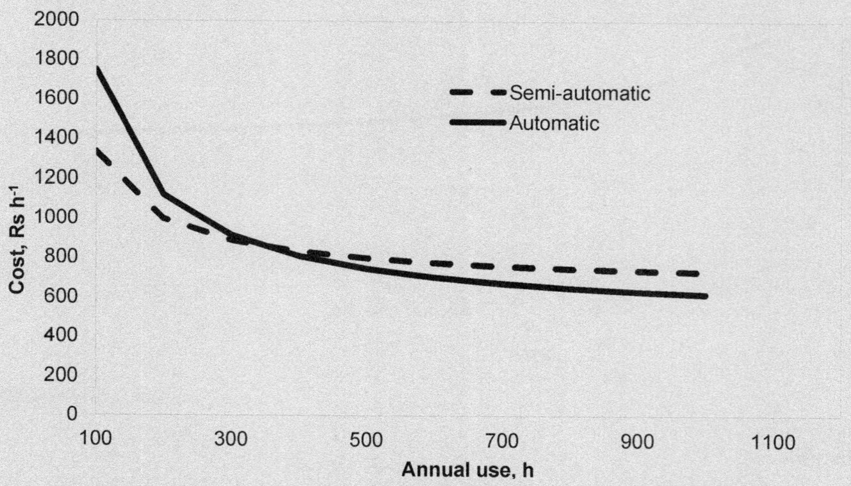

No Permission Required

Discussion:

The point at which the two curves intersect gives the minimum annual use in hours for the automatic transplanter, from which point onward the cost per hour of use falls below that of the semi-automatic machine. The operating cost for both machines is 800 Rs h^{-1} at the annual use of around 400 h Yr^{-1}. This implies that if the person making the purchase can ensure an annual hiring use of 400 h Yr^{-1}.

2. Similarly, calculate the total cost per hectare of each machine using the corresponding expression developed in the table above. Now calculate a series of annual area use values for the annual hour use based on the respective field capacities of each machine. Plot the curves using the appropriate cost per ha against the relevant annual area covered.

Continued

Box 11.3 Example 11.3—cont'd

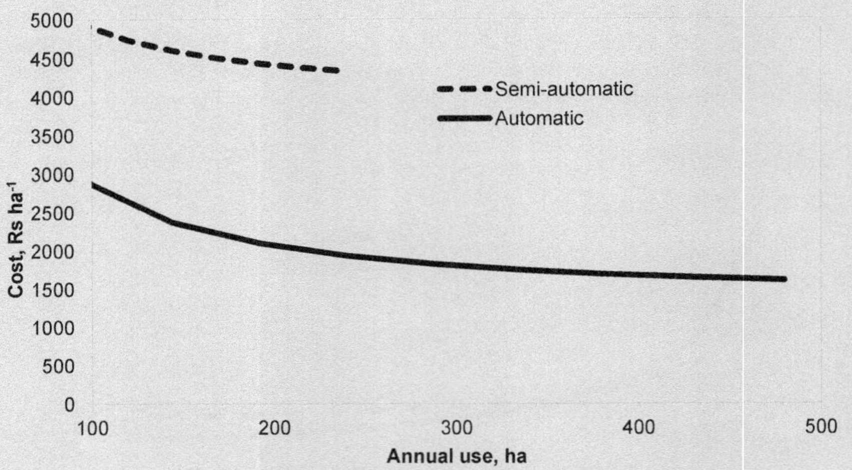

No Permission Required

Discussion:

However, the plot of cost per ha of the machines against annual area covered with the machines gives a completely different picture. Because of the relatively slower speed of the semi-automatic machine, the effective field capacity is half of that of the automatic machine. Owing to this reason, even at an annual use of 1000 h the machine yields only 250 ha of field coverage. This low capacity has again caused a higher operating cost per ha much above the automatic machine. Automatic machine can render 480 ha per ha for an annual use of 1000 h. It provides much lesser cost per ha even for low field coverage per year. If the buyer is able to hire out 480 ha area per year, the cost per ha will really be very cheap.

However, the farmer has an opportunity to purchase another machine for Rs 150,000 but with the same capacity. Over its three-year life, the machine will have reduced operating costs, of Rs 60,000 per year. It is estimated that the new machine can be sold for Rs 55,000 at the end of year three. Let us analyse whether a replacement is justified as of now using a cash flow diagram (Figure 11.3).

Option 1: Keep the defender machine. There is no additional cash expenditure. The annual operating cost for the next three years will be Rs 80,000 per year, and the machine's salvage value three years from today will be Rs 50,000.

Option 2: Replace the defender with the challenger (new machine). The defender can be sold for Rs 100,000 now. The cost of the challenger is Rs 150,000. Thus, the initial combined cash flow for this option is a cash flow of

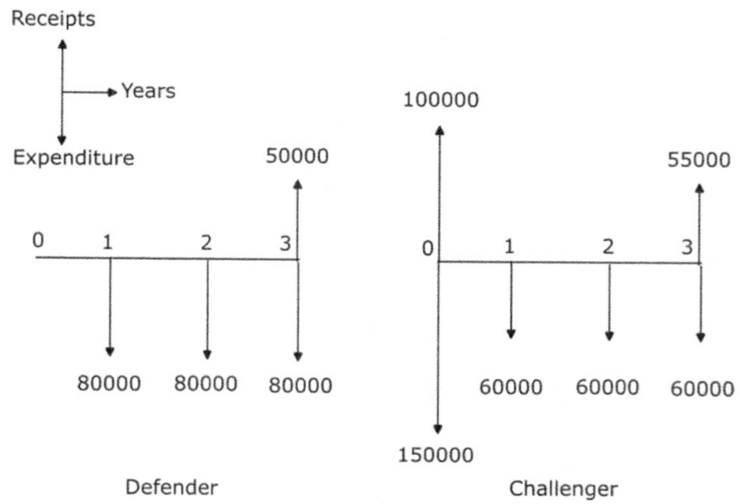

FIG. 11.3 Cash Flow Diagram for the Replacement Problem

No Permission Required

-150,000 + 100,000 = - Rs 50,000. The annual operating cost of the challenger is Rs 60,000. The salvage value of the challenger after three years is Rs 55,000. The cash flow diagrams for both options are shown below.

Before proceeding with the analysis, let us define some important economic terms.

Present worth value (PV) is the money value today for the amount available in the future (FV). Due to the time value of money, PV is always less than FV.

$$PV = FV\left(\frac{1}{(1+r)^n}\right)$$

Annual equivalent (AE), or annual worth, is the cost per year owning and operating the machine over its entire life. It is the uniform annual amount of money over a specific period that is economically equivalent to a given present value.

$$AE = PV\left(\frac{r(1+r)^n}{(1+r)^n - 1}\right)$$

where r is the interest rate prevailing on the amount.

We then use these cash flow values depicted in the problem (see Figure 11.3) to find the present worth of each option and then use the total value to find the annual equivalent value.

Let the interest on capital be r = 12 percent.

For the defender, we have:

The PV of the salvaged Rs 50,000 three years from now is:

$$PV = FV\left(\frac{1}{(1+r)^n}\right)$$

$$= 50,000\left(\frac{1}{(1+0.12)^3}\right) = \text{Rs } 35,589$$

The PV of Rs 80.000 spent annually is:

$$PV = 80,000\left[\left(\frac{1}{(1+0.12)^3}\right) + \left(\frac{1}{(1+0.12)^2}\right) + \left(\frac{1}{(1+0.12)^1}\right)\right]$$

i.e., for third year, second year, and next year, respectively:

$$= \text{Rs } 192,146$$

The present value of the future amount (PV) is easily computed by an Excel function.

PV (rate, nper, pmt, [fv]), *where the 'rate' is the interest rate, 'nper' is the number of years, and 'pmt' is the annual amount spent each year.*

In our problem, the total present worth (PV) of Rs 80,000 annually spent each year for the next three years is found using the function ***PV(0.12, 3, 80,000)*** *omitting [fv].*

And present value of the salvage value of Rs 50,000 got after 3 years is calculated using the function ***PV(0.12, 3, 50,000),*** *where 'pmt' is omitted and 'fv' the future value is given as 50,000 . (Note the two commas without any value in between)*

The total PV of the money as of today = Rs 35,589 (receipt) - Rs 192,146 (spent) = Rs 156,557 (spent).

Then, the annual equivalent worth (AE) of this present value split over the three years is:

$$AE = PV\left(\frac{r(1+r)^n}{(1+r)^n - 1}\right)$$

$$AE = 156,557\left(\frac{0.12(1+0.12)^3}{(1+0.12)^3 - 1}\right)$$

$$= \text{Rs } 65,182 \text{ (spent each year)}.$$

AE is easily computed by the Excel function PMT(rate, nper, pv), where 'rate' is the interest rate, 'nper' is the number of years of annual value, and 'pv' is the present value for which the AE is to be found. In our problem, the function used as PMT(0.12, 3, 156557) will compute the annual equivalent value.

For the challenger, we have:

The PV of the salvage value of Rs 55,000, three years from now is

$$= 55,000\left(\frac{1}{(1+0.12)^3}\right) = \text{Rs } 39,147$$

The PV of Rs 60,000 spent annually is:

$$PV = 60,000\left[\left(\frac{1}{(1+0.12)^3}\right) + \left(\frac{1}{(1+0.12)^2}\right) + \left(\frac{1}{(1+0.12)^1}\right)\right]$$

i.e., for the third year, second year, and next year, respectively"

$$= \text{Rs } 1,44,109$$

The cost of a new m/c after deducting the traded value of an old machine = 150,000 – 100,000 = Rs 50,000

The total PV as of today = Rs 39,147 (receipt) - Rs 144,109 (spent) - 50,000 (spent)

$$= \text{Rs } 154,962 \text{ (spent)}$$

Then, find the annual equivalent worth (AE) of this spent value split over the three years as:

$$AE = 154,962\left(\frac{0.12(1+0.12)^3}{(1+0.12)^3 - 1}\right)$$

$$= \text{Rs } 64,518 \text{ (spent each year)}$$

The annual equivalent expenditure of the challenger (replacement) is less than that of the defender (old machine); therefore, the replacement can be made, as planned. If it was found that the old machine should not be replaced now, then the question of whether the defender should be kept for one or two years (or longer) can be addressed before being replaced.

11.7 Storage building facilities for machines on a farm

The storage and maintenance of machines on a farm have many advantages apart from the protection they offer the machines. Farms that are in remote locations may need protection against the theft of these valuable machines. Some parts, such as rubber belts and hoses, are safeguarded from intense weather conditions when they are properly sheltered. The proper care and maintenance of high-cost machinery enhance trade-in values at the time of replacing the machinery with newer and more modern machines and, thus, worth the investment a farmer makes in building storage facilities. Planned facilities for central storage improves the parking, accessibility, and use of these machines on a farm. Machines that are used only in certain seasons are especially more benefited since maintenance is relatively easier. Many times, old farm buildings such as barns and unused livestock sheds can be remodelled into machine storage facilities and worth the investment made earlier.

The location of such a storage building should be chosen strategically such that it is not too near the farmhouse, to avoid noise from heavy machines, and, at the same time, not too far, which would hinder access in times of emergency. Security should be a major consideration in site selection and building orientation. Yard security cameras and surveillance systems are available that can monitor the storage area all the time. Most importantly, it should be located centrally such that it can be accessed from all the areas of the farm in approximately equal time. A maintenance shop can be built in the storage building to do minor repairs and servicing. The machinery storage building should be built on high ground such that rainwater drainage is not an issue. The storage building should not be near any fire-prone areas, such as hay and bale stacks, or moist environments, to avoid rusting of the machines.

The machinery that is to be stored is decided by taking into account the present availability as well as any expansion for future needs and purchases. The total floor area should be found for storing these machines with due consideration to space for manipulating and parking the machines judiciously in the available area. Simultaneously, the type of building, the superstructure, and the roofing should be decided based on the weather conditions of the region. The size and location of access doors in the storage building should be decided along with the pattern of moving in and out of the building. Alternate plans should also be considered to select the best layout.

11.7.1 Floor space calculations

The floor space required for each machine to be stored can be acquired from their specification sheets. Parsons *et al.* (1986) provided the space requirements for many farm machines commonly used in US farms. Table 11.1 presents some of that data for reference. Many machines, such as disc harrows, spring-toothed harrows, and field cultivators, have foldable sections that reduce the width of the machines considerably; this fact should be considered when calculating the space requirement. While calculating the space, future expansion should also be considered. After summing the total space requirement of the available machines and those that may be added in the future, a 15 percent increase could be added to allow space between machines, which is needed to manipulate the machines in the facility.

It is suggested that buildings that are about 16 m long are provided with doors on both ends, such that a straight drive through the building is possible, while allowing the machines on both sides of the central lane to be easily manoeuvred. The central lane is kept vacant for the machines to be moved and parked on the sides. Inclined parking spaces on the sides of the lane allow for easy hitching and moving in or out. Wider doors are provided on the wider sides so that bigger machines can be manoeuvred directly in the place allotted adjacent to the inside of the walls. The long-term storage of seasonally used machines should be allotted space near the extreme

Table 11.1 Floor area requirements of various farm machines

Equipment and size	Length, m	Width, m	Area, m^2	Height, m
Tractors				
2-wheel drive 4-5 plow	4.2	2.85	11.97	2.85
4-wheel drive < 300 hp	6	3.6	21.6	3.45
4-wheel drive > 300 hp	6.75	3.6	24.3	3.75
Tillage machinery				
Subsoiler (v-frame)				
3-13 shank (rear mounted)	0.6	1.2	7.2	
Mould board plough				
3-bottom	2.7	1.5	4.05	
8-bottom	9.3	4.2	39.06	
Tandem disk harrow (trailed)				
6-9 m (dual fold wings)	5.85	4.05	23.69	3.15
Spring-toothed harrow				
5-9 m (trailed, folding)	4.5	3.9	17.55	2.4
Rotary hoe (rear mounted)				
7-11 m (wings)	1.5	3.3-5.4	5-8.1	2.4
Row crop cultivator				
6-40 / 20-75 cm (rear-mounted folding)	2.4	3.45	8.28	3
Planting and seeding machinery				
Grain drill				
3.6-7 m. (rear mounted)	2.4	3.3-6	8-13	3.3
Row-crop planter (corn and bean)				
4-40 / 15-75 cm (rear-mounted toolbar)	2.4	3.9-4.6	9.5-11	2.7
Field application equipment				
Field sprayer (rear mounted)				
7-14 m. (fold-up boom)	2.1	2.55	5.35	
Fertilizer spreader (trailed)				
4 tonne (centrifugal)	3	1.8	5.4	
Manure spreader (trailed)				
6-10 cum rear discharge	7.05	2.55	17.9	
Grain harvesting machinery				
Combine (self-propelled)				
(base machine without header)				
6-row/ 8-row	7.8	3.9	30.42	3.75
8-row/12-row	7.8	4.35	33.93	3.9
Row crop header (corn and beans)	3	4.8	14.4	
Forage harvesting machinery				

Continued

Table 11.1 Floor area requirements of various farm machines—*cont'd*

Equipment and size	Length, m	Width, m	Area, m^2	Height, m
Mowers (rear mounted)				
2 m (cutter bar)	2.25	1.95	4.38	
Round baler				
295 kg bales	3	1.95	5.85	
400 kg bales	3.45	1.95	6.72	
Stack wagon				
1.5 tonne	5.55	3.15	17.48	3.9
3 tonne	6.9	3.6	24.84	4.5
Forage harvesters (self- propelled)				
(base machine without header)				
175-200 hp	4.2	2.4	10.08	3.15
Forage harvester (mounted)	3.6	2.55	9.18	3
Hauling equipment				
Forage wagon				
4.5 m box	5.1	2.4	12.24	3.45
5.1 m box	5.85	2.4	14.04	3.45
Trailer (cargo/implement)				
6 m	9	3	27	
Truck				
1.5 tonne	6.3	2.4	15.12	

No Permission Required

ends of the building so that other, more frequently used machines can be accessed easily year round.

Most of the frequently used doors are made 3 m high to allow passage of most of the farm equipment. For bigger machines, such as combines, higher doors are provided directly ahead of their allotted spaces. This is because higher-powered tractors, self-propelled combines, and large fold-up tillage toolbars and planters need more head space. Typically, higher-powered tractors and self-propelled combines are 4.5 m high. Sugar cane harvesters are more than 5 m high. Most importantly, wide, fold-up tillage or planting machines require a 6 m high door to enter.

A storage area in the building is allotted to the machines based on the space required for each machine; the geometry of the area; and the doors and alleys, their sizes, and locations. This can be planned on a graph sheet, affixing scaled cut-outs of the machines available. These cut-outs can be moved at will to check their ease of manoeuvrability inside the building. Or, one can use drawing software to do this job of allocation.

11.7.2 An example of storage allocation problem

Let us say a farmer owns the following farm equipment and desires to have a storage facility.

Equipment and size	Length, m	Width, m	Area, m^2	Height, m
Tractor 4-wheel-drive 200 hp	6.0	3.6	21.6	3.5
Mould board plough 5-bottom mounted	5.0	2.8	14.0	
Tandem disk harrow (trailed) 9 m (dual fold wings)	5.3	4.0	21.2	3.2
Row-crop cultivator (rear mounted folding)	2.4	4.0	9.6	3.0
Grain drill (rear mounted)	2.4	5.0	12.0	3.3
Row-crop planter (rear mounted toolbar)	2.4	4.0	9.6	2.7
Field sprayer (rear mounted) (fold-up boom)	2.0	2.5	5.0	
Combine (self-propelled) with header 8-row/12-row	9.0	4.8	43.2	4.0
Round baler 295 kg bales	3.0	2.0	6.0	
Truck 1.5 tonne	6.3	2.4	15.1	
Total		**35.1**	**157.3**	

Except for a trailed tandem disc harrow, a combine, and a truck, all the other pieces of equipment on the list are mounted ones and require a tractor to hitch them at the parking spot, move them forward by a distance of the overall length of the equipment and tractor, and then take a relatively sharp turn, as made by the tractor. The trailed and self-propelled pieces of equipment have their own turning radius and require space accordingly to move in or out of their allocated parking spot.

Let us first decide on the width and length of the storage facility for the machines listed above in the table. We also need to plan a path pattern (Figure 11.4) in and out as well as for letting the tractor manipulate the equipment in the storage building.

The total width of the facility can be approximately half of the total width of all equipment $= 35.1/2 = 18$ m appr. Adding a 15 percent space increase between the equipment, the width can be assumed as $18 \times 1.15 = 20.7 = 20$ m appr. As for the length of the facility, it should include the width of a pathway, in which the implement has to be drawn out to be taken out or parked. The total length would then be a sum of the width of this pathway in between the two rows of equipment and the maximum lengths of each implement row on either side. But, this length of the row should also be decided judiciously for very long machines such as a combine. Therefore, a hitched implement would need a pathway with a width at least 20 percent more than the overall length of the tractor and implement. This is so since the tractor can hitch the implement from the pathway, travel out, and turn into the path after clearing at least three-quarters of the length of the implement from the

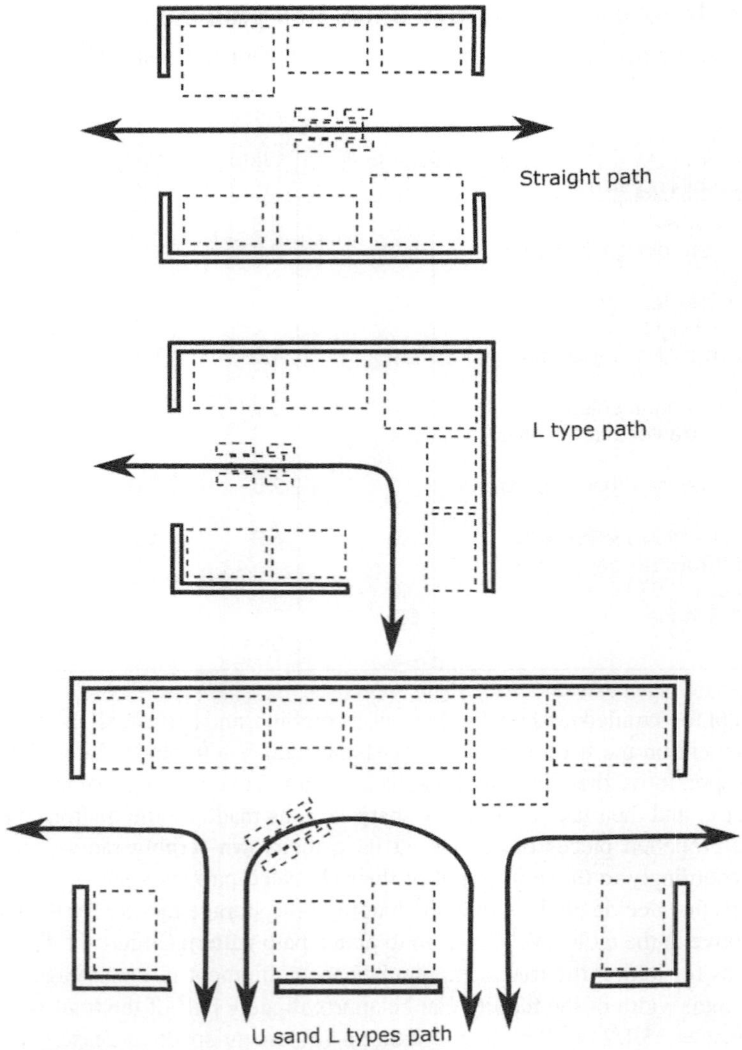

FIG. 11.4 Different Path Patterns in Storage Buildings

No Permission Required

row of implements. A self-propelled machine or a truck would need its whole length to be cleared to move out into the path, which would make it practically too big.

Here, in our case, the maximum overall length of the tractor and implement combo is that of the disc harrow, which is 6 + 5.3 = 11.3 m. Adding 15 percent to it is 13 m. Then, finding for the maximum width of the implement rows, the combine is 9.0 m and the truck 6.3 m long, respectively. Let us leave them out for now and look at the lengths of the rest. The maximum in that lot is 5.3 m, and this can be the width of one row of implements. The width of the other row shall be assumed as 3 m for now, since many of the rest are less than 3 m. Then, summing the assumed lengths, we get the overall length of the facility $= 13 + 5.3 + 3 = 21.3$ m $= 22$ m, approximately. The facility, as we have decided, is now of an area of 22 m × 20 m.

Let us now arrange the machines graphically on a sheet with grid lines (Figure 11.5). A word processor such as the MS Word can facilitate this since it allows arranging the machines as drawn rectangles. The doors in this example are allocated as to have an 'L'-type path in and out of the facility. The combine and the

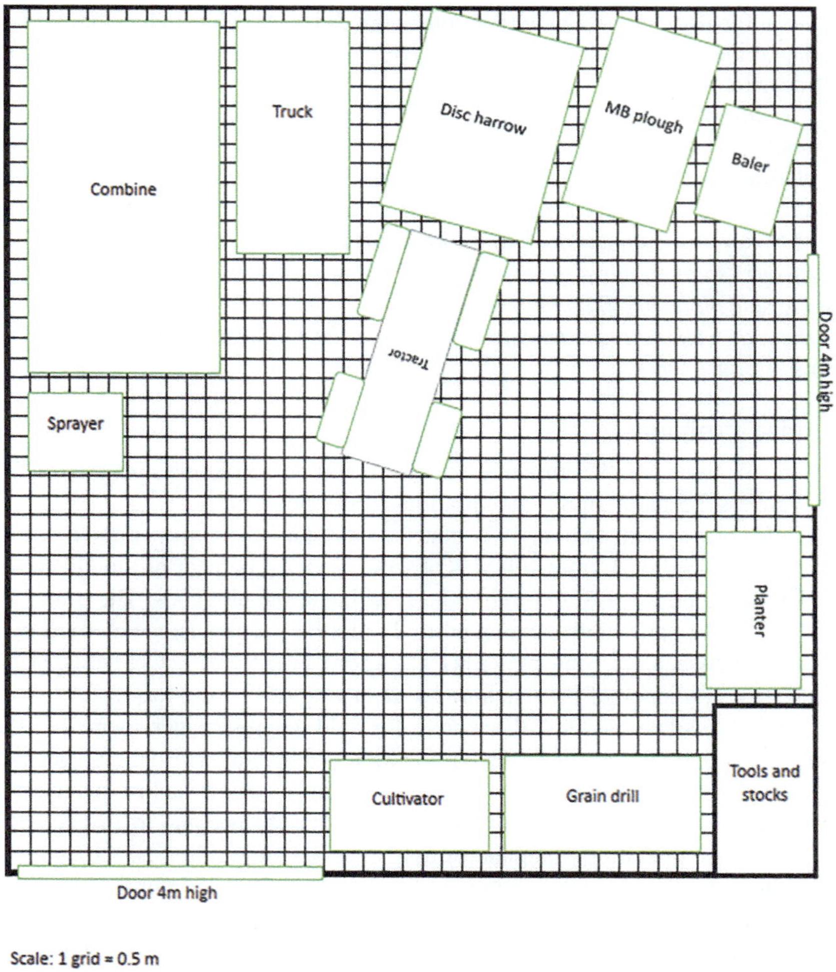

FIG. 11.5 Final Allocation of Spaces in the Facility

No Permission Required

truck, which are much longer, are parked along the side wall of the building so that they can be driven in or out straight through the door in the front. The sprayer is parked in front of the combine since the usage of the latter is seasonal and the sprayer can be moved away into the central path before taking it out. The truck, which may be used on a daily basis, has free access to the doorway in front of it. All the other items are on the sides that can be accessed easily. Some of the machines are parked at an angle to the pathway, such that the tractor has easy access from the pathway. The calculation of the dimensions so far has been adequate. If needed, the positions of the machines and doors on the WORD page can be rotated and moved at will to find more suitable allocations. The tractor can be parked on the pathway itself since it is used on a daily basis and is required to move most of the machinery. A shop area for storing tools and other consumables is also allocated inside the building. The dimensions of the facility can be varied to suit modifications, if any, until a final decision is made.

(A rudimentary game software developed by the author is furnished along with this book, which is self-instructive. A tractor with a set of implements parked on the sides are shown on the game's canvas. An enclosure of walls of desired dimensions can be drawn first at the available space on the canvas. The tractor can be moved by keyboard strokes in all directions. It can hitch and un-hitch the implements one by one, so as to move them to desired locations inside the enclosure. This can give the student an insight on how much space is required for storing the implements in a shed, as well the requirements of gates and their allocations for ease of moving them in or out).

Exercise Problems

1. A tractor-mounted, 40 cm, three-bottom mould board plough operating at 5 km h^{-1} covers 3.5 ha in an 8 h day. Find its field efficiency.
2. A field sprayer with a boom of 20 nozzles spaced at 45 cm is operated in a cotton crop at a speed of 4 km h^{-1}. It was observed that 14 percent of the effective operating time was lost to turning the machine at the plot ends. The time spent on filling the spray tanks amounted to 25 min ha^{-1}. The average amount of overlap between adjacent spray runs was 22 cm. Find the effective field capacity and field efficiency of the machine by deducting other time losses.
3. Find and tabulate the field efficiencies for field lengths in increments of 100 m onward, up to 1000 m, for a seven-tined inter-cultivator with 20 cm spacing between tines. The cultivator is operated at 5 km h^{-1}. The time lost due to minor repairs, adjustments, and miscellaneous interruptions is equal to 12 percent of the effective operating time, and turning at each end requires 20 sec.
4. We understand that, when casting or gathering a rectangular plot (Section 1.3.1), it is always better to do the ploughing along the length of the field, so that the length of turning runs in the headland will be less than that of the ploughing runs; therefore, the useful ploughing time is more than the time for idle runs in the headland. Derive an expression for finding the field efficiency in terms of the

ratio between the length and width of the field, so that the calculation can be simulated for various such ratios. Discuss the results.

5. A field plot of size 100 m by 50 m is being ploughed by a tractor-mounted, three-bottom plough of 1 m width. The plough is right-handed, and ploughing is done by gathering from the widthwise centre of the plot. A 5 m headland is marked around the plot, and ploughing is done along the length of the field by turning the tractor in the headland on the lengthwise ends. The headland is finally ploughed by going around the ploughed region. The minimum turning radius of the tractor is 3 m. The average speed of ploughing is 3 km h^{-1}, and the speed of idle travel and turning is 4 km h^{-1}. Calculate the field efficiency of the operation assuming that 2 percent of the actual ploughing duration was spent on adjustments and other activities not related to the area of coverage.
6. A 20 hp, self-propelled, four-row, semi-automatic vegetable transplanter is sold for Rs 400,000. A 30 hp, four-row, automatic vegetable transplanter is available for Rs 850,000. The row spacing is 40 cm for both machines. An entrepreneur plans to buy one of these machines to lease it out on a custom basis. The semi-automatic machine requires four labourers to sit on the machine for its operation. The automatic machine requires only one labourer to do the loading and feeding of nursery plug trays. The labour cost is Rs 100 per hour of work. The operating speed of the semi-automatic machine is 1.5 km h^{-1}, and that of the automatic machine is 3 km h^{-1}. The field efficiency is 70 percent and 80 percent, respectively. Find the minimum annual use and annual area to be covered that justifies the purchase of the automatic machine. Assume repair and lubrication costs as 15 percent of the fuel cost.
7. A tractor costing Rs 500,000 was purchased two years ago and is expected to have a 10-year life and a salvage value of 10 percent of the purchase value. The farmer expects operating costs to run at the rate of Rs 40,000 per year for three years and rise at the rate of 15 percent after that. The farmer finds that the current market value for the machine is Rs 300,000. The farmer can now replace the existing one with a new tractor of the same horsepower for Rs 600,000. The farmer expects it will reduce variable costs from Rs 40,000 to Rs 30,000. It is estimated that the new machine can be sold for Rs 400,000 at the end of year three. Determine whether this replacement decision is profitable to the farmer.
8. A farmer in India hires a wheat combine to harvest his crop in a season when an average 10 hours of working per day is possible with a probability of a working day being 0.9 as it is least affected by the weather. The effective field capacity of the machine he has hired is 1 ha h^{-1} and the minimum support price that he could get for his product is Rs 25 kg^{-1}. The crop yield that he expects is 3 tonnes ha^{-1}. He expects the machine to start work 3 days before the optimal day of harvest and work for 3 more days so that he can minimize his crop losses. But unfortunately, the hired combine arrives at his field to work only 1 day before the optimal day for harvest and completes the work 5 days after the optimal day. Calculate the timeliness cost. Assume K, the timeliness coefficient as 0.006 for the harvesting operation.

References

ASAE D497. (2006). *5 FEB2006, ASAE standards : Standards, engineering practices and data adopted by the American society of agricultural Engineers*. American Society of Agricultural Engineers.

Fairbanks, G. E., Larson, G. H., & Chung, D. S. (1971). Cost of using farm machinery. *Transactions of the ASAE, 14*, 98–101.

Hunt, D. (2008). *Farm power and machinery management*. Waveland press.

Kepner, R. A., Bainer, R., & Barger, E. L. (1978). *Original from the university of Michigan*. Avi Publishing Company.

Mayfield, W., Hines, G. S., & Roberts, L. (1981). A new method for estimating farm machinery costs. *Transactions of the ASAE,* 2151-0059. *24*(6), 1446–1448. https://doi.org/10.13031/2013.34470.

McKibben, E. G. (1930). Some fundamental factors determining the effective capacity of field machines. *Agricultural Engineering, 11*, 55–57.

Parsons, S. D., Strickland, M., Jones, D. D., & Friday, W. H. (1986). *Farm machinery management series*. AE-115 Planning Guide to Farm Machinery Storage.

Siemens, J., Hamburg, K., & Tyrrell, T. (1990). A farm machinery selection and management program. *Journal of Production Agriculture,* 08908524. *3*(2), 212–219. https://doi.org/10.2134/jpa1990.0212.

Srivastava, A. K., Goering, C. E., Rohrbach, R. P., Buckm, D. R., & ster. (2006). *Engineering principles of agricultural machines*. ASABE Publications.

Sullivan, W. G., Wicks, M., & Koelling, C. P. (2010). *Engineering economy*. Pearson publ.

CHAPTER

Earth-moving equipment

12

Chapter Outline

The correct choice of equipment and procedures (Nichols and Day, 2005; Peurifoy, Schexnayder, and Shapira, 2006) is critical for clearing land efficiently and preparing it for agricultural production. Virgin land may have a lot of trees and brushes, plain grass, or a different vegetative terrain. The type of terrain and soil may also vary for a given piece of land. With the right land-clearing equipment, vegetation including trees can be uprooted, hauled out, and levelled to till and prepare the soil for agricultural production. The excavation of drainage/irrigation facilities such as those for laying pipes and drainage tiles may also be desired. An engineer should be able to identify the right combination of machines and attachments for a given job.

Preparing land for agricultural production requires an assessment of the variety of operations to be done, such as cutting down large trees, knocking down trees, uprooting tree trunks and heavy bushes, removing heavy boulders, and, finally,

Machines of the Farm. https://doi.org/10.1016/B978-0-443-32906-7.00014-1

levelling and grading. The size of the area and production capacity of the machines to be used must be understood to select the right machinery. The type of terrain for machine mobility as well as the mode of disposing of the debris must also be considered. The major operations to be performed and machines to be used are mainly for pushing the soil, rubble, and vegetative debris; land clearing such as uprooting trees and vegetation; digging/excavating soil for the purposes of land forming, levelling, and grading; and, finally, loading and hauling material from the land or within the land area.

12.1 Pushing soil, rubble, and vegetative debris

Soil should be dug and pushed so that the basic work of the "cut and fill" process can be achieved to rough level the given land. Moreover, trees, heavy brush, cut and uprooted trees, and rubble need be pushed and moved short distances or heaped to be loaded and hauled from the land.

12.1.1 Crawler tractors

Because of the amount of heavy power required for land-clearing works, crawler tractors are usually desired over conventional agricultural tractors. These are tractors with tracks in place of wheels as traction elements. They are high-powered machines with greater than 1000 hp and usually weigh more than 4 tons. The track system is capable of generating more traction power than a wheeled system. Earth-moving machines, especially dozers (Figure 12.1) and excavators, are crawler tractor systems. A crawler tractor has a chassis that is comprised of a diesel engine, transmission, steering units, and a pair of track frames that supply traction and support. The drawbar horsepower is less than the engine's horsepower because of the losses mainly in the traction shoes interacting with the soil. The main clutch is usually controlled by a hand lever at the operator's left and is used to shift the gears in the power transmission. A torque converter provides a shock-absorbing effect in the power transmission between the engine and the tracks and allows gear shifting even when the crawler tractor is moving. A crawler tractor has a power take-off shaft similar to what wheeled tractors have.

The rear live axle drive drawing power from the transmission system transmits power through ring gears to the bull sprocket through the final drive and steering clutches on either side.

This pair of steering clutches and brakes is used for steering the tractor by stopping the drive on one track relative to the other. The bull sprockets on each side drive the track chain in turn and are big sprockets of very heavy construction. The track on each side consists of a roller chain with the traction shoes bolted to them serially over the complete roller chain. Many types of track shoes are available, two of which are shown in Figure 12.2. The typical type is a flat, steel plate with a single high cleat

FIG. 12.1 Crawler Tractor as a Dozer (CASE 850N dozer, CNH Industrial, USA)

or grouser across it, providing good traction in soil. Rubber-faced shoes or flat shoes are used on paved roads, but they cannot provide traction as the conventional ones do.

The track frame (Figure 12.3) and its mechanism are constructed in such a way that the frames rest on the small track rollers. The idlers, which are flanged wheels in the same width of the bull sprockets, are mounted on spring-mounted yokes on the frames. The rollers and idlers revolving on fixed axles are flanged in order to keep the track in line. The track and support rollers have outer flanges on each side of the track rail. The rollers are mounted on rugged taper roller bearings with mud-proof seals. A big idler held on a sliding yoke is at the front of each track, acting as the driven wheel of the track's roller chain. Hydraulic-type track tensioners are typical fool-proof devices, which are just grease-filled rams, the piston of which moves the main idler forward by filling grease in the cylinder. Similar to wheeled tractors, a crawler tractor has a heavy drawbar projecting to the rear.

12.1.2 Dozers

A bulldozer is basically a high-powered crawler tractor that has a dozing blade at its front. The blade is used to shear, cut, push, and roll the soil ahead of the dozer. Dozers have a heavy traction system, such as crawler tracks; this system allows the machine to push heavy loads at the front by using the track's traction. Dozers are effective and versatile earth-moving machines used for land clearing and land

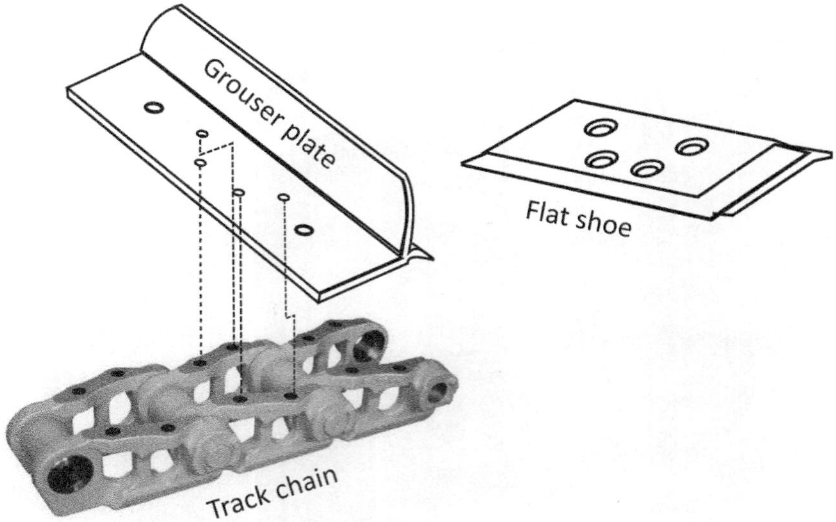

FIG. 12.2 Track Chain and Types of Track Shoes

No Permission Required

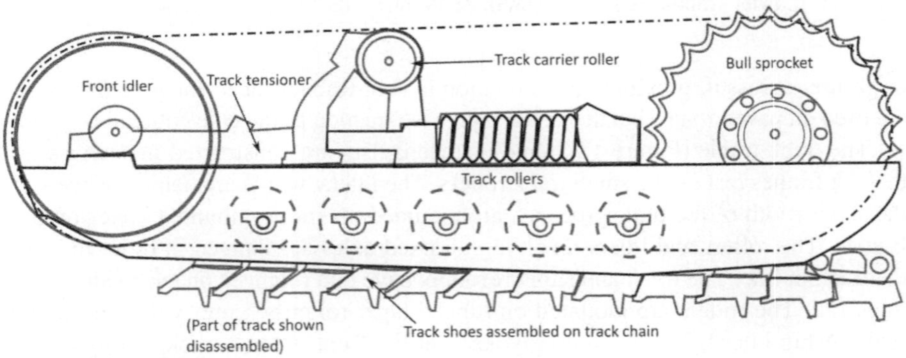

FIGURE 12.3 Track Assembly

No Permission Required

forming, leading to farmland development. They are very versatile and used for operations such as:

1. moving earth or rock for short distances;
2. spreading earth or rock fills over level ground;
3. backfilling trenches;
4. digging underground to create water bodies and basements for construction;
5. desilting dry water bodies;
6. clearing land of trees, stumps, and root mats.

12.1.2.1 Dozer blades

A dozer blade (Figure 12.1) is a curved, steel board with replaceable cutting edges at the bottom and sides. A C-frame pivoted to the dozer's sides supports the blade at the inner centre of the blade. In turn, the blade has hydraulically operated push-arm cylinders and tilt cylinders to raise and lower the blade as well as to tilt or "angle" it sideways to a certain extent. The blade is raised or lowered, strongly reinforced by the C-frame support. The top of the blade can be "pitched" forward or backward, varying the rake angle, which is the attack angle of the soil surface, to make a cut. Angling the blade in the direction of the dozer's travel causes the pushed material to roll off the trailing end of the blade. This process is known as side casting. The hardened steel-cutting edges at the bottom as well as the sides of the blade are made replaceable since they are subjected to abrasion, making it wear out rapidly. The various processes of which dozers are capable are:

1. Digging: This action includes moving the dozer forward, lowering the blade to make the cut, and spreading or transporting it. Digging collects the soil in the blade, and some soil flows to the side, making ridges. The depth of blade increases the soil slice cut and vice versa.
2. Transporting of soil: Transporting is usually performed by digging and moving repeatedly in the same path, so that the ridges on the sides built by previous passes do not allow the further flow of soil from the side of the blade.
3. Spreading of the soil: Spreading is performed by keeping the blade above the surface to which the spread is to be added so that the pushed soil slips under it in a smooth layer. Spreading is done by varying the travel path each time so that the side flowing ridges are not well formed.
4. Piling the soil: Piling is usually performed by first heaping a series of small piles and then riding over these piles to heap more on top so that the heap grows higher.
5. Breaking heaps: Breaking is done by traveling over the pile and then knocking down the soil layer by layer.
6. Back-blading: Back-blading is the operation in which the back of the blade is allowed to graze over the surface of the already levelled soil. This is performed by operating the dozer in reverse so that the small ridges formed in spreading as well as the track footprints are levelled.

12.1.2.2 Dozer production estimation

The amount of material a dozer moves, which is known as the blade capacity, is dependent upon the quantity that will remain in front of the blade during the push. The factors that influence the volume of material pushed are the following.

Blade Type

Straight blades, which have less curvature, roll material in front of the blade; universal blades control side spillage by holding the material within the blade. The latter forces the material to move to the centre and collects more material. Of course, the blade capacity depends upon the size of the blade that matches the bulldozer's tractive power.

Type and Condition of Material

Cohesive materials such as clay soils will form heaps when pushed. Cohesionless materials such as sand do not heap or swell in front of the blade; therefore, a smaller volume is collected.

The volumetric load of a blade can be estimated approximately as

$$V_s = 0.65\ WH^2\ m^3$$

where

V_s = the approximate capacity of a straight or angle blade, in I cycle;
W = the blade width, m;
H = the effective blade height, m.

Cycle Time

The cycle time is the sum of the time required to push the material forward, backtrack the dozer, and bring the dozer back into position to push. The time required to push and backtrack can be calculated for each dozing situation depending upon the speed and distance moved in each cycle. Dozing is usually done at slow speeds, of 3 km h^{-1} and less. Box 12.1 shows a worked-out example of calculating the volume of loose soil moved by a typical dozer in a cycle.

Box 12.1 Example 12.1

Example 12.1. A dozer can push an average blade load of 6.15 m^3 of loamy sand in one cycle. The average push distance is found to be 30 m in this situation. Assuming a dozing speed of 3 km h^{-1} and a backtracking speed of 4.5 km h^{-1}, calculate the loose soil moved in one hour.

Pushing time $= \frac{\text{Push distance}}{\text{Pushing speed}}$

$= 30\ m / 3\ km\ h^{-1} = 30\ m / (3000 / 60)\ m\ min^{-1}$

$= 0.6$ min

Backtracking time $= \frac{\text{Backtrack distance}}{\text{Backtrack speed}}$

$= 30\ m / 4.5\ km\ h^{-1} = 30\ m / (4500 / 60)\ m\ min^{-1}$

$= 0.4$ min

Manoeuvring time = 0.05 min (say)

Loose soil moved in one hour = 6.15 m^3 in one cycle / Total cycle time

$$= \frac{6.15}{(\text{Pushing time} + \text{Backtracking time} + \text{Maneuvering time})}$$

$$= \frac{6.15}{(0.6 + 0.4 + 0.05)} = 5.86\ m^3\ min^{-1} = 5.86\ m^3\ min^{-1} \times 60\ min\ h^{-1} = 352\ m^3\ h^{-1}$$

12.2 Land clearing

The clearing of vegetation and trees is necessary before undertaking subsequent earth-moving operations. Trees, brush, tall grass, and weeds need to be removed before moving the earth. Land clearing hence includes the removal of all trees and stumps and all vegetation above the surface and disposing of the removed vegetation by stacking and burning or transporting. Dozers with conventional earth-moving blades, special clearing blades, and clearing rakes are used for this purpose. Before felling large tree trunks, dozers with earth-moving blades excavate the earth around trees and cut the main roots. Dozers specially designed for use in felling trees are also used. Vertical and horizontal knives cut the tree in two planes for easy clearing. Bigger tree trunks are removed by stump removal practices such as uprooting them by pulling them with crawler tractors or dozers. Huge trunks are blasted by explosives before clearing them with dozers. Crawler tractors are fitted with rear-mounted rippers which are constructed more like wedge ploughs with minimal tines. A ripper is a relatively narrow, bladed implement. Because of the power and tractive force available with large crawler tractors, the penetration depth of a rear ripper on such machines is greater than 1.2 m; this machine can rip out rocks, clearing the land.

12.3 Excavation of pits and trenches

Excavators are used to dig pits and trenches in the process of land forming such as the construction of large irrigation and drainage channels, ponds, and farm roads. They have shovels that are used to dig into the ground or pile of soil or rubble, scoop it, and dump it as desired to another spot that is within its reach. They are used for scooping dirt and loading it into dump trucks that can haul the dirt or other material to another location.

An excavator consists of three structural components, namely, the top revolving unit, the unit for locomotion, and, finally, the arms and buckets that do the excavation work. Depending upon the types of mechanisms used on the arms to perform certain functions, they could be classified as front shovels, backhoes, draglines, clamshells, or cranes. The first two components, namely, the revolving unit and the locomotion unit, are almost similar in all these machines. The top-mounted unit has a vertically rotated drum, which is actuated by hydraulic motors and the allied controls. This controls the boom on which the digging unit is mounted and responsible for the up and down movement as well as the in and out function of digging and dumping. This whole assembly is mounted, in turn, on a horizontal drum that rotates it around a central axis so that the digging units can reach the desired direction. These drums are supported on heavy roller supports and driven by a pinion that meshes with the bull gear on the drum or turntable. Hydraulic motors allow the turntables to move in any direction as desired. This enables the digging elements to face any direction for digging, moving, and dumping the loads quickly anywhere within the excavator's reach. The locomotion unit is similar to that of a crawler tractor and generally has traction chains for rendering the derived locomotion and grip required in the terrain in which these machines are expected to work. Excavators are typically powered by a heavy diesel engine driving a central hydraulic pump that, in turn, supplies power to hydraulic motors performing various functions. The tracks are either directly driven by the engine or by hydraulic motors. Similar to crawler tractors, a torque converter is used to cushion the drives against shocks and to give additional power in hard digging. There may be a two-speed engine transmission, chiefly to allow faster travel. Excavators have shovels of a width ranging from 1 m to 3 m, with the latter weighing as much as 20 tons. The largest machines used for mining, such as draglines, are gigantically huge and weigh thousands of tons.

12.3.1 Revolving excavators

Excavators are either crawler or wheel-mounted machines. These machines are used for digging, transporting, and dumping material, using a steel bucket with sharp teeth on a digging edge. They are classified by the digging motion of the bucket. The machine, when its bucket is facing forward and digs by a forward/upward arc, is known as a "front shovel". When the bucket of the machine is facing backward and it digs and loads by a downward arc motion, the machine is known as a "backhoe".

As in all other excavators, these machines also typically consist of the same three major components, namely, the locomotion, revolving, and boom units. The locomotion unit either has a track system or a rugged set of wheels to carry and move the machine over the terrain in which it is expected to work. The revolving unit, mounted on the locomotion unit, has all the parts, such as the engine and drives, arranged such that their weight counterbalances the forces of the bucket operation as well as the overhanging weights of the bucket and the extending boom. In addition to this, a massive counterweight attached to the rear of the revolving deck outside helps in counterbalancing these eccentric loads on the revolving unit. The axis of rotation, too, is off-centre to the centre of the locomoting track to achieve balancing while the machine is functioning. The boom unit carries the digging bucket on a long boom extending from the revolving unit and has actuators to swivel the bucket into digging or dumping angles. The boom will also be in parts so that it can fold as desired to make the bucket reach up and down relative to the machine. The revolving unit rotates on a horizontal plane on the locomotion unit to make the bucket reach all 360 ° around the machine. Altogether, the machine's bucket can be moved three dimensionally to dig, transport, and dump the soil or material as desired. The different types of revolving excavators arise from the construction, mechanism of actuation, and purpose for which they are intended.

12.3.1.1 Backhoe

The boom assembly of a backhoe (Figure 12.4) is comprised of three rugged structural members, namely, the boom, the stick, and the bucket. The boom is hinged to the excavator's revolving deck at the front through massive trunnions so that it can vertically swing up and down. The other two units are hinged to each other and with the boom again. The swinging movement at each of these hinges is actuated by two-way remote hydraulic cylinders.

The boom has a concave bend toward the ground so that the bucket can work more closely to the machine and permit deeper digging without hitting the tracks below. The boom end has a huge bracket, on which the stick is hinged with heavy pins. The stick cylinder is mounted on the boom top between the bucket dump arms and the boom itself. It is usually one piece but may extend and retract by telescoping. The stick's connection to the boom is nearer the top of the stick. The bucket teeth will be moved by the stick cylinder four to eight times faster and farther than its piston moves, with one-fourth to one-eighth the force. The digging bucket is connected to the lower end of the stick by a hinge and to a triangular "Stick cylinder"

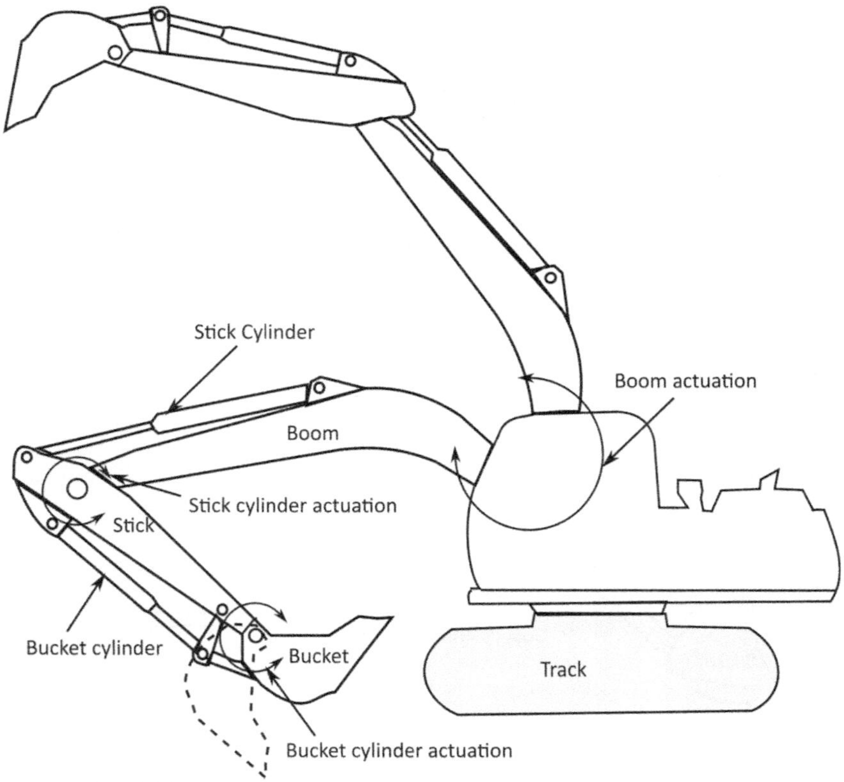

FIG. 12.4 Construction of a Backhoe

No Permission Required

in figure 12.4 set of paired dump arms so that the bucket can have an extended arc of rotary movement around the stick hinge. When the bucket cylinder connecting the stick and the bucket is extended, the bucket's teeth curl inward in a digging motion. When the cylinder is retracted, the bucket dumps by swinging back. Several sets of eccentricities are provided so that the bucket action can be changed by moving hinge pins. This can alter the bucket motion to the requirement. The primary use of a backhoe is to dig ditches, leading to excavation as desired. The bucket is usually slightly wider at the tooth end to avoid friction of its sides with the dug pit's walls. The digging edge has replaceable teeth. Backhoes are useful in digging pits and trenches below the ground surface on which the machine stands. But, they are also frequently used in loading dug soil and rubble into hauling trucks as well. Their positive bucket control, in comparison with other types of excavators such as draglines, makes them ideal for operating in areas close to the hauling trucks, near where the dug material will be loaded.

12.3.1.2 Front shovels

These excavators (Figure 12.5) are used for digging soil and loading it into hauling trucks. The loading of gravel and dug soil is a typical application. As previously explained, the bucket of this machine faces outward, and it digs and hauls using a forward arc motion of the bucket. The basic construction of a front shovel is similar to that of a backhoe except that the bucket is lowered to the ground with the teeth

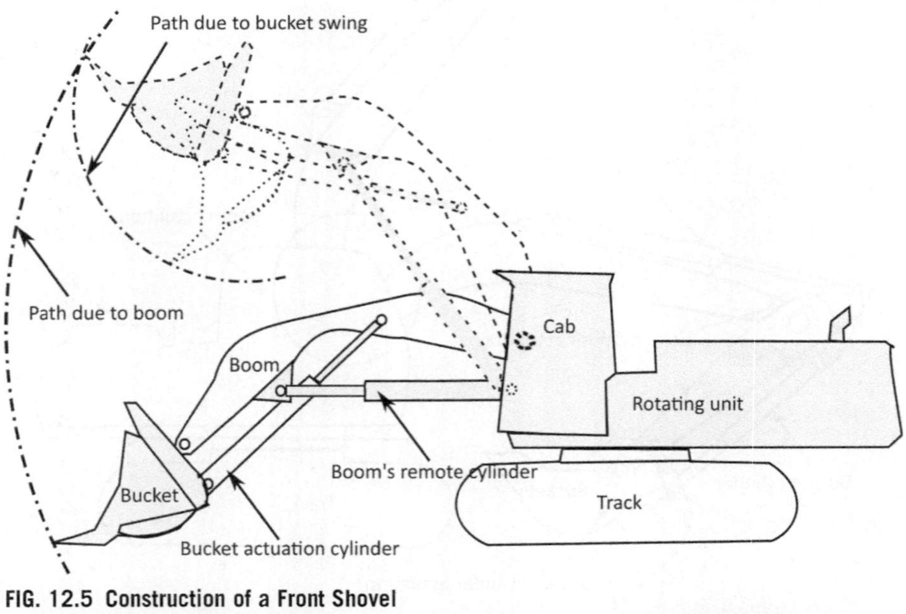

FIG. 12.5 Construction of a Front Shovel

No Permission Required

pointing outward. An actuation force is applied to the bucket cylinder, which pushes the bucket through the material to be loaded into the bucket. A fronthoe can also rotate the bucket downward, as shown, so as to dump the material at the desired height, as reached by the boom actuation.

12.3.1.3 Front loaders

Front loaders (Figure 12.6) are also called front-end tractor shovels, shovel dozers, or dozer shovels, and are similar in construction to front shovel excavators, providing a similar function. It is a short machine, used for digging, loading, rough grading, and short-distance hauling. A typical loader, as shown in the figure, has a pair of push-cum-lift arms, which act as a boom, hinged to the machine. A bucket, which is the same width as the tractor, is hinged to the front of the arms, and a pair of dump arms are hinged to both push arms and to the bucket. The actuation of all the arms and the bucket is performed by means of two-way remote hydraulic cylinders. This machine is used more like a dozer, so that the bucket can scoop up the material by moving the bucket forward by a dozing action, lift it, and dump the material from a height by tilting the bucket downward. Crawler and four-wheel drive traction is used for heavy work, and two-wheel drive tractions are used for lighter work.

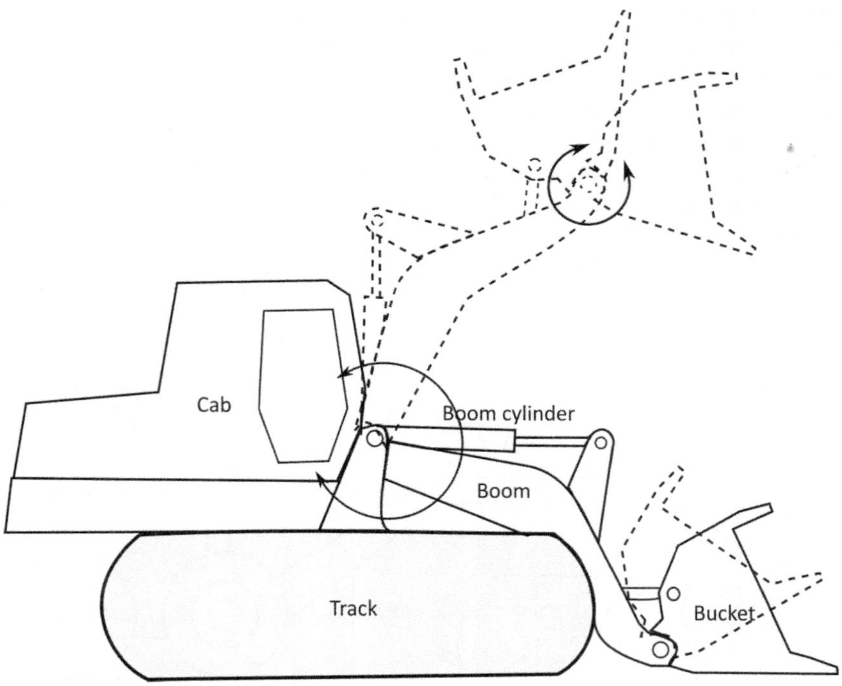

FIG. 12.6 Construction of a Front Loader

No Permission Required

12.4 Land-leveling and grading

Graders or levellers are machines that are used for the final shaping and finishing of land and are used after rough leveling is performed by dozers. Though motorized, heavy-duty graders of up to 600 hp exist for intensive construction work such as road laying and embankment works, light graders are available at powers as low as 35 hp. A self-propelled grader consists of a wide, controllable grading blade mounted at the centre of a wheeled machine with a long wheelbase. The blade is usually mounted at an angle to the travel direction; it makes a light grazing cut of the surface and side casts the cut soil.

Most heavy graders are tandem-drive units (Figure 12.7) with the engine mounted at the rear; the engine drives four single wheels in tandem pairs through gears and chains. The frame carrying the front axle is long and high, allowing for a long wheelbase and enough space to carry and manipulate the blade under the frame. This arrangement provides stability and an accuracy of the level for the blade's cutting edge. The blade is typically lifted and lowered by two remote hydraulic cylinders cradled on frame brackets. The blade's side shift or rotation is also controlled by hydraulic cylinders. The steering is hydraulically controlled because of the long base. Multi-directional control for the blade provides accuracy for its finishing ability. The operation of these machines requires a lot of expertise to bring about the required finish and level to the soil surface.

Drawbar graders are simple levelling blades mounted on the three-point linkage of the tractor's rear (Figure 12.8). The blade can be manually rotated and locked straight or at different rotational angles to either side. By raising or lowering the right lower link of the tractor, a cross slope is provided to the blade, too. The blade is pitched forward or backward to adjust the blade's rake, as in the case of a typical plough, thus controlling the penetration. As the blade is entirely behind the tractor, it lacks the stability of a self-propelled grader.

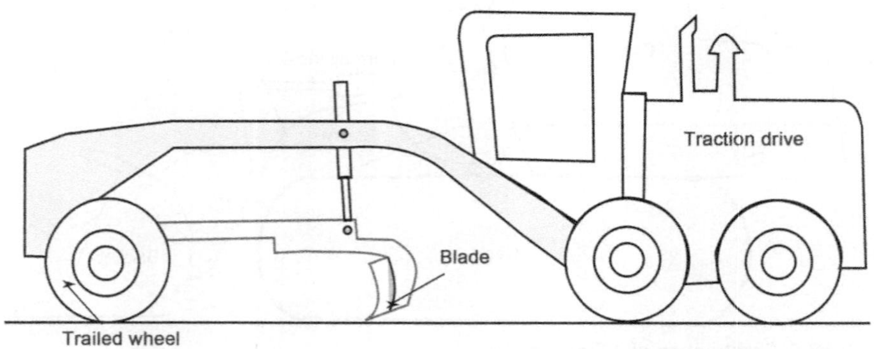

FIG. 12.7 Self-Propelled Grader

No Permission Required

FIG. 12.8 Tractor-Mounted Drawbar Grader/Levelling Board (India Agrovision Implements Pvt. Ltd)

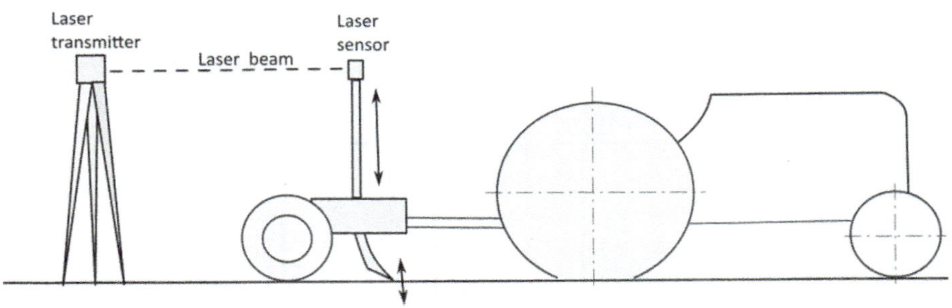

FIG. 12.9 Tractor Trailed Laser-Controlled Grader

No Permission Required

Machines that have automatic level control are also available now. One such technology allows for the grade to be controlled by laser guidance (Figure 12.9). A laser beam of an adequate intensity and pencil-thin thickness emanates from a ground tower installed in the land itself and beamed over 100 m in length. This beam rotates around the tower, thus forming a plane of laser at that height. The grader or leveller machine has a laser sensor installed on a mast which seeks to locate the circular plane of laser beam and accordingly raises the grading blade along with it. This motion is used to set the required level relative to the plane of laser and automate the grading operation. The laser-guided machine can hence move around the laser tower and grade the land within its range. The grading can be done in steps to refine the surface to the required level. The plane of laser can also be set to rotate at an inclination to provide a sloping grade desired for an area of land to be irrigated in a particular direction. A tractor-trailed grader mounted on wheels and that is laser-controlled is also available on the market.

12.4.1 Earth work calculations for a land-leveling job

Typically, any land-leveling job requires that the existing levels of the given land are measured first through standard surveying methods. A dumpy level is generally used to assess these levels with respect to a benchmark level. The levels are then used to draw the contour map of the land, which depicts the elevation of the land terrain. A profile level is the one that is desired and may consist of a flat level that is either flat all over or sloping gently in any desired direction. Based on the level differences, the earth-moving pieces of equipment are used to "cut" the soil above the profile level and "fill" the cut soil into the places that are below the profile level. Any land leveling is hence a job of cutting soil from elevated places and transferring it to the places to be raised to a specified level. Depending upon the profile level required and the existing elevations available, the job may result in an excess of soil or a lack of soil; then, soil will have to be hauled out or in, as may be demanded.

To understand the basis of calculating the soil to be moved by the equipment, let us consider a strip of land 250 m × 50 m. Assume that dumpy level measurements were taken on a grid along the length of the land on both of the 50 m sides at every 50 m space. Table 12.1(a) shows the measured levels, as from zero datum; Figure 12.10

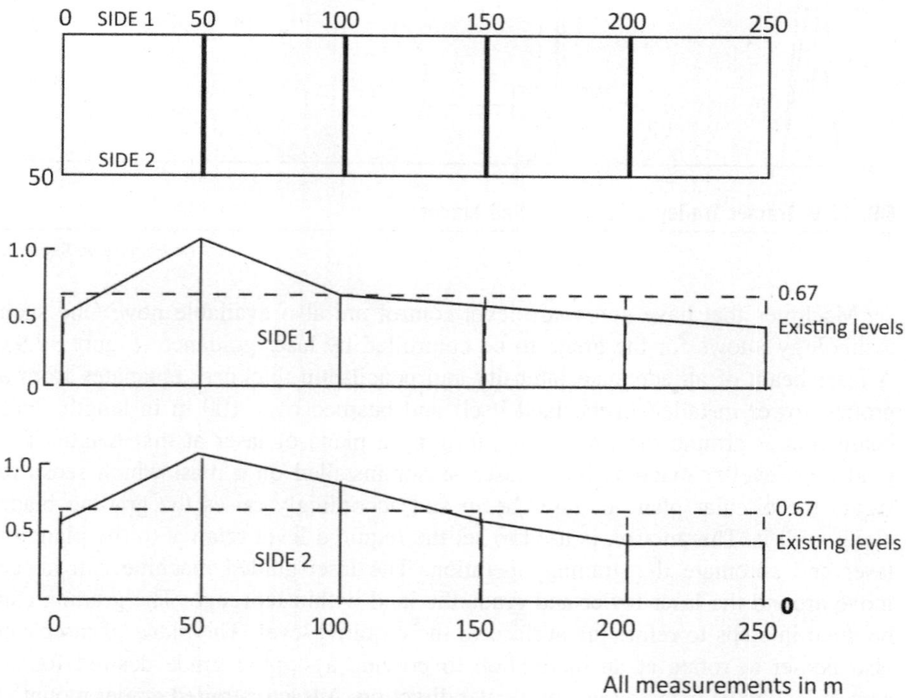

FIG. 12.10 Earth Work Calculations for Levelling

No Permission Required

Table 12.1 Stepwise earth work calculations

	Side 1 @ 0 m			Side 2 @ 50 m				
Stations at	**(a).** Measured levels at measuring stations							
0 m	0.50			0.58				
50 m	1.10			1.00				
100 m	0.67			0.90				
150 m	0.58			0.61				
200 m	0.50			0.43				
250 m	0.43			0.43				
Stations at	**(b)** Height of soil level above or below required profile							
0 m	0.50-0.67	=	-0.17	0.58-0.67	=	-0.09		
50 m	1.10-0.67	=	+0.43	1.00-0.67	=	+0.33		
100 m	0.67-0.67	=	+0.00	0.90-0.67	=	+0.23		
150 m	0.58-0.67	=	-0.09	0.61-0.67	=	-0.06		
200 m	0.50-0.67	=	-0.17	0.43-0.67	=	-0.24		
250 m	0.43-0.67	=	-0.24	0.43-0.67	=	-0.24		
Sections between	**(c)** Net sectional area of cut or fill, m^2						**(d)** Mean of sides 1 & 2, m^2	**(e)** Net Volume, m^3
0-50 m	(-0.5*0.17*15) +0.5*0.43*35	=	6.25	(-0.5*0.09*10) +0.5*0.33*40	=	6.15	6.20	310.00
50-100 m	0.5*0.43*50	=	10.75	0.5*(0.33 +0.23) *50	=	14.00	12.375	618.75
100-150 m	-0.5*0.09*50	=	-2.25	0.5*0.23*40 -0.5*0.06*10	=	4.25	1.0	50.00
150-200 m	-0.5*(0.09+0.17) *50	=	-6.50	-0.5*(0.06+0.24) *50	=	-7.50	-7.00/0.9	-389.00
200-250 m	-0.5*(0.17+0.24) *50	=	-10.25	-0.24*50	=	-12.0	-11.125/0.9	-618.00
				Total volume				-25

Negative values are "fills" and positive values are "cuts".

shows an illustration of these elevations on both sides with also the desired flat profile level, which is 0.67 m (say) from the datum level. Table 12.1(b) shows the elevations with respect to the profile levels by deducting 0.67 m from the measured values. The positive values derived indicate that the soil surface is above the profile level and should be "cut" by the earth mover; the negative values indicate that the elevation below the profile must be "filled".

Table 12.1(c) shows the net area of the cut or fill in each 50 m section along sides 1 and 2. The areas of "cuts" are positive and "fills" are negative. These areas are calculated using the trapezoidal rule as these sections are either triangular, trapezoidal, rectangular, or a combination of these. The lengthwise 50 m sections, having combination of both cuts and fills, must be summed to find the net area. The triangle of the cut and the triangle of the fill are similar triangles, and the base lengths are found accordingly by the principle of an equal ratio of similar sides. The mean of the net sectional areas on the two sides of each 50 m segment is calculated next (Table 12.1(c)). In Table 12.1(d), the mean net area of the cut or fill pertaining to each lengthwise segment of the land is multiplied by the length of 50 m to find the soil volume to be cut or filled. The compacted soil "cut" by the machine, when used for "fill"ing the lower elevations, needs to be compacted; hence the volume required for "fill"ing is to be accordingly more than the volume of the loose, "cut" soil. The "fill" volume is divided by the shrinkage index of 0.9 m to increase the loose volume to be compacted "fill" volume. The algebraic sum of the cut and fill volumes of the lengthwise segments lets us know whether the "cut" soil is in surplus of the "fill" requirement of the job or not. If the algebraic sum is negative, the cut soil is not adequate for the fill and soil has to be hauled in from elsewhere to finish the job. Here, in our example, the sum is negative; hence the cut soil is in a deficit and soil must be hauled in to fill the deficit.

Given a contour map of a strip, the cut and fill calculations can be performed as explained in the example worked out in Box 12.2.

In the example given in Box 12.3,the cut and fill calculations are performed on the contour map of a whole field plot. The grid levels are interpolated using supporting software and calculations automated on an Excel sheet. The logic of automation is explained so that the student can learn to analyze the problem.

12.5 Hauling of material

In many instances, soil and rubble must be hauled over long distances from inside the site or from an external site to the work spot. Trucks are the work horses of material hauling. Excavators, such as backhoes, are used to load the trucks to move soil

Box 12.2 Example 12.2

Example 12.2. The contour map of a 250 m × 250 m piece of land is given below. The land has to be levelled to a flat elevation of 0.6 m from the zero datum. Find the soil volume to be cut or filled in the first 50 m strip in an E-W direction.

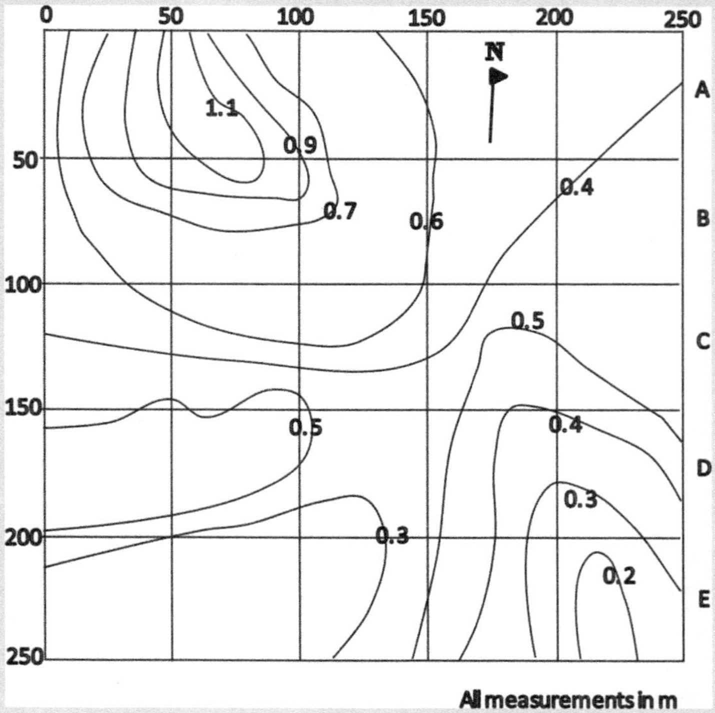

The measured elevations may first be derived for the grid points of the first 50 m strip in the E-W direction. Then, the procedure for soil volume calculation explained in Sec. 12.14.1 may be followed for that strip. The soil volume calculated for the lengthwise 250 m strip is finally summed.

A printout of the contour map is taken and the points of contours crossing the grid lines, such as a and b, are measured from the nearest grid points. These values may be used to interpolate the elevations at the grid points to solve the problem.

As an example, elevations at some grid points at the NW corner of the map, as shown below, are calculated.

Continued

Box 12.2 Example 12.2—cont'd

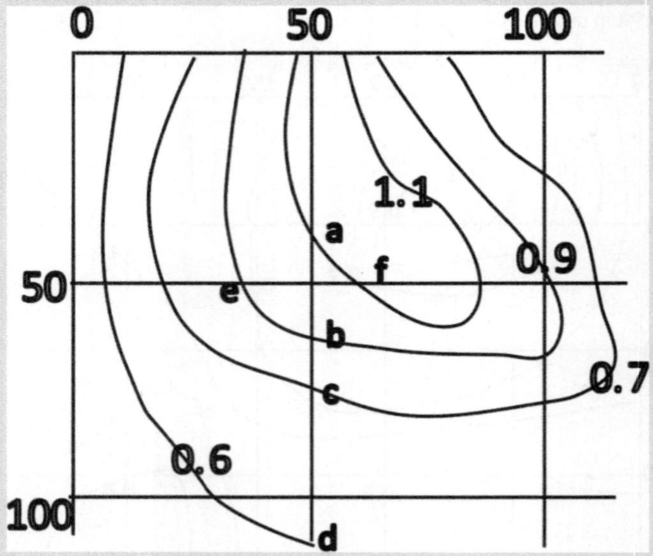

i. Contour of 1.1 m elevation at point a is 0.4 cm up from the 50×50 m grid point;

0.9 m contour at point b is 0.53 cm down from the 50×50 m grid point.

Interpolating the difference in elevation at the 50×50 m grid point between points a and b as

$= \frac{(1.1-0.9)\text{m}}{(0.4+0.53)\text{cm}} \times 0.4\text{cm} = 0.086\text{m},$

the elevation at the 50×50 m grid point is 1.1-0.086=1.014 m.

ii. 0.9 m contour at point e is 0.76 cm left from the 50×50 m grid point;

1.1 m contour at point f is 0.58 cm right from the 50×50 m grid point.

Interpolating the difference in elevation at a 50×50 m grid point between points e and f as (just for the verification of the last step)

$= \frac{(1.1-0.9)\text{m}}{(0.76+0.58)\text{cm}} \times 0.58\text{cm} = 0.086\text{m},$

the elevation at the 50×50 m grid point is 1.1-0.086=1.014m (verified).

iii. 0.7 m contour at point c is 1.42 cm up from the 50×100 m grid point;

0.6 m contour at point d is 0.28 cm down from the 50×100 m grid point.

Interpolating the difference in elevation at a 50×50 m grid point between points a and b

$= \frac{(1.1-0.9)\text{m}}{(1.42+0.28)\text{cm}} \times 0.28\text{cm} = 0.0164\text{m}.$

The elevation at the 50×50 m grid point is 0.6+0.0164 = 0.616 m.

iv. At a 100×50 m grid point, the 0.9 m contour is passing through; so

the elevation at the 100×50 m grid point is 0.9 m.

The elevations at the boundaries will be difficult to interpolate but can be approximated.

The following results of calculation using the interpolated levels at the grid points show the sequence of calculations similar to that shown in the work-out of Sec. 12.14.1 on cut–fill calculations .

Box 12.2 Example 12.2—cont'd

The final sum of soil volume is positive; hence the cut soil is in excess of what is required and must be hauled out to the other strips or out of the field as required.

	Interpolated levels at grid points, m					
	0	**50**	**100**	**150**	**200**	**250**
0	0.58	1.1	0.67	0.58	0.5	0.43
50	0.58	1	0.9	0.61	0.43	0.43

	Levels above or below required profile of 0.6 m					
	0	**50**	**100**	**150**	**200**	**250**
0	-0.02	0.5	0.07	-0.02	-0.1	-0.17
50	-0.02	0.414	0.3	0.01	-0.17	-0.17
			(-ves are fills and +ves are cuts)			

	Net sectional area of cut or fill, m^2				
	0-50	**50-100**	**100-150**	**150-200**	**200-250**
0	12	14.25	1.25	-3	-6.75
50	9.85	17.85	7.75	-4	-8.5

	Average volume of cut or fill at stations, m^3						
						Sum of cut and fills	
0-50	546.25	802.5	225	-194.444	-423.611	955.6944	Excess

Box 12.3 Example 12.3

Example 12.3. The 250 m × 250 m land given in Example 12.2 has to be levelled to a flat elevation of 0.5 m from the zero datum. Find the soil volume to be hauled out or in, as excess or deficit, respectively, from each 50 m E-W strip of the land and also the final volume that is to be hauled in from elsewhere to fill the deficit or hauled out as excess. (Use the software "contourgrid.exe" given with the book to find the levels at the grid points.)

Take an image scan of the contour map given and save it as a GIF file, sized to 590 x 590 pixels, or just use the image copy given along with the software. The software may be used to interpolate the levels of the land at each grid point. Some grid points will be difficult to assess and have only to be approximated by eyeballing. The interpolated levels at the grid junctions are saved in the image of the contour map, as below.

The following Excel sheet automates the calculations according to the level differences between the grid points in each E-W strip of the land. The study of such Excel solutions will lead the engineering student in understanding how such calculations can be automated by simple functions, which will prove to be of great value in engineering analysis.

The first part of the spreadsheet is fed with all the interpolated grid points as against the grid junctions. The second part of the sheet calculates the difference between the existing and desired levels at the grid points. This is performed by simply deducting the desired field level (0.6 in our case) from the interpolated values at each grid point. A very small value of 0.0001 m is added to these values to avoid zero values in the results that may cause "divide by zero" errors in automating the calculations. This addition of a penalty is negligible in the results. Positive and negative values indicate the required "fills" and "cuts", respectively.

As may be observed from Example Problem 12.1, the third part of calculating the sectional areas of each 50 m side segment of the E-W strip have two major possibilities, wherein:

1. the two values on the side of the 50 m side segment of the E-W strip are both positive or both negative; this means that the sectional area is in the shape of a trapezium, resulting in an area of positive (fill) or negative value (cut).
2. one side of the 50 m side segment of the E-W strip is positive and the other is negative, implying that the sectional area is of a shape having two right-angle triangles, one with a negative side and the other a positive side (Figure 12.10 may be referred).

The Excel expression for this part of the calculation is of the form:

```
IF(E15/F15<0,(0.5*E15*(ABS(E15)*$C$2/(ABS(E15)+ABS(F15))))+(0.5*F15*(ABS(F15)*
$C$2/(ABS(E15)+ABS(F15)))),0.5*$C$2*(F15+E15))
```

where E15 and E16 hold the two level differences. It is an IF function, which tests whether the two values are opposite in signs.

Box 12.3 Example 12.3—cont'd

Grid space, m		**50**					
Desired level, m		0.5					
		Interpolated levels at grid points, m					
		0	**50**	**100**	**150**	**200**	**250**
Part 1	**0**	0.45	1.1	0.66	0.57	0.5	0.42
	50	0.48	1.02	0.9	0.61	0.45	0.42
	100	0.41	0.63	0.65	0.6	0.46	0.46
	150	0.48	0.5	0.5	0.42	0.4	0.49
	200	0.45	0.3	0.3	0.46	0.25	0.36
	250	0.3	0.3	0.3	0.47	0.25	0.25
		Levels above or below desired profile, m					
		0	**50**	**100**	**150**	**200**	**250**
Part 2	**0**	-0.0499	0.6001	0.1601	0.0701	0.0001	-0.0799
	50	-0.0199	0.5201	0.4001	0.1101	-0.0499	-0.0799
	100	-0.0899	0.1301	0.1501	0.1001	-0.0399	-0.0399
	150	-0.0199	0.0001	0.0001	-0.0799	-0.0999	-0.0099
	200	-0.0499	-0.1999	-0.1999	-0.0399	-0.2499	-0.1399
	250	-0.1999	-0.1999	-0.1999	-0.0299	-0.2499	-0.2499
				(-ves are fills and +ves are cuts)			

Continued

Box 12.3 Example 12.3—cont'd

		Net sectional area of cut or fill, m²				
		0-50	**50-100**	**100-150**	**150-200**	**200-250**
	0	13.755	19.005	5.755	1.755	-1.995
	50	12.505	23.005	12.755	1.505	-3.245
	100	1.005	7.005	6.255	1.505	-1.995
Part 3	**150**	-0.495	0.005	-1.995	-4.495	-2.745
	200	-6.245	-9.995	-5.995	-7.245	-9.745
	250	-9.995	-9.995	-5.745	-6.995	-12.495

		Average volume of cut or fill at stations, m³					Sum of cut and fills	
	0-50	656.5	1050.25	462.75	81.5	-145.556	2105.444	Excess
	50-100	337.75	750.25	475.25	75.25	-145.556	1492.944	Excess
Part 4	**100-150**	12.75	175.25	106.5	-83.0556	-131.667	79.77778	Excess
	150-200	-187.222	-277.5	-221.944	-326.111	-346.944	-1359.72	Deficit
	200-250	-451.111	-555.278	-326.111	-395.556	-617.778	-2345.83	Deficit
					Grand total		-27.3889	Deficit

Box 12.3 Example 12.3—cont'd

If being TRUE, the areas of the two right-angle triangles are calculated and added. The ratio ABS(E15)/(ABS(E15)+ABS(F15)) is used to find the adjacent side of each triangle, applying the principle of similar triangles to the two opposing triangles. The opposite side of each triangle along with its sign, multiplied by half of the adjacent side, gives the area of each triangle with a proper sign by the expression:

0.5*E15*(ABS(E15)*$C$2/(ABS(E15)+ABS(F15))))+(0.5*F15*(ABS(F15)*$C$2/(ABS(E15)+ABS(F15)))). The ABS() function is used to get the absolute value of the level differences; $C$2 holds the grid interval (50 m in this case).

If the IF function results in "FALSE", it implies that the two values are both positive or both negative and the expression 0.5*$C$2*(F15+E15) calculates the trapezoidal area of the section. The signs of the two values add up to areas of corresponding signs of fill or cut.

The fourth part calculates the average area of the 50 m segment sides, which when multiplied by the grid spacing, gives the volume of cut or fill soil of each. As seen in Example12.1, the "fill" volume is divided by the shrinkage index of 0.9 m to increase the loose volume to the compacted "fill" volume. Accordingly, only the negative volumes are divided by 0.9 m, using an IF function. In the last column of the fourth part, the total volumes of the segments in each 50 m E-W strip are calculated. In the next column, each of these volumes are tested for whether the total soil volume is in excess or deficit, so that soil can be moved from one segment to another or not. The grand total of the volumes is calculated and found whether it is in excess or deficit to decide how much soil must be hauled in or hauled out.

In the example, the total volume is in deficit of 27.4 m^3. Try altering the value of the starting level of 0.5 m in the spreadsheet to find how its selection changes the total volume to be hauled out or in.

from one spot to another. Typically, trucks with a dumping facility are used for the dumping of the loaded soil on a desired spot. Trucks are classified according to:

1. the dumping mode they use, such as rear dumping, side dumping, or bottom dumping, of which rear-tipping dump trucks are the most popular;
2. the type of chassis, whether rigid or articulated, with the latter used for very heavy loads and difficult terrain;
3. the type of drive, such as four-wheeled or two-wheeled drives;
4. the number of wheels and load-bearing axles;
5. the load capacity.

Trucks are driven by heavy diesel engines and are usually wheeled vehicles. The wheels for light- to medium-capacity trucks are for both off-road and on-road hauling. Heavy trucks used in mining and difficult terrains have special treads on their wheels. Turbo chargers are used along with the engine for higher power delivery under shock loads. Trucks have suspension systems built on heavy spring leaves. Unlike tractors, they do not have final drives and are built like on-road automobiles that ply at relatively higher speeds. They have automobilelike steering systems, with a differential drive to the wheel axles. A hydraulic system is available that actuates the dumping system. Hydraulic remote cylinders are used to tip the dump trough to dump the load.

Exercise problems

1. The measured dumpy level readings of the 250 m strip given in Sec. 12.4.1 may be used again to calculate the total volume of soil to be hauled in or out of the strip, if a 0.2 percent downward slope is required in the direction along the strip starting at 0.6 m from zero datum at the left end.
2. The 250 m × 250 m land given in Example 12.3 has to be levelled for flood irrigation, requiring a downward slope of 0.25 % starting from an elevation of 0.6 m at the north end and going in a N-S direction. Find the total soil volume to be hauled out as excess or to be hauled in from elsewhere to fill a deficit.
3. The available bulldozer can push an average blade load of 6.0 m^3 of the given field soil in one cycle. The average push distance is found to be 30 m in this situation. Assuming a dozing speed of 2.5 km h^{-1} and a backtracking speed of 4.0 km h^{-1}, calculate how many days of eight hours of work are required to move the total volume of soil in the plot to bring about the required level in Exercise Problem 2.

References

Nichols, H. L., & Day, D. A. (2005). *Moving the earth – the workbook of excavation*. McGraw Hill companies.

Peurifoy, R. L., Schexnayder, A. J., & Shapira, A. (2006). *Construction planning, equipment and methods*. McGraw Hill publ.

CHAPTER

Testing of farm machines 13

Chapter Outline

Machines of the Farm. https://doi.org/10.1016/B978-0-443-32906-7.00013-X

"Testing" refers to an analysis of the machine behavior with reference to predefined standards under ideal conditions, and "evaluation" is to measure its performance under real farm conditions (Dixit, 2011; Smith, Sims and Neill, 1994). There is a need for testing the marketed agricultural machines for their quality and endurance. A reliable agency, if it tests, evaluates, and certifies the credibility of the marketed machines, that would enhance a machine's marketability. In many countries, such certification has been made mandatory, so that the manufacturing firms sell good-quality machines on the market. Based on such test reports, decisions could also be made by banks to extend credit to machinery manufacturers to set up or extend their facilities, to farmers to own machines for their use, and to entrepreneurs who wish to buy and hire them out to farmers.

Though the technological advancements in agricultural mechanization have benefitted the international farming industry in a big way, millions of poor farmers do not yet have access to these technical improvements and machines, which can potentially improve their land and labor productivity, thus raising their living standards. Research effort was directed at targeting these groups of farmers, and small- and medium-sized machines, emulating their bigger counterparts, were developed. These innovations are tested by such testing facilities, to see whether they are compatible to a user's farming system. Stringent testing standards evolve to assure the production and use of such innovations.

13.1 Objectives of testing and evaluating agricultural machinery

The objectives of testing and evaluating agricultural machinery, including prime movers such as tractors, engines, pumps, etc manufactured in the country, are to assess their performance and functional suitability under different agro regions. The test report will:

- Assist a government in deciding whether the type and model of machinery, manufactured in the country or abroad, are suitable for specified conditions and to permit their production and popularization.
- Help farmers and other prospective buyers of such machinery to understand their performance characteristics.
- Provide documented material to researchers, designers, and extension workers to improve their designs as well to help farmers select the right machinery for their requirements.
- Help develop standards for similar equipment for future use.
- Help financial institutions with the approval of financial assistance to dealers to set up dealerships and farmers to buy the tested equipment.

As previously stated, farm machinery manufacturers are particularly interested in having their machines tested by approved agencies since that will encourage farmers to buy their products without hesitation. The testing of farm machines is usually conducted by approved state and central agencies and institutions. The standard test code mandates a systematic series of tests concerning functional performance, power demand, durability, and wear performance. Hence, testing mandates manufacturing quality and suitability. It also helps manufacturers to publicize their products with much confidence, thus promoting sales. Manufacturers can improve the design of their products based on the test reports. The testing of farm machines helps farmers to select their machines suitably with the assistance of professionals, having access to the test reports of the concerned machines.

Different countries have various organizations and institutions approved to test farm machines brought to market. The government of India has established the Bureau of Indian Standards, which prepares the standard test codes for different types of farm machines. Other countries in the world also have similar organizations, such as the Nebraska Tractor Test Laboratory in the United States, British Standards Institutions in London in the United Kingdom, Organisation for Economic Co-operation and Development (OECD) in Paris, France, and the International Organization for Standardization (ISO). The ISO was formed in 1947, with headquarters in Geneva, Switzerland, to promote the development of standardization codes to facilitate the international exchange of goods and services and to develop cooperation in the sharing of intellectual, scientific, technological, and economic activities. The organization consists of a Technical Committee (TC) and Technical Advisory Group (TAG). The ISO comprises the National Standard Bodies of 91 countries including India, which is represented by the Bureau of Indian Standards (BIS).

In India, tests are conducted by the Farm Machinery Training and Testing Institutes under the jurisdiction of the Ministry of Agriculture, Government of India. Currently, there are four institutes located in Budni in Madhya Pradesh, Hissar in Haryana, Anantpur in Andhra Pradesh, and Vishwananth Chariali in Assam. The development of international standards for quality management systems led to the ISO 9000 series of standards in 1987. In all, the following six standards were issued:

ISO 8402: Quality Management and Quality Assurance – Vocabulary
ISO 9000: Quality Management and Quality Assurance – Guidelines for Selection and Usage
ISO 9001: Quality System Model for Quality Assurance in Design, Development, Production, Installation, and Servicing
ISO 9002: Quality System Model for Quality Assurance in Production, Installation, and Servicing
ISO 9003: Quality System Model for Quality Assurance in Final Inspection and Test
ISO 9004: Quality Management and Quality System Elements – Guidelines

The testing of farm machinery is about:

- Assessing the functional performance of the machine tested;
- Finding the power demand of the machine;
- Assessing its durability and endurance;
- Finding the wear performance of the components, particularly the soil working parts;
- Determining the adequacy of the machine structure to withstand higher stresses;
- Assessing the ease of making adjustments in the machine to suit the operating conditions and that in operating the machine itself.

Machines are tested at the written request of the manufacturer or accredited importers. All the tests are conducted as per IS Test Codes.

13.2 Methods of measuring soil, plant, and machine parameters

While testing farm implements, it is imperative that several of the soil, plant, and machine parameters are assessed. The following is a brief description of the methods by which each of these parameters is measured and presented. It is important that these methods are accurate enough to evaluate the performance of the implements in the field as well as simple enough to complete at the field level.

13.2.1 Assessment of soil parameters

13.2.1.1 Soil type and texture

Soil particle size or mechanical analysis helps us to determine the percentage of clay (below 0.002 mm), silt (0.002 mm to 0.06 mm), and sand (0.06 mm to 2.0 mm) in the soil, based on which its textural class can be determined. Soil properties, such as structure, soil water permeability, infiltration rate, runoff rate, fertility, etc, are influenced by the texture of the soil. Textural classification of soil is important in the

Table 13.1 Assessment of the Soil Moisture by Feeling the Texture by Hand

Soil type	Ease of forming it into a ball
Sand	Soil remains loose and cannot be formed into a ball at all.
Loamy sand	Sufficient silt and clay to become somewhat cohesive to form a ball but the ball easily breaks.
Silt loam	Soil can be shaped by rolling it into a short, thick cylinder.
Loam	Soil can be rolled into a 15 cm long cylinder, which breaks when bent.
Clay loam	Soil can be rolled into a 15 cm long cylinder that can be bent into a U without breaking it.
Light clay	Soil can be rolled into a 15 cm long cylinder that can be bent into a circle that shows cracks.
Heavy clay	Soil can be rolled into a 15 cm long cylinder that can be bent into a circle that does not show cracks

No Permission Required

context of testing machines that have soil working components. A mechanical analysis is accurately performed by the method of determining the sedimentation, which can be performed in a soil laboratory having accurate equipment.

In the absence of such an analysis, a method for field estimation of the texture can be used that is adequate for the testing of implements with soil working components. Soil samples, from steps of 10 cm depth zones of the ploughed/cultivated soil, are taken, starting from the surface to 10 cm below the intended working depth of the implement. About 200 g of soil from each depth zone is collected, and three such repeated samples in each depth zone are collected. Water is slowly added to the soil until a ball of 2.5 cm diameter can be formed from the collected soil as it approaches a point at which the soil just starts to stick to the hand. The extent to which the moist soil can be shaped by hand can approximately be found by its texture, as indicated in Table 13.1.

13.2.1.2 Bulk density of soil

The dry bulk density of soil is the mass per unit volume of dry soil in its undisturbed state. It is representative of its resistance to penetration by soil working tools or plant roots. The method of estimating the bulk density in a field is to use a coring cylinder to withdraw a core sample of soil. The coring cylinder is driven into the soil and then extracted out. Then, the soil core that is in the cylinder is taken out of the cylinder for weighing. The cylinder is generally 6 cm long with a 5 cm inner diameter. The sampled soil core is then weighed using a laboratory scale. After weighing, the sample is dried in a hot-air oven at 105 °C for 8 hours and then cooled. The core is again weighed after drying.

Dry bulk density in kg m^{-3} = weight of the dried sample/volume of the soil core

where the soil core volume = $\pi R^2 L$; R = the inner radius of the core sampler; and L = the length of the cylindrical soil core.

Soil core samples, from steps of 10 cm depth zones of the soil, are taken, starting from the surface to 10 cm below the intended working depth of the implement. Three such repeated samples in each depth zone are collected. After operating the implement, soil cores are drawn for bulk density measurements. The change in bulk density will give an idea of the degree of soil disturbance made by the implement being tested.

13.2.1.3 Soil moisture content

Soil moisture is that which is available as free moisture in soil pores that are between the soil particles. There are many methods available, both for laboratory and in situ assessments. Soil moisture detectors are also available for in situ measurements. But the most reliable and accurate method is the gravimetric method, which is done in a laboratory. It is recommended that the sampling plan for the determination of dry bulk density is used for soil moisture determination, too. Each soil sample is transferred to a soil moisture cup of known weight and weighed on a laboratory scale as W_1 immediately after sampling. W_1 is the weight of the soil excluding the weight of the cup. The sample is dried in a hot-air oven at 105 °C for 8 hours, cooled in a desiccator, and then re-weighed as W_2.

Soil moisture content, percent (dry basis) $= \{(W_2 - W_1) / W_2\} \times 100$

13.2.1.4 Mean clod diameter of soil

The amount of soil pulverization induced by a tillage or cultivating tool is measured by evaluating the mean clod diameter of the soil. The ploughed soil sample is passed through a set of special sieves, and the weight of the soil retained on each sieve is measured. The sieves that are usually used for this purpose have 10 mm, 20 mm, 30 mm, 40 mm, and 50 mm mesh sizes. While sieving the dry soil, the clods tend to break up; this method may not yield a very accurate assessment of the mean clod diameter. However, for the purpose of testing the implements, the approximate assessment is deemed adequate. The calculation of mean clod size is summarized in Table 13.2.

13.2.1.5 Soil cone index

The standard cone penetration apparatus (Figure 13.1) complies with ASAE standards, and the instrument is pressed into the soil by hand pressure so that the standard cone at its bottom end penetrates the soil. The cone resistances at each depth level as marked on the instrument is measured on the instrument panel and recorded as penetration progresses. Cone index is the average of penetration resistances recorded over a specified depth. It is an indication of the soil hardness and is expressed in kPa. The cone penetration resistance is less on the soil surface and rises substantially as the cone is pressed down deeper into the soil.

13.2.1.6 Soil inversion caused by tillage

Soil inversion, as imparted by a plough, is assessed from the ploughed field. The inversion is a measure of how effectively the soil in the top is buried and the soil

Table 13.2 Calculation of Mean Clod Size

Mesh sizes, mm	Soil clod diameters that do not pass through the larger mesh into the smaller one, mm	Average size of clods retained on the sieve, mm	Weight of soil, kg
50	>50	N	A
40	40-50	45	B
30	30-40	35	C
20	20-30	25	D
10	10-20	15	E
Pan without mesh	<10	5	F

Mean soil clod diameter, mm = (A×N + 45B + 35C + 25D + 15E + 5F) / (A+B+C+D+E+F) where N = the average diameter of soil clods retained on the largest mesh of 50 mm.

No Permission Required

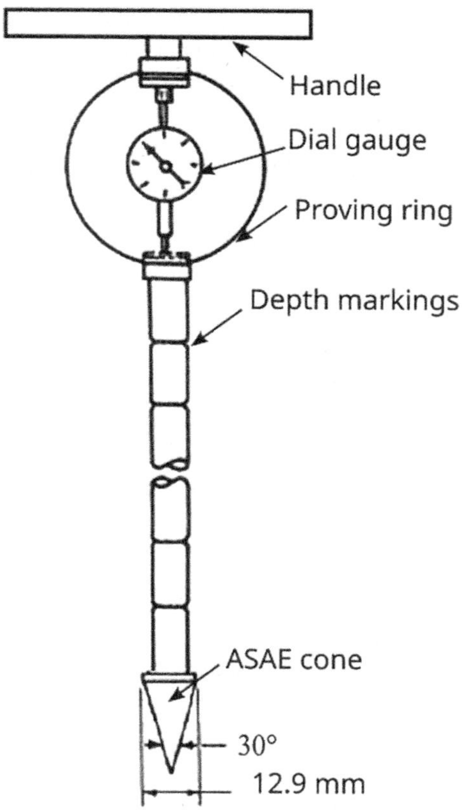

FIG. 13.1 Soil Cone Penetrometer

No Permission Required

in the bottom layers is brought to the top surface. An easier method of quantitatively assessing soil inversion is to count the number of weeds or crop stubble found on the surface before and after the ploughing operation. It is computed by the expression $F = \{(Wp\text{-}W_E) / W_p\} \times 100$

where F = soil inversion in percent,

which is representative of the weed and crop stubble being covered by the inverted soil.

W_p = No. of weeds or crop stubble before operation per unit area;

W_E = No. of weeds or stubble recorded on the surface after operation.

A light 1 m × 1 m square frame of wood or metal is used to sample the number of weeds or stubble in the field. The frame is randomly thrown on the soil surface and the number of weed or stubble roots in the 1 m square is counted and recorded. Several samples are recorded, and the value is averaged to acquire the percent of soil inversion.

13.2.2 Assessment of machine parameters

13.2.2.1 Theoretical working width of implements

The theoretical working width of implements is the calculated width of the implement accounting for the width of each soil working component multiplied by the number of such components.

13.2.2.2 Effective working width

The effective working width is obtained by dividing the width of the test plot by the number of passes of the implement.

13.2.2.3 Draft of farm implements

Draft, the horizontal component of the pull applied on an implement (Kepner et al., 1978), is measured by a dynamometer. The dynamometer could be a hydraulic unit or a strain gauge load cell. The hydraulic dynamometer is a simple hydraulic cylinder-and-piston assembly with hydraulic oil sealed inside the cylinder. The force to be measured is transmitted through the piston to the oil in the cylinder. The oil pressure, as sensed by a pressure gauge on the cylinder, shows the corresponding force on the gauge's calibrated dial. The strain gauge load cell is that which, when connected to its electronic indicator, shows the force applied on the load cell. The electrical signals corresponding to the load applied can be recorded as against time on a digital data logger to save and retrieve the force data later and compute the average force measured.

On a trailed implement, the dynamometer is connected in line with the tow line and measures the pull on the implement. The draft is calculated as the horizontal component of pull by measuring the angle of pull to ground. However, the draft of mounted or semi-mounted implements cannot be measured this way. Three-point dynamometers, which are basically six-cell measuring devices, as explained in Section 5. 2, can be used but are rarely available. So, a simple two-tractor method

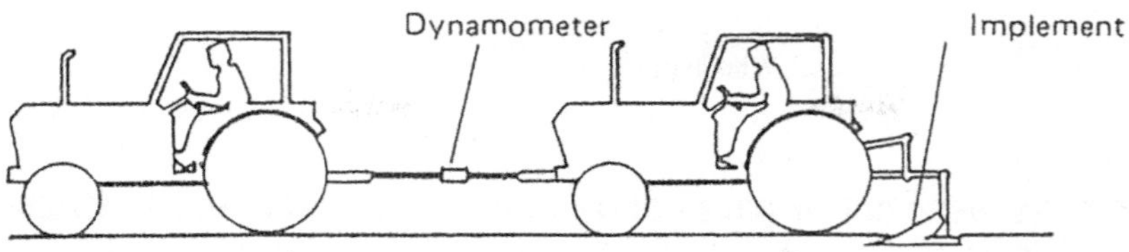

FIG. 13.2 Two-Tractor Method for Draft Measurement

No Permission Required

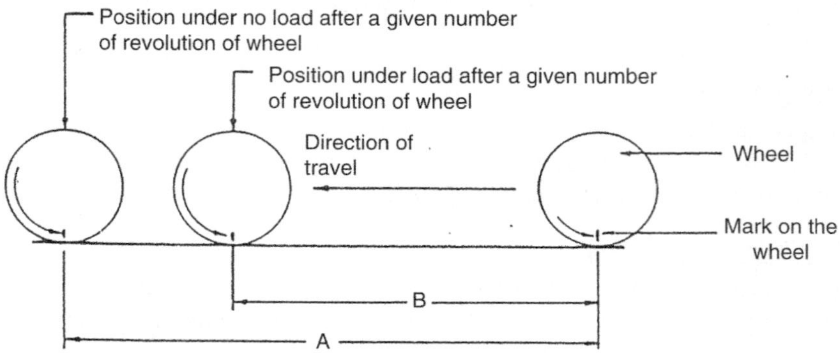

FIG. 13.3 Wheel Slip Measurements

No Permission Required

(Figure 13.2) is used wherein the dynamometer is mounted on the towing line between one tractor pulling another tractor that has the tested implement mounted to it. The tractor in the front pulls the implement-mounted tractor, thus transmitting the pull through the dynamometer. The pull is measured once when the implement is in operation and next by raising the implement out of the soil. The difference between the two measurements yields the pull on the implement. By keeping the tow line horizontal to the ground, the draft on the implement can be measured directly.

13.2.2.4 Wheel slip

In wheeled machines, such as the tractor, there will be wheel slip since the driving ground wheel may not be able to deliver the required traction for the locomotion of the machine. The distance that the machine moves forward for a given number of revolutions of the traction wheel decreases when the wheel slips. The traction wheel is marked at a place on the periphery. While the machine is operated without any tractive load, the distance it traverses (A) for a predefined number of revolutions of the traction wheel is measured first (Figure 13.3). Then, the distance the machine traverses is measured again (B) for the same number of revolutions of the wheel with a tractive load on the machine. Hence,

$$\text{Wheelslip}(\%) = \frac{(\text{A–B})}{A} \times 100$$

13.2.3 Assessment of parameters concerning seed and fertilizer devices

13.2.3.1 Seed weight

The seed weight is usually specified as the 1000-grain weight since the weight of a single seed is very minimal. The 1000-grain weight is found by weighing at least eight samples of 1000 grains. The average weight for 1000 grains is calculated thereon.

13.2.3.2 Seed size

The length, thickness, or diameter of a seed is measured as the average of at least 50 seeds. This is done in a laboratory under a magnifying microscope with an eyepiece reticle (a micron-graduated scale).

13.2.3.3 Seeding depth

Accurate seeding depth is measured by digging up plants after germination. But, for practical purposes, the seed furrow depth can be measured randomly while measuring the seed spacing in the row.

13.2.3.4 Metering performance of seed drills and planters

The metering performance of seed drills is evaluated following the method used for seed drill calibration (explained in detail in the chapter on crop planting). The weight of seeds metered out from each row is weighed and presented as a histogram, which will represent the variation of quantity between the rows. Since drills meter out seeds continuously without offering plant spacing in a row, this would be the only quantification of metering performance. These tests are done on all seeds that the manufacturer claims to be sown by a tested machine. As for the seed planter, the uniformity of seed spacing in a row is the true measure of its performance. Additionally, the quantity metered in each row is also tested, as in the case of seed drills. The variation of quantity of seeds metered between rows will be due to the presence of multiples and misses rendered by the metering devices. Graded seeds should be used in these tests so that the weight variations are not due to the differences in seed sizes.

13.2.3.5 Wheel skid

If a machine such as a seeder or planter is ground-wheel driven, there will be wheel skid since the ground wheel interacting with the loose soil may not be able to deliver the required torque to the metering drive. The number of revolutions of the ground wheel decreases when the wheel skids and will not correspond to the distance the machine moves forward. The ground wheel is first marked at a place on the periphery. The metering system is disengaged from the ground wheel, and the number of revolutions (A) of the ground wheel is counted for traversing a predefined distance the machine moves forward. Then, for the same distance moved by the machine, the number of ground wheel revolutions is counted with the ground wheel driving the seeding mechanism. Hence,

$$\text{Wheelskid}(\%) = \frac{(\text{A–B})}{A} \times 100$$

13.2.3.6 Uniformity of seed spacing

A planter's seed hopper is loaded with seed, and the planter is operated in a prepared seed bed. The furrow closer is removed from the planter so that the furrow is not closed after the seeds drop into the furrows. This allows for measuring the seed-to-seed spacing offered by the planter on each row of sowing. The spacing data are thus recorded randomly at several locations on different seed rows. These data are used to calculate the uniformity of seed spacing:

$$Uniformity\ of\ seed\ spacing = \frac{Standard\ deviation\ of\ seeds\ pacings}{Mean\ seed\ spacing}$$

13.2.3.7 Metering performance of fertilizer drills

Testing of the metering performance of fertilizer drills is very similar to testing the seed drill performance. The standard method of calibrating the seed drill is used in these machines to compute the variation of quantity of fertilizer metered in each row. The results of the fertilizer metered out is weighed and presented as a histogram, which will represent the variation of quantity between the rows.

13.2.3.8 Metering performance of fertilizer or seed distributors

The objective of testing a fertilizer distributor, especially the centrifugal broadcaster, is to assess whether it applies fertilizer or seed at a desired application rate (kg ha^{-1}) with a uniform application rate over the entire swath. The machine being tested is operated on level ground at a recommended speed and feed rate settings. The material broadcast is collected in collection trays placed in a square grid of multiple rows and columns at right angles to the line of travel (Figure 13. 4). After the operation, the contents of each tray are weighed and the quantity of fertilizer collected over the area of the collecting container is computed for each. The distribution pattern is plotted as histograms of distribution on each row and then averaged over the rows (Figure 13.4). The histograms are similar to the assessment of a spray distribution pattern of nozzles (explained in the chapter on plant protection equipment). The distribution histograms can also be used to find the overlaps required to bring about the desired application uniformity over the area. These assessments can be compared with the recommendations of the manufacturer. The overlapping of distribution histograms is also very similar to

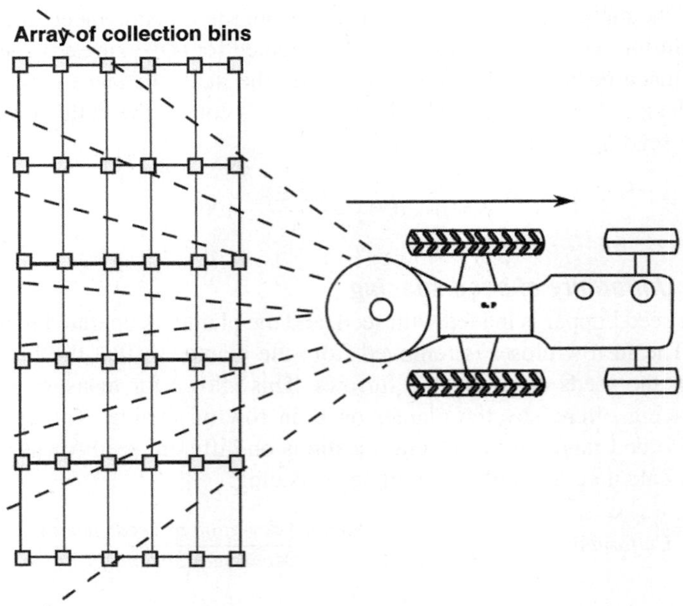

FIG. 13.4 Measurements on Fertilizer Distribution

No Permission Required

that used in boom sprayers (explained in the chapter on plant protection equipment).

13.3 Collection of machinery details from the manufacturer

As a first step in the testing of any implement, all the specifications of the implement including the type of implement, draft, power source, the make, model, the manufacturer's name and address, and the implement's overall dimensions are obtained from the manufacturer. The details of a complete machine, its specifications, and the working manual containing all the adjustments possible should be obtained and documented. A complete assembly and detailed drawing set of the implement may be collected from the manufacturer and studied in detail. All the parts such as the type, number of bottoms/rows, and working width of the implement should also be recorded. The type of mould board or the disc diameter and concavity, materials of the soil working parts, their hardness and thickness, and details such as animal-drawn plough suctions should be observed and compared with the manufacturer's specs. Details of the accessories supplied with an implement, such as a coulter, its type, size, range, and types of adjustments, are next observed. Details of implement wheels, if any, their adjustment ranges, and the details of the frame and its materials are recorded. In the case of seeders and fertilizer machines, the types and details of the metering devices and drives should be documented. Details such as the types of pumps and nozzles for sprayers should be assessed.

13.4 Testing primary tillage implements

13.4.1 Tested implement

A complete report, as per section 13.3, can be obtained from the manufacturer and documented along with a set of complete, detailed drawings of all the parts and an assembly drawing set.

13.4.2 Laboratory work

The main objective of this work in the laboratory is to study and confirm the specifications and essential components, comparing them to those given by the manufacturer. All the following aspects also should be studied and verified before proceeding to the field testing of an implement.

a. Range of adjustment of working width, depth, and level;
b. Type of supplied accessories such as a coulter;
c. Material of plough share, mould board, or disc;
d. Safety aspects of the implement including overload prevention;
e. Weight of soil working parts before and after a test;
f. Hitching arrangements, including their types and specifications.

13.4.3 Field work

13.4.3.1 Test conditions

13.4.3.1.1 Prime movers or draft animals

Tractors used for a test should be selected so as to be compatible with the implement under testing and should be driven by experienced operators. Draft animals and their handlers should be highly trained and, if possible, familiar with the type of implement under testing. The animals should be in good condition, and records of their health and feeding regimens should be available.

13.4.3.1.2 Fields

The performance of any implement varies and is influenced by the type of soil, its moisture content, weed growth, crop residue, and shape of fields. The following conditions should be clearly recorded:

a. Area and shape of test field;
b. Type and conditions of soil;
c. Last crop in the field;
d. Height of stubble of last crop;

e. Condition of weed and intensity of weed infestation.

13.4.3.2 Preliminary trials

Preliminary trials should be conducted on the implement in a plot adjacent to the test plots so that adjustments are made to the plough and testing instruments. It will also acclimatize the test engineers and operators to the implement's operation.

13.4.3.3 Performance tests

The main objectives of performance tests are to obtain reliable data on the implement such as capacity and quality of work, ease of operation, maintenance requirements, and adaptability to varied soil conditions. At least three series of tests should be made under different soil conditions. When the test plots have been marked out and before any tests being carried out, soil samples should be taken for measurement in a laboratory to assess the soil type, moisture content, bulk density, and mean clod diameter. Cone index, shear strength measurements, and weed counts should also be made.

Each plot should be completed without stopping unless necessary due to adjustments, breakdowns, or animal rest periods.

The system of operation being used for the test should be recorded meticulously along with the working depth, width, draft, speed, tractor wheel slip, and time taken to complete the operation. Based on these recorded values, the power and field efficiency can be calculated.

The following measurements should be recorded:

a. Width of ploughing;
b. Depth of ploughing;
c. Total area of ploughing;
d. Traveling speed;
e. Draft;
f. Wheel slip;
g. Time spent turning at headlands;
h. Time spent for any other reason;
i. Total operating time.

At the end of the test, measurement should again be made of weed cover and clod diameter.

During the test, the following observations should be made, along with relevant comments:

a. Ease of handling;
b. Ease of adjustment;
c. Maintenance of depth;
d. Adhesion of soil to shares, mouldboards, or discs;
e. Blocking with weeds or trash;
f. Visible deformation;

g. Wear of soil working parts.

13.4.3.4 Durability trials

To obtain measurements on the wear of working parts and the durability of the working parts and to identify problems of maintenance and operation, trials are made for longer periods of work, of about 100 hours. This is usually done in a farmer's field since the land area of trial plots will usually be smaller. All details of plot conditions and measurements specified in the performance tests are recorded.

Repairs and adjustments

The details of hitching of the implement are recorded, including whether it is trailed or mounted, if it is animal-drawn, the type of hitching, and the method. Constructional details, such as the category of three-point linkage, should be recorded. The range of adjustments in operating width, front furrow width, depth, level, and suction should be assessed. The recommended working speed and field capacity, as claimed by the manufacturer, are to be recorded.

Table 13.3 Sample Tables for Collection of Data

Test No.	1	2	3	4	5	6
Date						
Location						
Plot size (m x m)						
Topography						
Soil description						
Previous cultivation						
Previous crop residue						
Weed count (n/m^2)						
Cultivation after last harvest						
Soil moisture (% DB)						
Dry bulk density (g/cm^3)						
Penetrometer reading (kPa)						
Shear strength (kPa)						
Operational pattern						
Working depth (cm)						
Working width (cm)						
Draft (or N)						
Speed (m/s)						
Power (W)						
Tractor wheel slip (%)						
Time taken to complete operation (min)						
Field efficiency (%)						
Soil inversion (%)						

No Permission Required

A sample set of tables is given below to provide an idea of the test parameters being assessed during the testing of primary tillage implements. Table 13.3 shows the summary of test conditions and results.

Comments on aspects of performance

a. Ease of handling;
b. Ease of adjustment;
c. Maintenance of depth;
d. Adhesion of soil to shares, mouldboards, or discs;
e. Blocking with weeds or trash;
f. Visible deformation;
g. Wear of soil working parts;
h. Results of durability trials.

Test conditions

a. Test No & Date;
b. Implement;
c. Make;
d. Site location;
e. Topography;
f. Soil description;
g. Previous cultivation;
h. Previous crop;
i. Crop residue;
j. Plot size.

The following sample table (Table 13.4) is the field work sheet.

A sample table (Table 13.5) of a calculation sheet is furnished for reference.

Table 13.4 Field Work Sheets

Weeds in m^2 area before test	1	2	3	4	5	6
Weeds in m^2 area after test						
Cone penetrometer reading (kPa)						
Soil shear strength (kPa)						
Bulk density samples						
Soil moisture samples						
Make, type, and size of tractor						
Distance for five revolutions of tractor wheel unloaded:m						
Time at start of test						
Time at completion of test						
Number of passes in plot width						
Time for stoppages						

No Permission Required

Table 13.5 Test Calculations' Sheet

Parameters	Unit	Value
Plot size completed (area)	m^2	
Cone index	kPa	
Shear strength	kPa	
Bulk density	g cm^{-3}	
Soil moisture (D.B.)	%	
Working speed	m s^{-1}	
Working width	cm	
Working depth	cm	
Total time to complete area m	hr	
Effective field capacity	ha h^{-1}	
Theoretical field capacity	ha h^{-1}	
Field efficiency	%	
Distance for 5 revs of driving wheel without load	m	
Distance for 5 revs of driving wheel when loaded	m	
Wheel slip	%	
Average turning and stoppage time per row	sec	
Measured draft	N	
Power input	W	
Cross-section area cut on each pass	cm^2	
Maximum depth	cm	
Weeds before test	Nos /m^2	
Weeds after test	Nos /m^2	
Weeding efficiency	%	

No Permission Required

13.5 Testing of secondary tillage implements

The testing of secondary tillage implements is very similar to that of the primary tillage implements in workflow and the data to be recorded. However, much emphasis should be given to soil manipulation, especially the tilth, soil inversion, and soil mixing. Clod size analysis on the tilled soil and stubble inversion should be meticulously recorded in respect to cultivators and harrows. Implements such as land-forming tools like a bund former, ridger, and leveller should be assessed by way of uniformity in the dimensions of the landform as well the levelness created by a leveller. As for the field condition and soil condition of a test plot, the clod size analysis before and after testing should be recorded. Other important parameters are the accumulation and clogging of openers in the case of cultivators. These need be quantified depending on the type and density of stubble or weeds present in the field. Row crop cultivators should be evaluated in terms of their manoeuvrability and transit between row crops and at the field ends. The time lost and the field efficiency thereon are to be recorded carefully.

Measurements to be recorded meticulously are:

- **a.** Width of cultivation;
- **b.** Depth of cultivation;
- **c.** Total area of cultivation;
- **d.** Traveling speed;
- **e.** Draft and pull angle;
- **f.** Wheel slip;
- **g.** Time spent turning at headlands;
- **h.** Time spent for any other reason;
- **i.** Total operating time.

At the end of the test, measurements will again be made of weed cover and clod diameter.

During the trials, observations such as ease of handling and adjustment, maintenance of depth, adhesion of soil to tines or discs, clogging of openers by weeds or trash, visible deformation of the implement parts, and wear of soil working parts are also to be recorded.

13.6 Test procedure for seeders, planters, and fertilizer application equipment

The effectiveness of a seeder or planter will depend not only on its mechanical design but also on the type and condition of the seed, the condition of the soil, and field topography, which should all be adequately described in the test report. The test parameters to be measured are similar to those for tillage implements, but additional parameters specifically assessing the sowing performance are also meticulously collected. Theoretical width, which is simply the product of row spacing and number of furrow openers used on a seeder, the effective or actual width, forward speed, and time for coverage should be measured, as in the case of tillage implements. Other than these, parameters specific to the seeding performance, as previously explained in Section 13.2.3, should also be assessed; these include the seed size, seed weight, the seeding depth, wheel skid, and uniformity of seed spacing.

Testing and the data to be recorded are very similar to that of other field machinery in the work flow. But, much emphasis should be given to the seeding performance, especially the seeding depth, uniformity of spacing, and the seed rate rendered. As always, a manufacturer provides the details of a machine completely, with the specified machine in working order for testing. Specifications concerning materials, construction, range of adjustments, and expected performance with various types of seeds are all documented. A complete specification with a set of detailed and assembly drawings are provided and will be added in the final test report.

13.6.1 Laboratory tests

In the laboratory, the specifications and range of adjustments, as given by the manufacturer, are confirmed. Details to be examined include:

a. Metering mechanism and method of changing rate of seed delivery;
b. Type of furrow opener and its adjustments;
c. Type of seed closing devices;
d. Type of hill dropping mechanism, if any;
e. Type of drive from ground wheel to feed shaft;
f. Depth control system and its adjustments;
g. Types of row markers.

Laboratory tests on the seed-metering system are performed and are very similar to the seeding device calibration explained in the chapter on sowing machinery. The machine is jacked-up and the ground wheel is rotated to ascertain the seed rate given at each setting for each type of seed. Tests should be done using at least three different types (sizes) of seeds for which the machine is suitable, as claimed by the manufacturer. After each of the test runs, random samples of metered seeds are collected and damaged seeds are weighed and expressed as a percentage of the total sample. Thus, the seed rate rendered and the seed damage are to be recorded. A histogram of seed quantity metered in individual rows is presented to understand the variability between individual metering devices.

Seed-spacing performance is tested in the laboratory by operating the machine without the furrow closers on a small test plot with a flat soft surface. Test runs are made for each type of seed, and the seed-to-seed spacing is measured and recorded. In the case of drills, this test is not required since they do not offer any seed spacing in a row.

13.6.2 Field testing

The effective performance of seeders and planters can only be determined by trials in the field. Reliable data can be obtained on how the seed rate and spacing of seeds are affected by movement and vibration and on ease of operation and maintenance and adjustment requirements to varied seed type and soil conditions.

Test conditions, as recorded for other field machinery, including the condition of prime movers or animals, should be recorded. As for the test field conditions, the following parameters should be recorded:

a. Area and shape of test field;
b. Type and character of soil;
c. Topography;

d. Type of prior cultivation (if any);
e. Soil conditions as assessed by the soil moisture, cloudiness, structure ,and topography.

Field tests are done on at least three types of seeds in three different soil conditions. The laboratory tests help the engineer to adjust the seeder to the desired seed rate of the seed used in each test. During the trial run, it should be ensured that there is no blockage in the seed tubes. The seed hoppers are emptied, and the remaining seed is weighed after each trial run. The measurements taken are:

a. Number of passes;
b. Number of rows per pass;
c. Row spacing;
d. Depth of farrow openers;
e. Forward speed;
f. Pull and pull angle;
g. Wheel skid;
h. Time spent on turning;
i. Time spent for other reasons;
j. Total operating time.

Similar to the testing of other field machines, observations are made on the ease of handling, adjustments, and working depth maintenance.

13.6.3 Seed placement performance

Measurements are done on seed placement during these trials. Some of the passes are operated after removing the soil-covering devices and the seed spacing is measured on the row. The average seed spacing and the standard deviation of this spacing are calculated and recorded later. Based on the weight of seeds totally sown in the test plot area, the seed rate rendered by the machine is calculated later.

After the seeds germinate in the test plot, the plant spacing, number of plants per unit area, and the depth of placement are measured randomly in the rows and documented. Plant emergence is an important measure of the seeding performance, as influenced by the seed placement depth, soil–seed contact, soil cover of the seed, and seeds damaged by the metering mechanism. Since emergence is influenced by other factors, too, apart from these machine parameters, germination of the tested seeds is tested in a small patch of the trial plot as sown on the same day of testing. Seed germination is expressed as the percentage of germinated seeds out of the total seeds planted in the test patch. A sample table (Table 13.6) for testing a seed metering system is given for reference.

Most seed drills are also fitted with fertilizer-metering devices. While testing the seed drills, fertilizer-metering devices should also be tested. Although fertilizer application is done simultaneously by the machine while sowing, the testing of fertilizer metering is done separately, operating only that part of the machine. The test

Table 13.6 Sample Test Data on Seed-Metering Performance

	Level of seed rate set								
	Maximum			Intermediate			Minimum		
	Quantity of seed in hopper			Quantity of seed in hopper			Quantity of seed in hopper		
	1/4	1/2	*Full*	1/4	1/2	*Full*	1/4	1/2	*Full*
Seed rate set, kg ha^{-1} at row spacing, cm									
Rate of damaged seed by metering mechanism, %									
Effective rolling diameter of ground wheel, m									
Revolution of ground wheel for measuring delivery									
Seed delivery for the above revolutions, kg									
Seed rate, kg ha^{-1}									
Mean seed spacing, mm									
Seed-spacing evenness									

No Permission Required

procedure is exactly the same as the testing of bulk-flow metering devices, such as seed drills. The quantity of fertilizer applied is measured both in laboratory and field tests and compared with that of the settings made on the machine for a particular application rate in kg ha^{-1}. From the laboratory data, a histogram of fertilizer quantity metered in individual rows is also presented to understand the variability between individual metering devices. The uniformity index or the coefficient of variation is calculated for each row of the tray rows and averaged.

13.7 Test procedure for boom sprayers

Testing is very similar to other field machinery in workflow and the data to be recorded. However, much emphasis is given to the spraying performance with reference to the spray application rate and the uniformity of spray on the plant target, as measured from the spray pattern rendered by the nozzles and the boom. The manufacturer provides all the detailed specifications of the machine completely, along with a machine in working order for testing. Specifications concerning range of adjustments and expected performance on various crops are all documented. A complete set of detailed and assembly drawings of the machine can be obtained from the manufacturer and will be added in the final test report.

13.7.1 Laboratory work

The specifications and details of settings given by the manufacturer should be confirmed. The following specifications should be examined and recorded:

a. Type of machine: mounted, trailed, or self-propelled;
b. Power source for pump;
c. Capacity of sprayer tank;
d. Length of boom;
e. Types of nozzles supplied;
f. Nozzle spacing adjustments available;
g. Size and weight of machine.

Other items are listed in the specification form.

13.7.1.1 Pump test

The discharge of the pump, in l min^{-1}, is measured at the recommended pump speed at the maximum and minimum discharges. The discharge pressure is varied by controlling the flow back of the spray fluid to the tank. The chapter on plant protection equipment may be referenced for how and why.

13.7.1.2 Spray nozzle tests

This information is in accordance with what is explained in the chapter on plant protection equipment. Usually, sprayer manufacturers purchase nozzles supplied by other firms. If so, the specifications and performance characteristics of those nozzles supplied with a machine should be provided by the manufacturer for comparison of performance. Typically, the discharge rendered by the nozzle at the optimal pressure setting advised should be measured and compared to the specifications. A spray patternator, explained in the chapter on plant protecting machinery, is used to assess the spray pattern at the recommended spray height. The histogram of the pattern is checked with the manufacturer's specifications. The tests are to be replicated with different nozzles of the same type, and the patterns are to be compared for systematic variation and underperformance. Spray boom discharges of the complete boom with nozzles at various spacings are measured from the spray tank level before and after spraying for a given period, and the application rate of the sprayer boom should be verified.

13.7.2 Field trials

The effective performance of field sprayers is also determined in the field on various types of crops. These trials will allow testing the machine on aspects of workability, control, adjustments, and maintenance. The recommended nozzle type, discharge pressure, boom height, and forward speed should be used for each type and size of crop.

The following observations should be recorded clearly for each test plot:

a. Size of plot;
b. Type and height of crop;
c. Weather conditions such as wind speed and temperature.

The sprayer should be tested in at least five different plots and crop conditions, and the following measurements should be made:

a. Actual field coverage along with time spent on turning, refilling, minor repairs, and idling;
b. Application rate rendered by the sprayer as measured from the spray tank level;
c. Forward speed;
d. Boom height setting;
e. Ease of filling, adjustment, and maintenance;
f. Ease of control;
g. Repairs of leakages, if any;
h. Ease of aligning passes as set by the row markers.

References

Dixit, A. (2011). *Course notes on Farm Machinery and Equipment-II, Lesson 28. Introduction to farm machinery testing, types of tests, testing parameters*. e_Krishi Shiksha. http://ecoursesonline.iasri.res.in/mod/page/view.php?id=2265.

Smith, D. W., Sims, B. G., & Neill, D. H. (1994). *Testing and evaluation of agricultural machinery and equipment - principles and practices, 110*. FAO Publications, FAO Agricultural Services Bulletin.

Kepner, R. A., Bainer, R., & Barger, E. L. (1978). *Original from the university of Michigan*. Avi Publishing Company.

CHAPTER

Automation and precision agriculture

14

Chapter Outline

14.1 Mechatronics and its application in agricultural machines

14.1.1 Mechatronics as a science

Advances in technology, particularly in micro-electronics and computer engineering, have made almost all products in use today dependent upon these sciences in terms of their functions and controls in our day-to-day life. Computing power in the form of a microcontroller has made tremendous progress in its application in these devices. The microcontrollers are shrinking in size, but their speed and number crunching power are growing exponentially. More importantly, their ability to interface with other engineering systems, such as mechanical, electrical, and fluid power elements, has made them much more versatile in every area of control systems in machines and processes. As with every other machine in every other arena of application, agricultural machines have also started using the enormous power of computation and control. Mechatronics (Onwubolu, 2005) is the science of

Machines of the Farm. https://doi.org/10.1016/B978-0-443-32906-7.00035-9

synergistic integration of mechanical, electronic, and control engineering principles in the development of machines and processes. In the past, the actuation and control of machines and processes were entirely based on mechanical elements, which were quite bulky and cumbersome. But, mechatronics has made it possible to design machine actuation and control much more simplified. Above all, designing such systems to be integrated into agricultural machines has become easier because microcontrollers and their allied boards are easier to program and implement without a steep learning curve. Agricultural engineers, like the author of this book, are able to build their own machine systems with the application of mechatronics. These boards can be approached as black boxes that do what the engineer wants them to do. The interfacing parts of these boards and controllers require a rudimentary knowledge of electronics especially about the types, working, sourcing, and synching of driving elements; a basic know-how of fluid power systems such as the hydraulic and pneumatic circuits and their elements; and, of course, an engineering knowledge of how machine elements are driven by the actuators and drives. As with learning any other science, acquiring more knowledge of electronics, mechanical principles, and fluid power engineering will always yield better results in designing such systems. The availability of common programming languages for coding these electronic systems has made the implementation of such systems much simpler for engineers of all disciplines. Complex problems that require much more expertise in any of the components of the mechatronic system can always be addressed by teams consisting of cross-disciplines, and a lead engineer can make it possible to build the machine as desired. Control systems engineering is another key component of mechatronics since, ultimately, the machine elements need to be controlled as desired. As a whole, mechatronics focuses more on the interaction and synergy between the various engineering systems to do a desired job.

14.1.1.1 Micro-electronic systems

Let us learn about a basic control board available on the market known as Arduino Uno@ to understand the use of micro-electronics in a rudimentary mechatronic system that we may desire to build. Figure 14.1 shows the outline of the UNO board. This board has an ATmega328, a microcontroller, as the central brain. The active and passive components built around the chip support the interfacing of this chip to external electrical circuits. The board can read electrical inputs, work on them, and send out signals that can control other devices. The whole process of interacting with the external world is performed by means of a user-written program or code uploaded to the chip.

The Arduino UNO has 14 output ports labelled as D0 to D13 that are basically digital inputs/outputs. Each of these ports can handle a binary-level signal and are known as digital pins. If they are assigned as inputs, they can read in the level of signal as 'HIGH' corresponding to 1 and 'LOW' corresponding to 0. When they are assigned as outputs, these pins output either a HIGH or LOW voltage corresponding to 1 and 0, respectively. In digital electronics, a HIGH means 5V and a LOW means 0V. The assigning of these pins is performed by the program that is

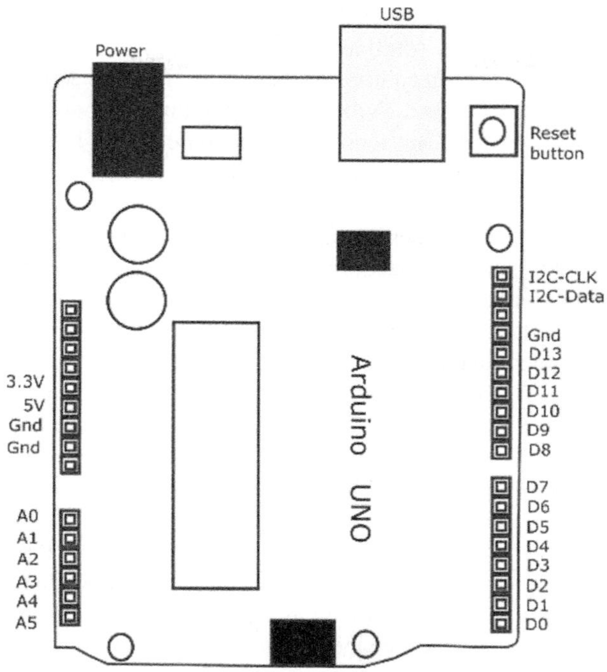

FIG. 14.1 An Outline View of an Arduino Uno Microcontroller Board.

stored in the microcontroller. These digital pins can act as a parallel channel or bus of eight bits or individually as input or output pins. Some of these pins such as D3, D5, D6, D9, D10, and D11 can also be assigned for outputting a pulse-width modulation (PWM) signal using the Arduino command *analogWrite()*. Pulse-width modulation (PWM) is a modulation technique that generates variable-width pulses to represent the amplitude of an analogue signal. This implies that values between 0 and 255 can be sent out as modulated analogue signals through these pins. An explanation of PWM beyond this point is beyond the scope of this text. Many of these pins have dual functions for specific operations such as handshaking serially with other devices. For instance, D0 and D1 can act as transmit and receive pins for serial communication with other electronic devices. A light-emitting diode (LED) on the board itself is connected by default to D13 and used directly for generic purposes and indications.

Apart from these digital pins, ANALOGIN pins labelled from A0 to A5 are available on the board to read in analogue signals. Analogue signals are not binary-state signals having only HIGH and LOW but can continuously vary from 0 V to 5V. This range of voltage is converted to a corresponding decimal value of 0 to 1023. These pins are used to read analogue sensors such as temperature or level detectors. For instance, a liquid level sensor may produce a voltage varying continuously from

0 to 5V depending on the liquid level. As the level rises from empty to full, it varies from 0 to 5V correspondingly. A full level is detected as 1023 on the analogue input by the board and empty as 0. In addition to these pins, there are some pins available, marked as 0V (GND), 3.3V, and 5V, known for tapping out power for driving the external circuits and sensors. But these pins can source only 40 mA of maximum current, above which the board will be overloaded. The GND pin acts as a common ground for the external circuit. A RESET button is provided on the board to stop and start the program flow from the beginning. The board has a 5V power port for powering the board and components including the microcontroller chip and other ICs that support serial communication. The board has a USB-driven port to link it to a computer's serial port so that the board can be uploaded with the program we desire to run in it. There are more details about the board, such as several functions for pulse-width modulation (PWM), which are beyond the scope of this text.

There are several ways of writing the code and running it on the board to do a job. The easiest is to use the "Arduino IDE", a free software developed by the maker of these boards. It is an integrated development environment (IDE) for writing the code in an Arduino C++ language with specific Arduino commands to interface the board to external elements and also to assign and control the pins and functions of the board. The language is easy to learn if one has a basic knowledge of programming logic and is familiar with a programming language like C++ or Python. The IDE is run on the computer that is connected to the Arduino UNO board. It identifies the type of Arduino board and interfaces with it. The user first writes the program on the IDE installed on the computer and then verifies it for any syntax errors. Then, the code is uploaded to the Arduino board connected to the computer. The IDE converts the code into a machine language code and uploads it to the board's microcontroller. Immediately upon uploading, the code starts to run on the board and performs the sequence of jobs commanded to be performed. It talks to the external devices connected to the board, takes in signals, and sends out commands, as desired. This is done after connecting the board completely to the intended devices. Once it is verified that the board is able to function as desired, the board can be separated from the computer to continue functioning independently. There is an extensive amount of free documentation and tutorials available on the Internet to learn the basics of programming these types of boards. Year by year, Arduino is releasing more advanced models with different names like MEGA and GIGA that are capable of more powerful functions such as WiFi, Bluetooth, etc. The UNO board has a program memory of 32K, which is quite sufficient for running a mediocre function. Advanced boards can handle more memory and enhanced functions.

Let us now look into how a simple program can control the onboard LED to blink at a desired delay. This code is given with the Arduino IDE for a beginner to learn. Many more such tutorial codes are available on it for learning. In-depth learning of the whole programming of Arduino cannot be the objective of this book. Let us understand some important commands with this simple code to pave way for the student to learn more in the future. The code 'Blink' is furnished below to start with.

```
/*
This example code is in the public domain.
Blink - Turns an LED on for one second, then off for one second, repeatedly.
https://www.arduino.cc/en/Tutorial/BuiltInExamples/Blink
*/
// the setup function runs once when you press reset or power the board
void setup() {
// initialize digital pin LED_BUILTIN as an output.
pinMode(LED_BUILTIN, OUTPUT);
}
// the loop function runs over and over again forever
void loop() {
digitalWrite(LED_BUILTIN, HIGH); // turn the LED on (HIGH is the voltage
level)
delay(1000); // wait for a second
digitalWrite(LED_BUILTIN, LOW); // turn the LED off by making the voltage
LOW
delay(1000); // wait for a second
}
```

Let us dissect the simple code above.

It may be noted that the program follows the syntax of the typical C language. The "definition of functions", "for" and "if" commands, all follow the syntactical rules of C. The comment of a line is done by a double slash, and a block of comments is started by "%*" and closed by "*%". As always, all the steps are commented elaborately so that any person reading the code can understand what the code is doing. The function *setup()* has all the definitions and commands that are to be performed only once by the board. All the assignments are performed under this function. Inside this function,

pinMode(LED_BUILTIN, OUTPUT); is a command that assigns D13 as an output. *LED_BUILTIN* is a built-in definition of D13. The term "*pinMode*" is the keyword for assigning a digital pin as input or output. If we need another pin D3 as input, then *pinMode(D3, INPUT)* will be the command.

The function *loop()* is where all the lines of code that are to be implemented are to be repeated again and again by the program without stop.

digitalWrite(LED_BUILTIN, HIGH); writes the digital pin D13 with a HIGH (5V) output, which turns on the built-in LED connected to the D13.

delay(1000); is the command that delays the running of the program by 1000 milliseconds, i.e., one second.

After the delay, the command *digitalWrite(LED_BUILTIN, LOW);* turns off the built-in LED by sending out a LOW (0V).

Then, the next command, *delay(1000);* introduces a one-second delay yet again.

The *loop()* function with the above steps repeats again and again, making the built-in LED blink with a one-second delay.

What we have learnt is to drive an in-built LED on the board. We can now do the same by connecting an external LED to D3, as shown in Figure 14.2. A resistor in series with the LED is connected between D3 and the ground. The resistor is needed to limit the current flowing through the LED. It may be noted that the LED's anode is connected to D3. Of course, this can be experimented only with a real board, which the student is encouraged to do.

There are diverse types of sensors that can detect various parameters in a physical environment to feed in the desired data to the board.

A simple microswitch or push button is a simple digital sensor for sensing a machine function. A limit switch is a microswitch that can be used to detect, for instance, the end of a reciprocating machine member at its stroke end. The element, when it contacts the switch, toggles it and sends a digital input to the board connected to it. Figure 14.2 shows how a push button is connected to the board's D4 assigned as an input. When the switch is closed, the board's pin is connected to the ground, making the input LOW (0V). When it is open, the pin will be connected to 5V through the resistor connected in series with the switch. This will make the input HIGH. A photoelectric sensor is a digital sensor that can detect the presence of an element by sensing the light falling on it. Hall sensors and counters are available to detect the vicinity of an element to another as well as to count gear teeth by

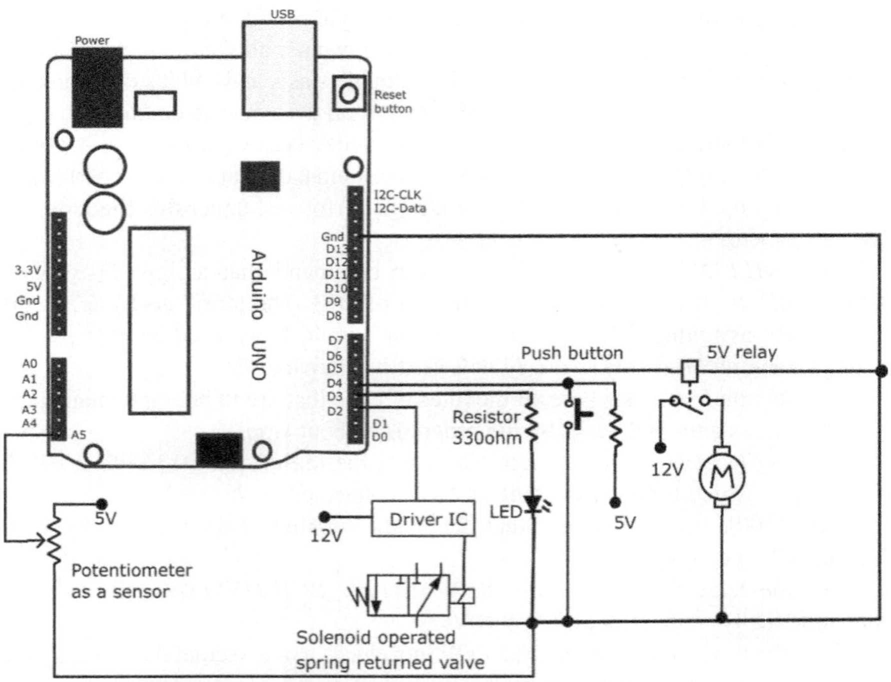

FIG. 14.2 Some Input and Driven Devices Connected to the Board.

digital outputs. Digital encoders provide digital signals corresponding to the angular position their axis takes. They can also be used to measure the rotational speed of a shaft.

Similarly, sensors that output analogue voltage according to the parameter they sense are available. Some of the types are explained below.

A simple linear potentiometer to sophisticated laser sensor is used to detect the linear or rotary motion of any element in a machine or process. For instance, Figure 14.2 shows a linear potentiometer or variable resistor connected as a voltage divider between 5V supply and the ground. The moving contact of the potentiometer is connected to the analogue pin, A0. When the potentiometer's axis is given an angular deflection, the resistance changes and, thereby, the voltage between the pin and the ground varies accordingly. This analogue input available on the pin A0 can be read by the board to detect the angular displacement of the axis. Motion sensors can well be used as sensors to detect liquid levels and displacements.

Simple solid-state humidity and temperature sensor ICs are available on the market that, when are connected to the Arduino board at a defined configuration, can input an analogue signal corresponding to the temperature or humidity in the environment.

Accelerometers can detect the acceleration of a machine element and its orientation at any instant of time.

Capacitive and resistive moisture sensors are available to monitor grain and soil moisture.

The controlling board can also send an output to the actuator that converts it to an appropriate electrical action in the process that we want to control. An example is a servo motor or another type of electrical motor through the necessary driving circuits or devices. It could be a solenoid valve that is driven to open a valve or move a plunger, etc. Generally, controller boards such as the Arduino UNO cannot source the amount of power that is required to drive an electrical device directly. Typically, driver circuits or relays are used for this purpose. Figure 14.2 shows an example of a typical magnetic relay connected directly to the pin D5 configured as an output. On the board's HIGH output to the relay, it closes the relay points, which connect the DC motor to a 12V voltage source. Here, the relay requiring much less current to be driven by the board drives the motor demanding a higher current. Solid-state driver ICs are also available to perform this job. Figure 14.2 shows an example of the board's pin D2 driving a solenoid-operated hydraulic valve through such a driver IC. The driver IC is one that boosts the power output of the board to drive the device requiring more power. On a HIGH output from the board, the electrical solenoid is energized, which actuates the valve to open.

14.1.2 Fluid power systems

Fluid power (Merritt, 1979) is one of the components of systems controlling many agricultural machines. Fluid power is either hydraulic, using pressurized oil, or pneumatic, using pressurized air. In many ways, the two are analogous; the main difference is the compressibility of oil and air, causing a quicker response in the latter and more power rendered in the former. Fluid power systems in full detail, their logic, and their design are beyond the scope of this text; hence, let us focus on the rudiments of this important area from the known example of a boom sprayer that the student has already learnt in the chapter on plant protection machinery. The hydraulic power sprayer basically has a hydraulic circuit in which the pressurized fluid is the medium. The only difference is that it is an open-to-atmosphere circuit that sprays the liquid off the nozzles or orifices. The circuitry of the sprayer can be used to study the components and their symbols in a hydraulic circuit. The physical hydraulic circuit is symbolically shown in its hydraulic circuit in Figure 14.3.

The pump, as shown, is the heart of any fluid power system; it pumps the hydraulic fluid (spray fluid here) from a reservoir or tank. A filter element is added in the suction line to filter away dust and impurities. The pump used is necessarily a positive displacement pump such as a piston pump, gear pump, or a vane pump. The characteristics of such a pump have been learned in the chapter on plant protection. The pump drives the fluid into the circuit through a pressure-relief valve. The construction and function of an adjustable pressure-relief valve was also explained in that chapter. Whenever the system pressure on the outlet of the valve exceeds a preset level, it opens and returns the fluid to the reservoir, thus safeguarding the pump from overload. It controls the discharge pressure to a desired level to suit the nozzles.

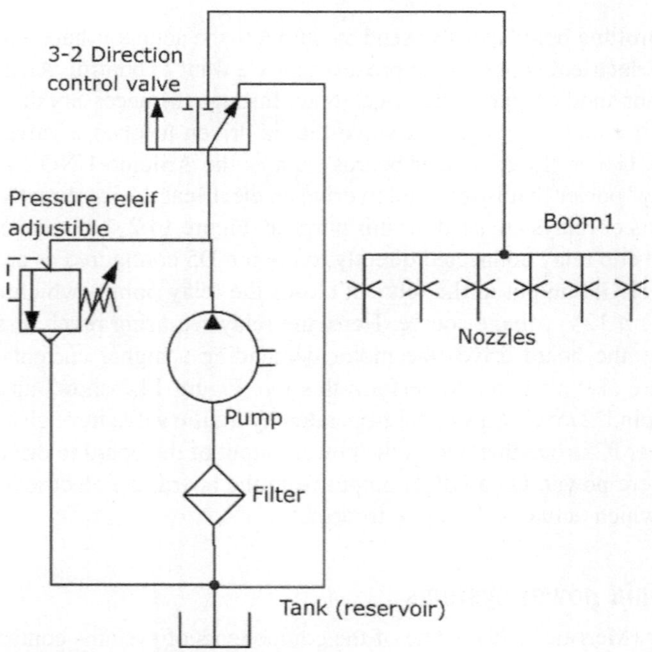

FIG. 14.3 Hydraulic Sprayer's Symbolic Circuit.

In closed fluid control circuits, this valve is mainly for the prevention of pump overloading. The pressurized fluid leaving the relief valve is controlled by a direction control valve (DCV) in the sprayer circuit. Here, it is used for shutting down the spray liquid moving to the boom and the nozzles by returning the fluid to the reservoir. The valve shown in the circuit here is a two-way DCV. It may be noted that the symbol for the two-way valve has two boxes, one for each position of the valve. In the position shown, the fluid is diverted from the inlet to the second outlet connected to the reservoir, and the first outlet connected to the boom is blocked. When the DCV valve is manually actuated, visualize the second box now connected to the same inlet and outlet ways. Now, it may be noted that the fluid flows through the inlet port and goes into the boom through the first outlet port, whereas the second outlet port is found blocked. Depending upon the number of ports present on the DCV, the number of positions the valve has, and the method of actuating them, there is a plethora of valve types available. Let us learn about some of the available types and actuations, as shown in Figure 14.4.

The first example is a four-port, two-position DCV, which is solenoid actuated against a spring. The normal position of the valve directs fluid straight to the opposing port, and the actuated position swaps the inlets to the other outlets. On applying a voltage to the solenoid, the valve is actuated; switching the voltage off returns the valve to its normal position. The solenoids can be actuated by either AC or DC voltage and are available at different voltage values. The second example, as the name implies, has only three ports and three positions. The mid position blocks all the ports, and the other two positions direct the fluid, as shown in their respective positions. The valve actuation is performed by a manual lever against a

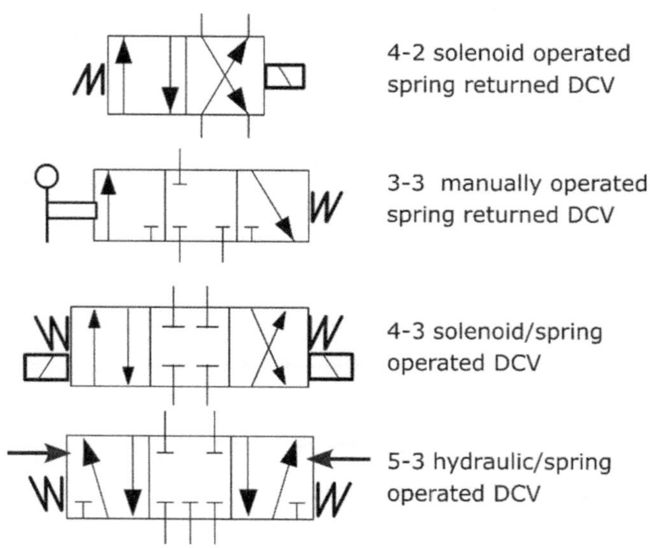

FIG. 14.4 Some Examples of DCVs and Their Hydraulic Symbols.

spring. Throwing the lever can position the valve in any desired position. The next example is a four-port, three-position solenoid-actuated DCV with a return spring both ways. The mid position that blocks all four ports is the neutral one. Upon actuation of the solenoid on any one side, the valve moves to that direction and directs the fluid accordingly, as shown. The final example is similar to the previous one, but the actuation is done by hydraulic pressure. These can be considered as slave DCVs to the main control valves in a circuit. But, here there are five ports available in the DCV configuration, and, as usual, the mid position is the neutral or off position of the three positions. The shown valves are only a very few examples, and the varieties available for use are very extensive.

The other important component of a hydraulic circuit is the actuator (Figure 14.5), which converts the fluid power of the pressurized oil into mechanical work in terms of a linear or rotational movement.

A hydraulic motor (Figure 14.5) converts the fluid power to a rotary motion and can easily be visualized as an inversion of a hydraulic pump. This implies that the pressurized oil is sent into the motor's ports and rotates the vaned rotor or actuates the plungers in the motor's sealed housing. After performing the work, the oil returns back to the reservoir after expending its pressure. The fluid power supplied to the motor is expressed as $P = \frac{Q(\Delta P)}{60}$ where Q is the flow rate of pressurized oil into the motor, *l*pm, and ΔP is the pressure drop across the motor, which is the difference between the inlet and exhaust pressures used by the motor. The speed of rotation of the motor's shaft is expressed as $N = \frac{Q}{V_r}$ where V_r is the volume of oil required by the motor for one revolution of its shaft, $l\ \mathrm{rev}^{-1}$. The torque developed by the motor is $T = \frac{\Delta P V_r}{2\pi}$. The actual power of the hydraulic motor is the product of the fluid power and the overall efficiency of the motor, the components of which are the mechanical and hydraulic efficiencies.

A hydraulic cylinder (Figure 14.5) is one that provides a linear displacement when oil is sent into a piston and cylinder assembly. The hydraulic cylinder can be a single-acting unit in which pressurized oil is supplied only on one side to move the piston, thus lifting or forcing the load to act against it. The return stroke of the piston happens by the counteracting force of the weight being lifted. Many old models of tractors had a single-acting cylinder for its three-point hitch; in these cases, the implement was lifted by hydraulic pressure but dropped by the weight of

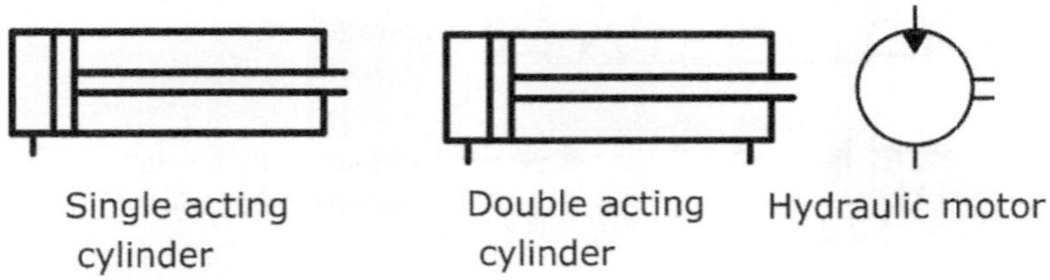

FIG. 14.5 Hydraulic Actuators.

the implement. In a double-acting cylinder, the pessurized oil is supplied on any one side to move the plunger to one direction causing the evacuation of unpressurized oil on the other side of the plunger. An opposite movement is obtained by supplying pressurized oil on the opposite side and emptying the other side of the cylinder. This allows the cylinder to work with a load on it in both directions. This is the cylinder commonly used in many hydraulic circuits for lifting and moving machine parts against a load. A typical example is the pair of cylinders used to lift and lower the blade of a bulldozer. The force of displacement is the product of the oil pressure and the plunger's effective area of action. In a double-acting cylinder, the force is correspondingly reduced on one side because of the presence of the piston rod that reduces the effective area of the plunger. If A_1 is the area of plunger's face on one side; A_2 the effective plunger area on the other side = A_1 – the area of cross section of the rod; then the net force the cylinder exerts = (P_1A_1 tilde P_2A_2), with P_1 and P_2 being the pressures on each side respectively. If Q is the flow rate of pressurized oil into the cylinder in *l*pm, then the speed of plunger, in m s^{-1}, *is* $v = \frac{Q}{6A}$ where A is the effective area of plunger in cm^2.

Figure 14.6 illustrates a simple, open-centre hydraulic circuit with no pressure-compensated hydraulic pumps and load-sensing elements in it. The pump supplies pressurized oil from the reservoir through a filter, with a pressure-relief valve to prevent any overloading of the pump. The supplied oil reaches the 4-3 DCV, which is manually operated and has three positions. The central position has all the ports

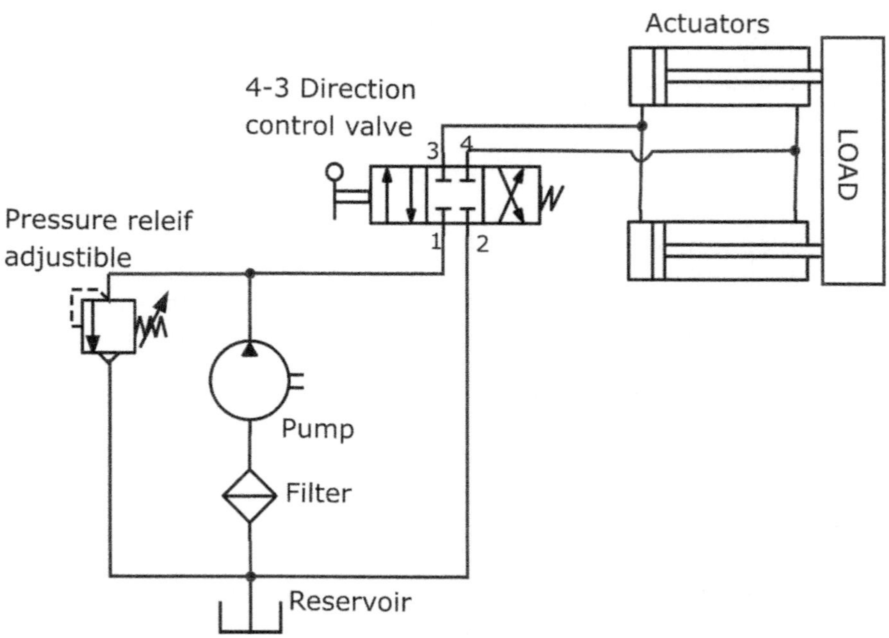

FIG. 14.6 A Basic Hydraulic Circuit Controlling Actuator Cylinders.

blocked, and the pressurized oil is relieved by the relief valve back into the reservoir. When the valve is thrown to one side (LHS) (*visualize the DCV's left-side box moved to the right and to the connecting lines at the centre*), the pressurized oil is directed through port 1 and through the opposite port 3 to the LHS of both cylinders simultaneously. This moves the cylinder rods to the right, thereby pushing the load. The oil on the other side of the two cylinders connected to port 4 is exhausted through port 2 back to the reservoir. Then, when the DCV is thrown to the right side (*visualize the DCV's right-side box moved to the left and to the connecting lines at the centre*), the pressurized oil is directed through port 1 and through port 4 to the RHS of both cylinders simultaneously. This moves the cylinder rods to the left, pulling the load back. The oil on the other side of the two cylinders connected to port 3 is exhausted through port 2 back to the reservoir. This is a very rudimentary circuit, which has many drawbacks but explains the principle.

14.1.3 An example of a mechatronic implementation

To keep things simple and explain an example system that can apply mechatronics, let us say we desire to build a variable-rate fertilizer application machine that should vary its fertilizer application rate as per the greenness of the plant canopy of a row crop. The greenness is measured by a green seeker (sensor) device that measures the Normalized Difference Vegetation Index (NDVI) of the crop leaves. NDVI is used to quantify the greenness of vegetation to assess the plant health and is calculated as a ratio between the red (R) and near-infrared (NIR) values. It has been found by research that fertilizer application rates can be varied based on the NDVI of the crop, maximizing the crop yield. A green sensor can capture the NDVI in situ and send it as an output voltage signal to a control board. The control board can run a program that converts the NDVI values to a corresponding fertilizer application rate using a predefined relationship between the two parameters. The board can then send out a signal to a motor drive to change its speed according to the calculated fertilizer rate. The motor driving the fertilizer-metering device through an appropriate reduction gear box can render the desired application rate at the location where the NDVI is measured. The green seeker can be fitted to the machine itself to measure the NDVI on the go and, thus, the system can change the fertilizer application rate accordingly.

The following is only intended for explaining the principles of building the machine. The data used, especially for relating the NDVI and fertilizer rate is not experimentally proven. We assume the relationship as shown in Figure 14.7

The circuit required for implementing the control is shown in Figure 14.8. The output from the green seeker is connected to the analogue input pin A0. The green sensor is supplied a voltage of 5V and also connected to the board's ground. We know that the digital pin D3 is also a PWM pin on the UNO board and can output an analogue signal to any device using the command *'analogWrite()'*. The D3 pin is connected to a NMOSFET transistor's 'gate' input that can control a DC motor's speed. The motor's armature is connected between the transistor's 'drain' terminal

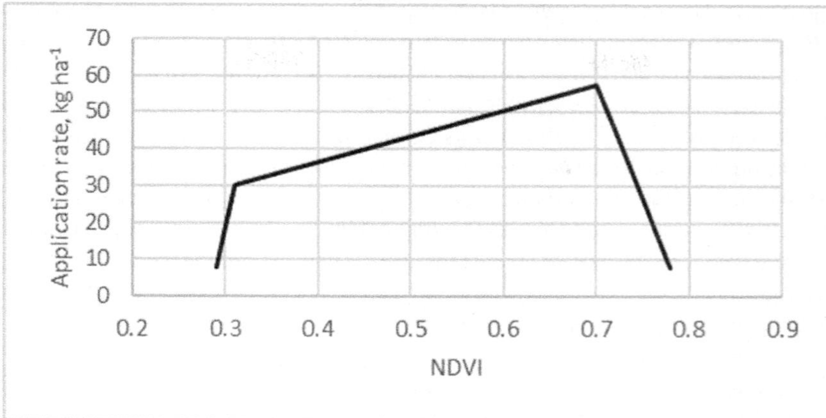

FIG. 14.7 Assumed Relationship between Fertilizer Application Rate and NDVI.

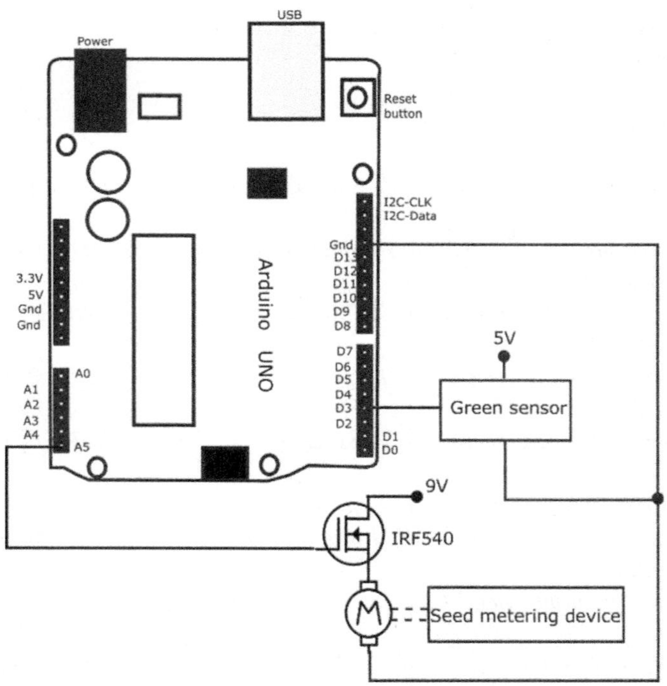

FIG. 14.8 Controller Circuit of a Variable-Rate Application Machine.

and ground, as shown. (*Since the basics of NMOS devices is beyond the scope of this text, the reader is advised to learn the basics of NMOS devices and DC motor control from appropriate reading material.*) The transistor's 'source' terminal is connected to the positive terminal of a 9V external supply because the motor needs a separate high-power supply. The DC motor thus conneced drives the fertilizer-metering devices that are similar to the fluted roller device. The metering devices one on each row of the sowing machine are driven together through a common worm gear drive.

Let us now write a code for the Arduino Uno, as below, and learn what it is doing step by step. Many of the steps are explained in the code's comments themselves.

```
// Initialize some variables for our use
int gsensor; // Integer variable for receiving the Green sensor signal
//Assuming the sensor is sending out an analog voltage 0 to 5 V corresponding to 0 to 1 NDVI
float fertrate; //Float variable for calculating the fertilizer rate
int motspeed; //Integer variable for sending out control signal to the motor driver (NMOS)
void setup()
{
// There is no need for assigning the pins, because we are not using any pin as digital in or output
}
void loop()
{
gsensor = analogRead(A0); // Read the Green sensor output into the variable gsensor
// Apply the prescription relationship between green value (NDVI) and fertilizer application rate
if(gsensor > 295 && gsensor < 317) fertrate =7.87 + (30.28-7.87)*(gsensor-296)/(317-296);
if(gsensor > 316 && gsensor < 716) fertrate =30.28 + (57.53-30.28)*(gsensor-317)/(716-317);
if(gsensor > 715 && gsensor < 798) fertrate =57.53 - (57.53-7.87)*(gsensor-715)/(797-715);
if(gsensor < 296 || gsensor > 797) fertrate =7.87;
motspeed = int(fertrate*100); // Multiply the fertilizer rate by 100 and convert decimal to integer
motspeed = map(motspeed, 0, 5753, 0, 255); // Map the range between 0 to 5753 to 0 to 255
analogWrite(3, motspeed);//Write the PWM signal corresponding to motspeed to D3
}
```

The line '*gsensor* = *analogRead(A0)*;' inputs the NDVI value of 0 to 1 as a corresponding digital value ranging from 0 to 1023. For instance, NDVI value of 0.7 is input as 0.7 × 1023 = 716 and so on.

Since the relationship between the NDVI and fertilizer rate has been assumed as a piecewise linear curve (Figure 14.7) on which the slope is changing in three different steps, we calculate the fertilizer rate as in:

if(gsensor > 295 && gsensor < 317) fertrate =7.87 + (30.28-7.87)(gsensor-296)/(317-296).*

Each linear segment of the relation is used to calculate the fertilizer application rate to the variable *'fertrate'*, thus covering the overall range of the NDVIs.

Subsequently, in the step *motspeed* = *int(fertrate*100);*, the calculated fertilizer rate is converted to an integer value, eliminating the decimal places, and stored in the integer variable *'motspeed'* using the int() function.

motspeed = *map(motspeed, 0, 5753, 0, 255);* is an expression that maps all the 'motspeed' values from a range between 0 and 5753 (the maximum of fertilizer rate values) to another range between 0 and 255. The *map()* function of Arduino performs this interpolation value to value. This implies that 5753 is converted to 255 and the values between 0 and 5753 are linearly interpolated to a value between 0 to 255. This is done to make the output compatible for writing these values to the pin D3, which can accept only values between 0 and 255.

The value in *'motspeed'* is now sent out to the NMOS IC by the statement *analogWrite(3, motspeed),* which controls the motor speed accordingly.

> *There are many Arduino simulators available on the Internet that can be used to simulate the circuit as well as to write the code. These are great sites for learning the basics of Arduino programming. One such example is the AUTCAD's TinkerCAD site: https://www.tinkercad.com/dashboard. The circuit can be built piece by piece, connected, and run after writing the code. The reader is advised to check on such simulators.*

Now that the motor speed is set to change according to the fertilizer rate, we need to find the speed ratio between the metering-feed shaft and the motor, as follows.

Say that the fertilizer application device has a fluted roller type of metering device, the row spacing is set at 45 cm, and the machine is traveling at a speed of 3 km h^{-1}.

Then, the area covered by the metering device on each row = (0.45 × 3000/60) = 22.5 $m^2 min^{-1}$.

The maximum fertilizer rate required from the data = 57.53 kg ha^{-1} = 5.753 g m^{-2}.

The maximum flow rate from the metering device = 22.5 $m^2 min^{-1}$ × 5.753 g m^{-2} = 129.44 g min^{-1}.

Say the metering roller is set to output 15 g per revolution. Then, the maximum rpm of the roller for rendering 129.44 g min^{-1} = 129.44/15 = 8.62 rpm.

Say the driving 9V DC motor is a brushless motor, the maximum speed of which is 17,550 rpm. Then, the speed ratio between the feed shaft and the motor is 17550/8.62 = 2034 rendered by a worm gear box.

This implies that the motor will revolve at 17,550 rpm to meter the maximum fertilizer rate of 57.53 $kgha^{-1}$ and proportionately for other rates according to the desired relationship.

The above is a very rudimentary control logic for the said purpose and the code is written to explain the basics. In fact, a reliable machine has to have provisions for choosing any kind of relationship between the fertilizer rate and the NDVI. It should have a detector to measure the offered fertilizer rate in real time and feed it back to the control system for corrections. A precise machine would be much more complex than the basic one we have learnt.

Now that the reader has an introduction to the fluid power systems too, he can easily visualize that such control boards can interact with the fluid power systems by sending out signals to solenoid-controlled valves to actuate hydraulic cylinders and hydraulic motors to control the machine's working. Analogous solenoid-controlled valves are available for pneumatic circuits, too, and can be controlled by the control board. The use of mechatronics in modern agricultural machines is already very prevalent. The automation and control, as implemented on a basic board such as the UNO, are very limited. When such automation is coupled to on-board computers, as may be found in today's agricultural machines, their potential is huge. Mechatronic systems are provided with increasing frequence in agricultural machines, giving the operator in the cab an overall control of all the required parameters under the operator's command. An era of autonomous agricultural robots is already underway, which may radically change agricultural mechanization altogether.

14.2 Global positioning system

A global positioning system, or GPS, is basically a satellite-based navigation system developed by the United States Department of Defense in the 1970s, mainly for military and space applications and expanded to public use later. The Global Navigation Satellite Systems, or GNSS, describes every satellite constellation in orbit around the globe. GPS is one of several GNSS constellations, including Russia's GLONASS, Europe's European Satellite Navigation System (GALILEO), China's COMPASS/Bei-Dou, India's Regional Navigation Satellite System (IRNSS), and Japan's Quasi-Zenith Satellite System (QZSS). The advancement in technology has given the GNSS advanced capabilities, many of which are proprietary to each country. Apart from their use in the military, GPS is used extensively in almost all areas of activities including agriculture, fisheries, transportation, aviation, sea navigation, etc.

This technology offers the measurement of Position (geographical location by way of latitude, longitude, and altitude), the rate of movement for use in Navigation, and Time (UTC); and so is abbreviated as a PNT (position, navigation, time) service. There are now 31 active GPS satellites in the US constellation. These satellites are in high orbit, at 20,000 km, and are going around the Earth in different orbital paths, thus making at least four satellites available for a ground location (receiver) anywhere and anytime on the globe. Each GPS satellite continually broadcasts information on its position by time signals, which a receiver can decode. The GPS receiver measures the time signals of at least four satellites at a time; then, by the principle of triangulation, detects the three-dimensional position (latitude, longitude, and altitude) of a receiver's location. The resolution of the GPS, depending on the receiving system, has improved to be closer to sub-centimetre in terms of detecting the position. Using the time signals continuously emitted by the satellites, the receiver's algorithm can also calculate the time rate of movement of the receiver. This aids in navigational applications that is vital for aviation, naval navigation, and ground transports.

GPS is provided in two classes of precision, namely, for a standard mode and a precision mode. The former is used by general public, and the latter is used for military and space applications. GPS satellites now have included a broadcast frequency called L5, which eliminates many setbacks of the earlier bands of frequencies. GPS receivers that use the L5 band have 30 cm resolution or positional accuracy, whereas land surveying systems have 2 cm resolution. Even sub-millimetre accuracy is feasible in highly accurate systems.

GPS consists of three main segments, namely, the space segment, the control segment, and the user segment. The space segment of GPS has about 31 satellites and is used mainly for highly precise satellite positioning. The GPS control segment under the US government's space force control has many ground-monitoring stations located worldwide. These stations monitor the health of GPS satellites in terms of their positions, signals, and accuracy. The user segment includes all receivers extending over a wide range, from hand receivers to highly precise aviation and navigational systems used for PNT measurements.

Real-time kinematic positioning (RTK) is an augmentation to GPS that corrects errors in the current GNSS systems. It relies on a reference station to provide real-time corrections that can enhance the positional accuracy to a centimetre level. RTK uses a fixed-base station and a rover to reduce the rover's positional error. The base station transmits the correction data to the rover. RTK systems use a single base station receiver and a number of rover or mobile units. A radio modem working in the UHF band is used for communication between the base and rover, constantly sharing correction data. This allows the units to calculate their relative positions to within millimetres. The resolution of these systems can, hence, be 1 cm horizontally and 2 cm vertically. The base station is located at a surveyed benchmark, and the rovers produce a highly accurate map relative to the base. RTK is mainly applied in land surveying, autopilot systems such as drone guidance, precision farming, and machine control systems. RTK networks, their density, and accuracy determine the use of RTK-assisted receivers. Moreover, a Continuously Operating Reference Station (CORS) network is a system of RTK base stations that broadcast corrections, usually over an Internet connection. CORS enhances accuracy many folds since RTK measurements can then use more than one base station.

14.3 Precision agriculture

Geo-referenced data are spatial data of an area, such as a field plot that has geographical location information (geo-reference) attached to it. The global positioning system, which has grown to have a sub-centimetre resolution, can assist in creating these data and, hence, is the main driving force in the use of such data. All the desired information about soil, plants, and the growth environment can be sensed and geo-referenced over the entire field. This can be done by measuring these data in real time by sensors mounted on agricultural machinery or proprietary sensor systems and stored or transmitted digitally. These voluminous geo-referenced data can be analysed for their variability — inherent, temporal, or caused by some factor. Based on this information, prescriptions for agricultural inputs, such as fertilizer, pesticides, seeds, and water, can be made at each localized point of the field using a geographical information system (GIS). The phenomenal rise of computing power has the capability of making such decisions by superimposing relevant data layers one over the other at geo-reference points on a GIS. The access and analysis of geo-referenced data to aim for higher productivity in a crop has brought out an information-based management method called "precision agriculture." This technology is also known as precision farming or site-specific crop management. It encompasses all the influencing factors of crop production including soil manipulation, planting, weed control, input applications, and harvesting. Profitability is improved by increasing yields, lowering input costs, and enhancing crop quality.

14.3.1 Geographic Information System (GIS)

A geographic information system (GIS) is a vital tool for all precision farming decisions since it is capable of preparing maps based on geo-references and performing map calculations at each location to create prescription maps for the inputs. This will form the basis for the decision-making part of the system. It can layer different property maps and yield maps, one over the other, to perform this job. Any parameter or piece of information that is geo-referenced on a map is a candidate to be a layer (map). As may be noted, the values of the parameter over the whole field are available at geo-referenced locations all over the field but are spaced apart according to the grid of measurement locations used over the area. The values in between these locations are interpolated using a process known as 'Kriging' so that the values all over the area are obtained and maps are made on the spatial distribution of the parameters.

Figure 14.9 shows a typical application wherein the topography of the field, soil electrical conductivity (EC), soil fertility, and yield map of the previous season are layered to produce a prescription map of seed rate based on a proven relationship between the prescribed seed rate and the influencing parameters, namely, the topography, soil EC, and the yield data. The overlaying of maps of different parameters influencing the independent variable is done in the GIS platform, and map calculations are made to generate the prescription map of the result (independent variable). For simplicity, say in our example above, a reliable expression in the form of:

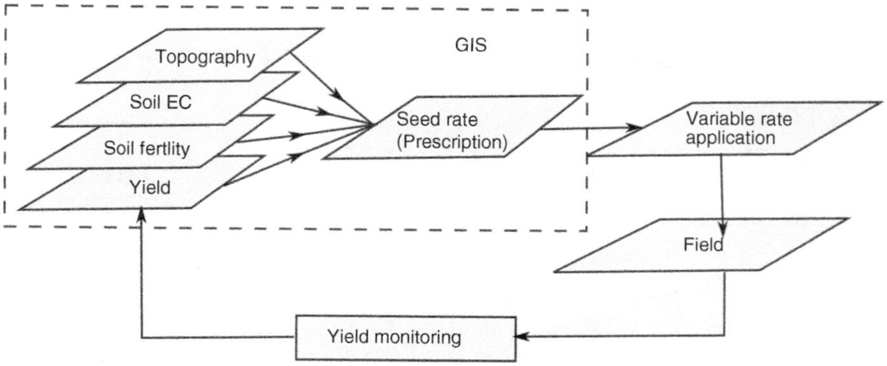

FIG. 14.9 GIS Overlays in Precision Agriculture.

seed rate = Fn(topography, soil EC, soil fertility, yield of previous season)

is available. Then, the seed rate is calculated by GIS using the above expression and a seed rate map is generated.

Powerful commercial GIS software, such as ArcGIS and IDRISI to name a few, is available for use. Equally powerful openware, such as QGIS and GRASS, too, are available absolutely free. Since GIS is mainly for academic work and cannot be easily used by end users such as farmers, customized platforms specifically for mapping and prescribing inputs are available on the market.

14.3.2 Yield monitoring and mapping

Two applications in precision agriculture are yield monitoring and mapping. A yield monitor, as fitted on a grain combine or a harvesting machine, measures the weight of a harvested crop on the go, point by point, and tags it with the corresponding geo-reference. These accumulated data are used to prepare a yield map using a GIS, which shows the locationwise variability in the field. These yield maps are recorded season after season, and these temporal yield maps are overlaid on each other in GIS to prepare prescriptions for site-specific inputs such as seed and fertilizer rates. This would help the spots that yielded less in earlier seasons yield more. Soil fertility and hydrologic properties of the field are factors influencing the yield, and correlations have been successfully made. Combines retrofitted with yield-monitoring units use yield monitoring more prevalently because the steady flow of grain through the machine makes it easy to measure yield on the go. Grain yield-monitoring systems use grain flow sensors to measure the weight of the grain, grain moisture sensors to apply conversions to a specific moisture percent, and ground speed sensors to correct the calculated grain yield in-situ. These pieces of yield information are spatially tagged for their geo-locations, as measured by the GPS. Yield monitors are now available for the mechanized harvesting of potatoes, tomatoes, sugar beets, peanuts, and cotton. Figure 14.10 shows the components of a yield-monitoring system, and Figure 14.11 shows an example of a yield map.

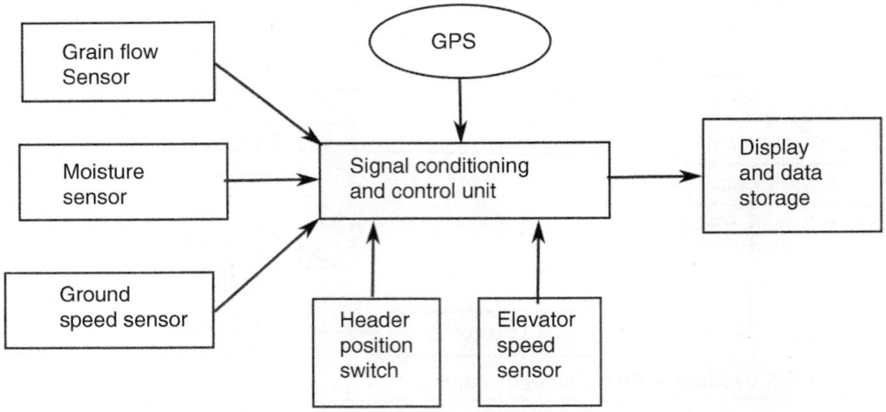

FIG. 14.10 Yield-Monitoring System Fitted on a Grain Combine.

The yield data are stored as ASCII files with each geo-reference tagged to the complete data acquired. These files can be imported into the GIS for the preparation of prescriptions.

14.3.3 Sensors for collecting spatial soil and plant data

Apart from using only the temporal yield maps to analyse and prescribe site-specific agricultural inputs, spatial data on relevant soil and plant data can be mapped and used with earlier yield maps to prescribe actions for enhancing the crop yield (see Figure 14.10). Maps on field boundary, soil survey, and chemical and physical properties such as electrical conductivity and pH are prepared and used along with yield maps to prepare the prescriptions on seed or fertilizer rate. Some of the maps, such as soil survey and boundary maps, are readily available from external sources. Other relevant soil fertility and physical maps are prepared by measuring their values in situ using machine-mounted sensors. Several sensor systems provide information on such soil and crop parameters.

Electromagnetic induction or electrical resistance is used for measuring the electrical conductivity (EC) of soil on the go. The Veris® EC Probe (Veris Technologies, Inc., Salina, Kansas, United States) is one such machine that measures EC by sending out an electrical current through the soil using a set of coulter electrodes mounted on a trailed machine and computes the EC by the measured voltage drop. Figure 14.11 shows an EC map prepared from data measured by such a sensor. Infrared reflectance or capacitance measurements can be used to determine soil moisture on the go. Optoelectronic soil organic matter sensors are also available; they use multiple-wavelength sensors. Sensors that determine mineral nitrogen rapidly in the field are helpful in prescribing the correct doses of fertilizer on a site-specific basis. Optical sensors for the on-site sensing of photosynthetic activity in plants are available for these applications.

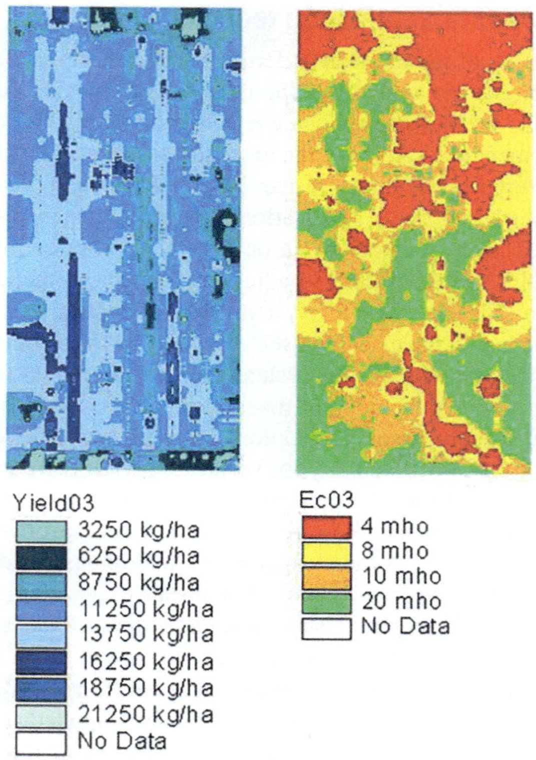

FIG. 14.11 Examples of Yield and Electrical Conductivity (EC) Maps (Ehsani & Durairaj, (Ehsani *et al.*, 2005)).

However, unless definite theories and related algorithms are developed, these maps may not be useful in predicting and prescribing the inputs. This is a researchable area in which many people are contributing. Though the technology of precision agriculture is viable and used popularly, it is limited by the following in many instances.

- lack of rapid, on-the-go and economically feasible sensing systems for assessing soil, crop, and environment data;
- appropriate models of adequate scientific validity for determining prescriptions based on the localized input data; and
- techniques for a more precise application of input. Decision support systems are now being used extensively for analysing and preparing prescriptions. Since the use of a GIS platform may not be ideal for end users to prepare prescription maps, customized, stand-alone systems with decision support are implemented to analyse and prepare the prescriptions.

14.3.4 Variable-rate application technology

Once the prescription of inputs, such as fertilizer or seed rate, is derived, machines and methods are needed to apply these inputs at the site-specific geo-referenced locations in the field. Variable-rate application technology (VRT) is one that is used by agricultural equipment that can apply the prescribed amounts of materials across the field at the geo-referenced locations. A map-based VRT system (Figure 14.9) adjusts the rate of product based on an application or prescription map, which has been created and placed in the memory of an on-board computer-cum-controller of the machine (Figure 14.2). It uses GPS to adjust the amount at the desired locations. On the contrary, a sensor-based VRT system accesses real-time data from soil and plant sensors mounted on the machine itself and controls the amount of input instantaneously at the spot. Obviously, this system does not need GPS for its operation.

A VRT system carried onboard VRT machinery, such as a variable-rate seed drill, has an application rate processor, application controller, and other on-board sensors such as dopplers to assess the velocity for verification. The current location and velocity are now integrated into the prescription map fed into the system, and commands are issued to the application rate controller. The controller in turn actuates the application drives, such as the stepper and synchronous motors, to output the desired quantity specific to that location. For instance, a variable-rate seeder has stepper motor-driven seed-metering drives that meter the seed rate specific to a given location based on the prescription map and the travel speed of the machine. Speeds of travel need be taken into account while applying the variable rate of inputs, since lags or leads in application influence the correct quantity applied at a given location. Feedback control architecture is used in many machines to monitor the applied quantity precisely. This implies that the quantity of agricultural input flowing from the machine is measured in real time and fed back to the controller. VRT machines for applying fertilizer and seeds are commercially available and used.

Precision is a means of applying precisely the correct amount of inputs like fertilizer and pesticides at the correct time to the crop and, hence, significantly reduces their excessive application spoiling the environment. This is in addition to substantial savings a farmer can realize for these inputs.

In addition to crop production activities, precision livestock farming is also possible; this allows for the continuous monitoring and control of animal productivity, environmental impacts, and health parameters. For example, animals can be tagged with electronic identification that allows a milking robot to identify and access that cow's coordinates for milking.

Automatic off-road vehicle guidance and driverless machines are also technologies in the offing due to the higher resolution of GPS available for use. Agricultural robots for planting, weeding, and other farm jobs can use these technologies. Spraying drones are already using GPS navigation for planning and controlling the path they take during their spraying flight. Artificial intelligence, machine learning, and smartphone integration are being used in many precision farming applications and control systems.

14.3.5 Automation and information solutions in management of agricultural machines

Commercial manufacturers of agricultural machines are striving hard to render as much efficiency and simplicity in the management of field operations. Today, the machine system in agriculture is being revolutionized by increased automation, leading to the integration and centralization of control of machine operations in the field. Software and hardware solutions are provided right from the tractor's cabin to control the operations of different systems of the machine a tractor is pulling. Sophisticated information channels are being utilized through the Internet and cloud computing to receive inputs from sensors and other types of hardware as well as to output control signals to the machine sections. In addition to these, GPS tagging and variable-rate technology are also manifested through these channels.

Telematics is also used to implement these channels by some firms, offering simple solutions. Automated sections of a machine respond to commands from the central unit in a tractor. Ultimately, the whole operation is well planned, sequenced, and controlled systematically by the tractor operator. Moreover, all such tractor outfits being operated in the field can also be controlled and guided from a central ground station such as a farm office through this technology. This leads to a substantial savings in time and in the operations yielding greater productivity. Even very big farms can be automated and controlled through this technology.

Large control terminals (Figure 14.12) mounted inside a tractor with bright screens that can be seen even in dark conditions are available. The software applications being run on these central units are so designed that many complex actions are possible. The terminals provide interfacial ports such as USB and RS232 and other ports for camera and sensor inputs. Touchscreen functions are offered for the operator to interact with the machine quite easily. Typically, large machines, such as harrows, seeders, and sprayers, have foldable sections. Individual sections are controlled from the tractor control terminal toselectively shut down sections, avoiding the overlap of machine passes and shutting down sections outside the field boundaries and in headlands. These operations are either automated or manually controlled on the go.

Variable-rate technology is also available in these systems to control the application rate of inputs such as seed, fertilizer, and pesticide according to pre-fed, GPS-tagged prescription maps. These vary the application rates as desired to improve the productivity of the crop being raised. Possibilities exist to log GPS-tagged data from the field, as provided by sensor inputs from the machine sections in real time and that are readily available for use in precision agriculture. Moreover, the systems of operations, routing, vehicle guidance, and pass manipulations are made as per previous input plans to make the field operation more field efficient. Field boundaries and headlands are automatically detected and dealt with according to the plan. Manual overrides are always available for the tractor operator. GPS antenna modules (inset of Figure 14.12) with a DGPS (differential GPS) accuracy of 30 cm minimum are supplied with these automation solutions such that the best precision possible is

FIG. 14.12 Touchscreen Panel of a Machine Control Unit and GPS Receiver (Isomatch Tellus Pro and Global-3, Kverneland Group, Norway).

obtained for machine guidance, automatic section control, variable-rate control, and input tagging and logging of data.

All the details of finished operations are recorded by the control terminal. One can view and analyse them as well as use them in the future as a decision support for future operations. Machine management modules are provided along with these systems for the operator to make decisions on choosing the steps and sequence of operations much more efficiently. Automatic machine learning and AI may be used in the future to make such series of decisions.

14.4 Robots in agriculture

Today, agriculture has proven to be energy hungry, demanding timely operations with huge machines that cause undue soil compaction. This has started a vicious cycle in which the compacted soil needs to be loosened for cultivation, thus unduly disturbing its structure and organic nutrient balances. The promulgation of the use of large machines has killed the possibility of managing small farms. Big machines are often underutilized in many operations, wasting energy and capital investment. Moreover, pesticides and inorganic inputs are being used indiscriminately, with huge quantities of these chemicals not being utilized for their specific intention. Maximizing production has been the only focus, leading to the indiscriminate use of energy and other inputs accordingly. This has created havoc on the environment. Future systems should be able to respond to real-time changes in weather, economics, and environment and should naturally seek information intensively.

A need for precise soil manipulation and the application of inputs is now being felt internationally. This could be achieved only through appropriate automation with relatively smaller and lighter machines. Unmanned machines and robots are the answer. Precision agriculture has come to stay because of its potential; subsequently, robotic agriculture is on its way, compounding the principles learnt already. In this context, it is now realized that medium-sized fields have a greater potential for better production management.

Robotic agriculture was attempted first by trying autonomous/guided tractors that can move in a field precisely and perform most of the operations. Advances in software, hardware, and sensors have expanded the yearning for the use of robots on farms.

Let us now explore some recent examples (Abrams, 2023) that have been successfully experimented and, in some instances, marketed.

A robotic seeder that is quite light and does not cause soil compaction performs the punch planting of seeds without disturbing the soil. The place and depth of planting is precisely controlled year after year, so that the soil nutrients are utilized efficiently. A light, robotic dibber can enter a field in any weather condition and do the planting.

Crop scouting is already being performed by drones that provide the farmer with vital information sensed at each geographical location of the cropped area. But, a robotic scout will be able to access these data at a very proximate range in real time. Variables such as the crop cover, rate of plant growth, presence of pests and diseases, weed intensity, and soil parameters including moisture and nutrient status can be monitored, researched, and remedied. The NDVI and thermal sensors are some examples that give the plant nutrient status. A crop-scouting robot built by the University of Athens has a thermal camera for measuring the soil moisture to schedule irrigation as well as a multi-spectral camera for assessing the nutrient status. A LIDAR (light detecting and ranging) sensor is used additionally to measure the canopy density and extent.

It is already known that only a meagre 1 percent of the pesticide sprayed by conventional sprayers reaches the target pest, causing mortality. Blanket spraying is, hence, known to be grossly inefficient. A robotic sprayer can apply the spray precisely on the target plant in real time.

Similarly, a robotic weeder can replace weedicide applications and remove young weeds efficiently. Laser weeding is being explored and used (Figure 14.13) with such robots. A robot uses computer vision and artificial intelligence deep-learning models to identify the weeds in real time, differentiate the crops from weeds, and kill the weeds with high-powered lasers of 30×150W power. Numerous such laser gun arrays fire every 50 milliseconds at sub-millimetre accuracy and kill the detected weeds at an enormous rate. The robot operates even at night and on any type of soil. Laser weeding is a form of organic weeding that does not use soil-harming chemical herbicides and does not disturb the soil.

The selective picking of fruits, vegetables, and flowers is another prime area in which picking robots can be used effectively. Though the work rate of such machines will be low, the quality of products, eliminating undue loss, is very high. The

FIG. 14.13 Laser Weeding Robot (LaserWeeder, Carbon Robotics, USA).

selection of picked products can be based on a variety of parameters such as the size, ripeness, firmness, nutrients, etc. For instance, selective picking or cutting of flowers based on desired colour can be done. Post-harvest grading, sorting, and packing can be done in situ. The availability of better sensor and image acquisition technologies has improved their possible uses. Soft grippers on multiple picking arms pick the selected products efficiently.

Robotic pollinators are yet another potential application for robotics in agriculture. Self-pollinating crops, such as tomato and the blueberry, have male and female parts on each flower, requiring a gentle vibration to self-pollinate. Robots can identify the flowers, aid in their pollination, and thus not depend on help from insects to do so. These robots can identify prime flowers ready for such pollination in any temperature and lighting and produce outstanding fruit quality.

Robots that can perform multiple operations such as seeding, weeding, crop protection, and harvesting are also being researched.

References

Abrams, M. (2023). *Agricultural robots bringing food to the table*. The American Society of Mechanical Engineers. https://www.asme.org/topics-resources/content/5-agricultural-robots-bringing-food-to-the-table.

Ehsani, M. R., Durairaj, C. D., Woods, S., & Sullivan, M. (2005). Potential application of electrical conductivity (EC) map for variable rate seeding. *Agricultural Engineering International: The CIGR Ejournal, VII*.

Merritt, H. E. (1979). *Hydraulic control systems*. John Wiley & Sona, Inc.

Onwubolu, G. (2005). *Mechatronics – principles and applications*. Elsevier. Publ.

Srivastava, A. K., Goering, C. E., Rohrbach, R. P., & Buckmster, D. R. (2006). *Engineering principles of agricultural machines*. ASABE Publications.

Index

Note: Page numbers followed by *b* indicate boxes, *f* indicate figures and *t* indicate tables.

D

E

F

G

H

I

K

L

S

T

U

V

W

Y